METHODS IN MOLECULAR BIOLOGY

For further volumes:
http://www.springer.com/series/7651

Homing Endonucleases

Methods and Protocols

Edited by

David R. Edgell

Department of Biochemistry, Schulich School of Medicine and Dentistry, Western University, London, ON, Canada

Humana Press

Editor
David R. Edgell
Department of Biochemistry
Schulich School of Medicine and Dentistry
Western University
London, ON, Canada

ISSN 1064-3745 ISSN 1940-6029 (electronic)
ISBN 978-1-62703-967-3 ISBN 978-1-62703-968-0 (eBook)
DOI 10.1007/978-1-62703-968-0
Springer New York Heidelberg Dordrecht London

Library of Congress Control Number: 2013957689

Printed on acid-free paper

Humana Press is a brand of Springer
Springer is part of Springer Science+Business Media (www.springer.com)

Preface

Homing endonucleases are site-specific DNA endonucleases that primarily function as mobile genetic elements, promoting their spread by recombination-based pathways. Homing endonucleases are easily distinguished from other site-specific DNA endonucleases by their lengthy recognition sequences (~14–40 bp) and by their tolerance to nucleotide substitutions within their recognition sites. The biotechnological potential of homing endonucleases as rare-cutting endonucleases was recognized soon after their discovery. Detailed structural, biochemical, and bioinformatic studies on homing endonuclease–DNA interactions has led to the realization that the specificity of homing endonucleases, especially the LALIGDADG family members, can be reprogrammed to target desired sequences. Engineering of designer homing endonucleases has set the stage for genome editing of complex eukaryotic genomes, with a broad range of potential applications including targeted gene knockouts in model organisms and gene therapy in humans. This volume is aimed at providing molecular biologists with a comprehensive resource to identify and characterize homing endonucleases from genomic sequence, to deduce the biological basis of binding and cleavage specificity, as well as to provide protocols to redesign endonuclease target specificity for genome-editing applications.

London, ON, Canada *David R. Edgell*

Contents

Contributors

DAVID BAKER • *Department of Biochemistry, Institute for Protein Design, University of Washington, Seattle, WA, USA*

SARAH K. BAXTER • *Medical Scientist Training Program, University of Washington, Seattle, WA, USA; Center for Immunity and Immunotherapies, Seattle Children's Research Institute, Seattle, WA, USA; Northwest Genome Engineering Consortium, Seattle, WA, USA*

MARLENE BELFORT • *Department of Biological Sciences and RNA Institute, University at Albany, Albany, NY, USA*

RICHARD P. BONOCORA • *Wadsworth Center, New York State Department of Health, Albany, NY, USA*

RUSSELL J. DICKSON • *Department of Biochemistry, University Of Western Ontario, London, ON, Canada*

DAVID R. EDGELL • *Department of Biochemistry, Schulich School of Medicine and Dentistry, Western University, London, ON, Canada*

JEAN-CHARLES EPINAT • *Cellectis SA, Paris, France*

JOSHUA I. FRIEDMAN • *Departments of Biochemistry and Pathology, University of Washington, Seattle, WA, USA*

EWAN A. GIBB • *Genome Sciences Centre, BC Cancer Agency, Vancouver, BC, Canada; Medical Genetics, University of British Columbia, Vancouver, BC, Canada*

GREGORY B. GLOOR • *Department of Biochemistry, University Of Western Ontario, London, ON, Canada*

TUHIN KUMAR GUHA • *Department of Microbiology, University of Manitoba, Winnipeg, MB, Canada*

MOHAMED HAFEZ • *Department of Biochemistry, Université de Montréal, Montréal, QC, Canada*

GEORG HAUSNER • *Department of Microbiology, University of Manitoba, Winnipeg, MB, Canada*

DAVID L. HERRIN • *Section of Molecular Cell and Developmental Biology, Institute for Cellular and Molecular Biology, School of Biological Sciences, University of Texas at Austin, Austin, TX, USA*

KYLE JACOBY • *Program in Molecular and Cellular Biology and Department of Immunology, University of Washington, Seattle, WA, USA; Center of Immunity and Immunotherapies, Seattle Children's Research Institute, Seattle, WA, USA*

BENJAMIN P. KLEINSTIVER • *Department of Biochemistry, Schulich School of Medicine and Dentistry, Western University, London, ON, Canada*

TAEGUN KWON • *Section of Molecular Cell and Developmental Biology, Institute for Cellular and Molecular Biology, School of Biological Sciences, University of Texas at Austin, Austin, TX, USA*

Abigail R. Lambert • *Center for Immunity and Immunotherapies, Seattle Children's Research Institute, Seattle, WA, USA; Northwest Genome Engineering Consortium, Seattle, WA, USA*

Hui Li • *Department of Pathology, University of Washington, Seattle, WA, USA*

Raymond J. Monnat Jr. • *Department of Pathology, University of Washington, Seattle, WA, USA; Department Genome Sciences, University of Washington, Seattle, WA, USA*

Obed W. Odom • *Section of Molecular Cell and Developmental Biology, Institute for Cellular and Molecular Biology, School of Biological Sciences, University of Texas at Austin, Austin, TX, USA*

Stefan Pellenz • *Department of Pathology, University of Washington, Seattle, WA, USA*

Eyal Privman • *Institute of Evolution and Department of Evolutionary and Environmental Biology, University of Haifa, Israel*

Weihua Qiu • *Section of Molecular Cell and Developmental Biology, Institute for Cellular and Molecular Biology, School of Biological Sciences, University of Texas at Austin, Austin, TX, USA*

Andrew M. Scharenberg • *Program in Molecular and Cellular Biology and Department of Immunology, University of Washington, Seattle, WA, USA; Center of Immunity and Immunotherapies, Seattle Children's Research Institute, Seattle, WA, USA; Northwest Genome Engineering Consortium, Seattle, WA, USA*

Jyothi Sethuraman • *Department of Microbiology, University of Manitoba, Winnipeg, MB, Canada*

Chen Shen • *Department of Microbiology, University of Manitoba, Winnipeg, MB, Canada*

Barry L. Stoddard • *Division of Basic Sciences, Fred Hutchinson Cancer Research Center, Seattle, WA, USA*

Ning Sun • *Departments of Chemical and Biomolecular Engineering, Chemistry, Biochemistry, and Bioengineering, Institute for Genomic Biology, Center for Biophysic and Computational, Biology, University of Illinois at Urbana-Champaign, Urbana, IL, USA*

Summer Thyme • *Department of Biological Sciences, University of Washington, Seattle, WA, USA*

Jason M. Wolfs • *Department of Biochemistry, Schulich School of Medicine and Dentistry, Western University, London, ON, Canada*

Huimin Zhao • *Departments of Chemical and Biomolecular Engineering, Chemistry, Biochemistry, and Bioengineering, Institute for Genomic Biology, Center for Biophysic and Computational, Biology, University of Illinois at Urbana-Champaign, Urbana, IL, USA*

Lei Zhao • *Residency Program in Pathology, University of Chicago, Chicago, IL, USA*

Chapter 1

Homing Endonucleases: From Genetic Anomalies to Programmable Genomic Clippers

Marlene Belfort and Richard P. Bonocora

Abstract

Homing endonucleases are strong drivers of genetic exchange and horizontal transfer of both their own genes and their local genetic environment. The mechanisms that govern the function and evolution of these genetic oddities have been well documented over the past few decades at the genetic, biochemical, and structural levels. This wealth of information has led to the manipulation and reprogramming of the endonucleases and to their exploitation in genome editing for use as therapeutic agents, for insect vector control and in agriculture. In this chapter we summarize the molecular properties of homing endonucleases and discuss their strengths and weaknesses in genome editing as compared to other site-specific nucleases such as zinc finger endonucleases, TALEN, and CRISPR-derived endonucleases.

Key words Homing endonucleases, Genome editing, Altered specificity endonucleases, Biotechnology, Endonuclease applications

1 History and Evolution

1.1 Discovery of the First Homing Endonuclease

In the early 1970s at the Institut Pasteur in Paris, researchers discovered a phenomenon that seemed to throw the principles of genetic inheritance described by Mendel out the window. While performing crosses with yeast strains, they discovered that a particular mitochondrial allele, known as omega (ω), was unidirectionally inherited at what was later determined to be the gene for the large ribosomal RNA subunit (LSrRNA) [1, 2]. The ω allele was shown to contain a LSrRNA gene disrupted by an intervening sequence known as a group I intron [3, 4]. At that time introns were thought to be purely "junk" DNA that were removed from a gene posttranscriptionally, leaving the mature RNA intact, fully functional, and without a discernible phenotype. How then could this junk exert such a strong influence on its own inheritance? The mystery was finally solved when, surprisingly, the

David R. Edgell (ed.), *Homing Endonucleases: Methods and Protocols*, Methods in Molecular Biology, vol. 1123, DOI 10.1007/978-1-62703-968-0_1,

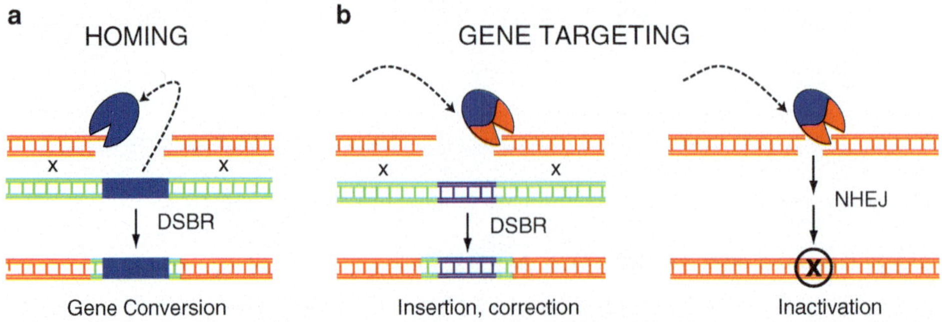

Fig. 1 Endonuclease-mediated DNA repair. (**a**) HE-induced DSB mediates homing. The HEG (*blue pacman symbol*), often encoded by an intron or intein (*blue bar*), cleaves an uninterrupted homing site. The DNA ends of the cleaved recipient (*red*) engage in double-strand-break repair (DSBR), such that the cut allele is repaired by gene conversion with homologous intact DNA. (**b**) Gene targeting by an exogenous endonuclease. An engineered endonuclease (*blue* and *red*) cleaves a desired target, which can be repaired with an allele that inserts or corrects DNA using a homologous template (*left*), or is repaired by nonhomologous end joining (NHEJ), an error-prone process, that inactivates the cleaved gene

ω intron was found to contain a separate gene. That gene encodes a DNA endonuclease that recognizes and introduces a double-strand break (DSB) in LSU rRNA genes lacking the intron [5].

The endonuclease-dependent mechanism for the "super" inheritance of ω was simple and elegant and has since been shown to be a general phenomenon known as "homing" [6]. By introducing site-specific DNA breaks that then are repaired, these homing endonucleases (HEs) stimulate the cellular recombination and DNA repair processes to fix the break by simply copying the gene encoding them (the homing endonuclease gene or HEG) and flanking DNA into the broken chromosome (Fig. 1a). This gene conversion event is termed double-stranded break repair (DSBR) [7]. The end result is the proliferation of the HEG by lateral transfer.

1.2 Evolutionary Considerations

Since the discovery of ω, evidence has accumulated that the HEG is the actual mobile element and that the intron provides safe haven in the form of a phenotypically silent locus, while the intron enjoys the ride [8]. First, HEGs have been found in different genetic contexts such as different classes of intron [9], protein-splicing elements called inteins [10–12] and as free-standing genes [13]. Second, phylogenetic analysis of HEGs and the splicing elements indicated that they have different evolutionary histories [12]. Third, similar HEGs are located in different positions of similar introns [14]. Fourth, similar introns have been invaded by different HEGs [15]. Finally, disruption of the HEG abrogates homing [5, 16–18].

One might then ask whence the HEs arose. In addition to their function in self-promoting their lateral transfer, their catalytic scaffolds have been found in genes involved in general housekeeping processes such as recombination, repair, gene regulation, RNA folding, genome stability, and restriction of foreign DNA [19]. In most cases it is unclear if the chicken or the egg, namely the HEGs or the housekeeping gene, came first. Although in some cases the order is defined, as for the derivation of the HO endonuclease, which promotes yeast mating-type switching, from the intein-encoded VDE endonuclease [20, 21], most origins are ambiguous. What is clear is that these enzymes live a dynamic lifestyle that must adapt to the host gene and genome that they invade to minimize their impact on the fitness of the organism [22, 23].

HEGs within splicing elements have been proposed to undergo a lifecycle characterized by rapid invasion to fixation of a gene, slow decay of the endonuclease activity, eventual complete loss of the element and subsequent reinvasion [24]. Amendments to this model include the opportunity of the HEG to develop new functions that benefit the host organism, as for the HO endonuclease and other examples described below (Table 1), thereby preventing loss of the element, or for the HEG to transpose to a new favorable location [25, 26].

Regardless of the genetic environment in which a HEG exists, a key aspect of the homing mechanism is that the converted chromosome is now protected from cleavage by the HE. For intron- and intein-encoded HEs the associated intervening sequence lies within the sequence recognized by the HE. Once the intron/intein is copied into the new chromosome the recognition site is split by the intervening sequence thereby preventing self-cleavage. The process for free-standing HEGs is similar, but with a specific problem: the HE recognizes a target site that can be separated by several hundred base pairs from the HEG, typically located in an adjacent gene. The mode of protection can vary here; often the recognition site contains sequence changes between the sensitive chromosome without the HEG (recipient) and the refractory chromosome with the HEG (donor), which make the chromosome with the HEG refractory to HE cleavage (known as intronless homing) [27]. The sequence change to prevent cleavage can even be an unrelated intron or other genetic element inserted within the HE recognition site (termed collaborative homing) [28, 29].

A characteristic shared by introns and HEs may have been key to the origin of mobile introns; both prefer conserved, functionally significant sequences such as those that encode enzyme active sites [30–32]. Extant splicing elements have been retained in these locations presumably because their loss has to be precise or the coding sequence will be altered and the gene function impaired. Precise loss, on the other hand, would lead to reinvasion. HEs benefit from recognizing a target site that is well conserved. This similarity in sites suggests that free-standing HEGs can be "preadapted" to recognize intron insertion sites.

Table 1
Homing endonucleases described in this chapter

Family	Member	Characteristics	Origin	Notable features	Related host functions
LAGLIDADG	ω I-SceI	Monomer	*Saccharomyces cerevisiae* mitochondria	• First HE to be discovered • Widely used in genome engineering including mosquitoes	• HO mating-type endonuclease • Intron-encoded maturases • Bacterial transcriptional regulators
	HO	Monomer	*Cerevisiae* nucleus	Mating-type switch	
	I-CreI	Homodimer	*Chlamydomonas reinhardtii* chloroplast	• First LHE structure solved • Scaffold for retargeting	
	I-MsoI	Homodimer	*Monomastix* chloroplast	Scaffold for retargeting	
	I-AniI	Monomer	*Aspergillus nidulars* mitochondria	Nicking variant isolated	
	I-OnuI	Homodimer	*Ophiostoma novo-ulmi*	Family of closely related enzymes	
GIY-YIG	I-TevI–	Modular DNA binding and catalytic domains	Coliphage T4	• First GIY-YIG structure solved • Catalytic domain used in fusions	• Restriction enzymes • Penelope retroelements • Eukaryotic flap endonuclease • Eukaryotic S1x1–S1x4 resolvase • Bacterial UvrC excision repair
	I-BmoI	Modular	*Bacillus mojavensis*	Similar catalytic domain to I-TevI	

PD(D/E)xK	I-Ssp6803I	Tetramer	*Synechocystis*	Motif also in recombinases, resolvases and DNA repair enzymes	• Restriction enzyme • Enzymes of recombination, tRNA splicing, DNA resolution
His-Cys Box	I-PpoI	Bends DNA	*Physarum polycephaleum* nucleus	Colicin-like	
HNH	I-HmuI	Nicking enzyme	*B. subtilis* phage SP01	Recognizes intron and intronless targets	• Group II intron endonuclease
	I-HmuII	Nicking enzyme	*B. subtilis* phage SP82	Excludes SPO1 DNA polymerase locus in mixed infection	• Cas9 CRISPR nuclease
	I-TevIII	Makes DSB	Coliphage RB3	Makes DSB by dimerization	• Bacterial colicins
	F-TslI	Nicking enzyme	Coliphage T3	Freestanding, downstream of gene *5*	• Restriction enzymes
	I-TslI	Nicking enzyme	Coliphage ΦI	In intron within gene *5*	• Resolvase

Such a scenario is likely to have occurred at the DNA polymerase gene (gene *5*) of T3-like phages. Phage T3 contains a free-standing HE F-TslI (encoded by gene *5.3*), immediately downstream of gene *5*. F-TslI recognizes and cleaves a site within gene *5* of related phages, within the enzyme's catalytic center. As expected, the corresponding sequence in T3 is not cleaved. The related phage ΦI lacks a HEG downstream of gene *5*, but instead has a group I intron inserted one nucleotide away from the F-TslI cleavage site. This intron contains a HEG that is similar to F-TslI and encodes a functional HE named I-TslI. Both of these HEs cleave intronless gene *5* alleles at precisely the same location. Thus F-TslI exemplifies a HE preadapted for an intron insertion site that has since invaded an intron [17].

Since both the HEGs and splicing elements converge on the same sequences, is there an advantage to their forming a composite element? Free-standing HEs generally cleave far from their insertion sites. As a result, transfer of the cleavage-resistant allele from the donor genome can occur without cotransfer of the HEG [17, 27]. The result is an increase of resistant alleles and therefore a concomitant reduction in homing opportunities and pressure to retain the HEG. The HEG solves this problem by coupling with the resistance element (a group I intron disrupting the HE recognition site) thereby ensuring the transfer of both. The intron also benefits as it is now intimately linked to a mobile element and will persist in the population.

1.3 HEs from Then Till Now

In the more than 40 years since the observation of unidirectional inheritance of ω that led to the discovery of intron homing, much has been learned about the recombination process and the HEs responsible. Although the biological role of HEGs remains elusive, the usefulness of HEs as tools in biotechnology, medicine, agriculture, and possibly population control of disease vectors is becoming increasingly clear. In this chapter we will provide an overview of the biochemistry and structure of HEs and how HEs can be tailored for the various applications. We further compare these enzymes to other agents of gene targeting.

2 General Properties of HEs

HEs are small proteins (<300 amino acids) found in bacteria, archaea, and in unicellular eukaryotes (reviewed by Stoddard [33]). A distinguishing characteristic of HEs is that they recognize relatively long sequences (14–40 bp) compared to other site-specific endonucleases such as restriction enzymes (4–8 bp). These lengthy recognition sites, and the name of the first such known enzyme, ω (also known as I-SceI), have given rise to the term "meganuclease" [34]. Another feature that sets HEs apart from restriction endonucleases

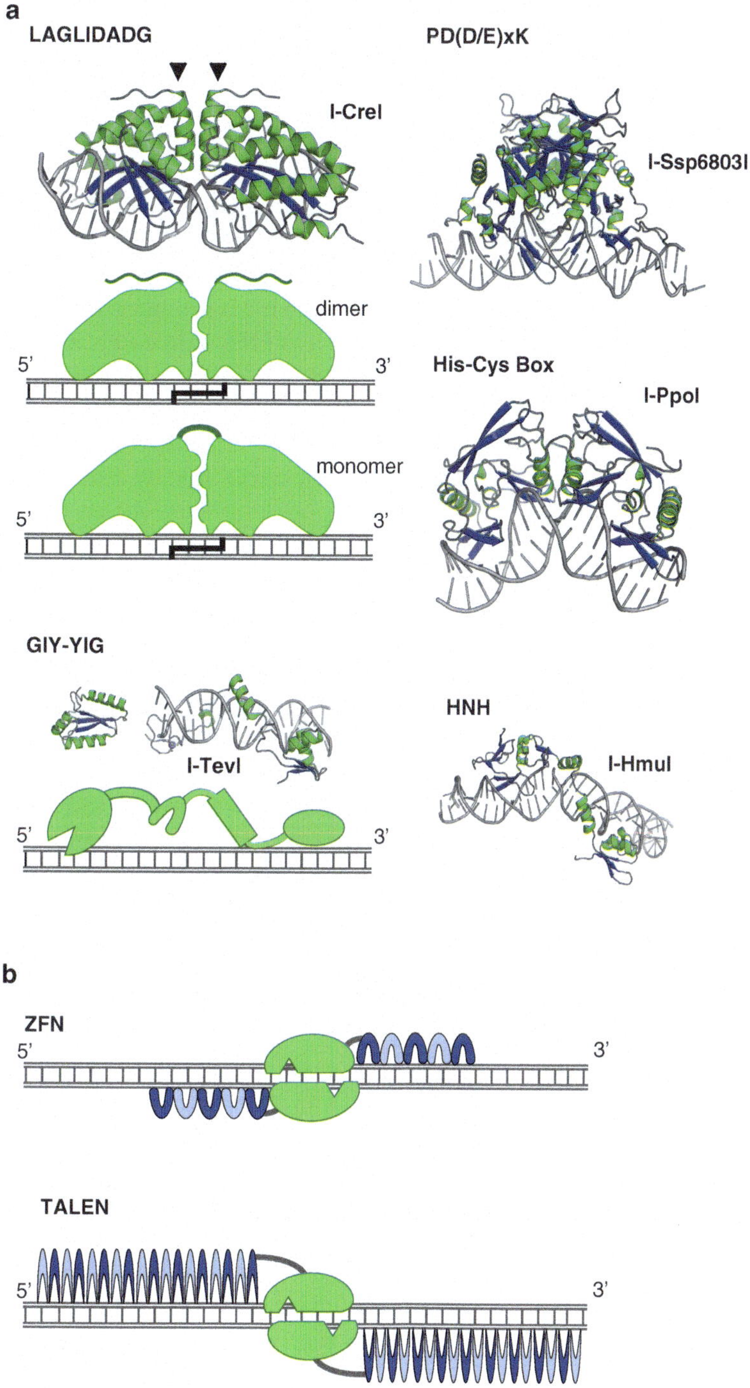

Fig. 2 Endonuclease–DNA interactions. (**a**) Five families of HEs are shown with examples indicated in parenthesis: LAGLIDADG (I-CreI), GIY-YIG (I-TevI), HNH (I-HmuI), His-Cys Box (I-PpoI) and PD(D/E)xK (I-Ssp68031). I-CreI binds DNA as a homodimer, while other examples of the family are monomeric (*cartoons below*). *Arrowheads* point to the LAGLIDADG helixes. The I-TevI binding domain is shown on DNA, whereas the catalytic domain is freestanding. A model of full-length GIY-YIG HE binding is shown in *cartoon form below the structure*. (**b**) Synthetic endonucleases, ZFNs and TALENs, are fusions between the respective repeated binding units and the FokI catalytic domain, which dimerizes on the DNA target

is their lack of absolute sequence specificity. Whereas restriction enzyme binding and/or cleavage depend on a perfect match to the recognition sequence, HEs are less discriminating, often tolerating multiple sequence changes within their recognition site [35, 36]. This is apparent at the structural level where there is a great disparity between the number of contacts made by restriction endonucleases and HEs. Restriction endonucleases exploit most of the potential hydrogen bonds between the proteins and their target sites [37] whereas HEs utilize only a fraction of the possible hydrogen bonds [38–40]. The positions that are tolerated by HEs are often those at third positions of codons, which vary naturally between organisms. Such tolerance allows homing into new sites. Despite the imperfect fidelity, the lengthy recognition sites can make HEs highly specific, often cutting large genomes only once. This attribute makes the HEs amenable to genome editing, where spurious off-site cleavages are detrimental.

HEs have been historically categorized by small conserved amino acid motifs. At least five such families have been identified: LAGLIDADG; GIY-YIG; HNH; His-Cys Box and PD-(D/E)xK, which are related to EDxHD enzymes and are considered by some as a separate family (Table 1, Fig. 2a). At a structural level, the HNH and His-Cys Box share a common fold (designated ββα-metal) as do the PD-(D/E)xK and EDxHD enzymes. The catalytic and DNA recognition strategies for each of the families vary and lend themselves to different degrees to engineering for a variety of applications.

3 HE Families

3.1 LAGLIDADG Endonucleases

The LAGLIDADG endonucleases (LHE) make up a large, well-described family of HEs identified largely in archaea and organelles of lower eukaryotes. LHE genes exist in a variety of genomic environments, being encoded within group I introns, archaeal rRNA introns, as in frame fusions with inteins, and as free-standing genes.

LHEs have either one or two copies of the signature LAGLIDADG motif and leave 4 nt 3′- extensions. Crystallographic studies with both dual and single motif LHEs, show that the motif is present in an α-helix that is responsible for mediating protein–protein interactions [39, 41–44] (Fig. 2a). The LAGLIDADG helices act as an intramolecular interface for the dual motif endonucleases, or as the site of dimerization between identical monomers for single-motif endonucleases. These helices are also responsible for correctly positioning the active site residues. Biochemical and structural evidence indicates that the second conserved acidic residue of the motifs (LAGLIDA$\underline{D}$G) acts in metal ion coordination critical for the catalytic activity of the endonuclease

[39, 45, 46]. I-CreI catalysis involves hydrolysis of the scissile phosphates following a canonical two-metal mechanism, with a single metal in each active site [47, 48]. Target recognition is mediated by interactions between a four strand anti-parallel β-sheet from each monomer/domain [39] (Fig. 2a). LHEs with a single motif recognize largely palindromic sequences whereas the two-motif nucleases need not be symmetric.

In some cases LHEs have taken on secondary roles, as for example the aforementioned HO endonuclease [21], which initiates a gene conversion event that results in mating-type switching [20, 49]. Additionally, several LAGLIDADG proteins encoded within group I introns of yeast act as RNA maturases instead of DNA endonucleases, assisting in the folding of their cognate intron into its catalytically active conformation. Some intron-encoded proteins can function as both maturase and endonuclease [50–53] while others can be converted from a maturase to an endonuclease by mutation [54]. I-AniI, a group I intron-encoded LAGLIDADG endonuclease/maturase, has the DNA- and RNA-binding sites located at independent surfaces on the protein [50, 55, 56], suggesting that maturase activity is an adaptation independent of endonuclease function. Finally, in an extreme example of LHE adaptation, the DUF199 family of bacterial transcriptional regulators resemble dual motif LHEs lacking catalytic aspartate residues, fused at their C-terminus to a helix-turn-helix DNA-binding domain [57–59]. Although the LAGLIDADG domain alone has lost its DNA-binding capacity, it improves binding of the full length protein about fourfold over the HTH domain alone, providing an example of adaptation of both its catalytic and DNA-binding functions [19, 59].

3.2 GIY-YIG Endonucleases

GIY-YIG endonucleases are modular proteins that exist in all three domains of life. Computational studies coupled with extensive genetic, biochemical, and structural analyses of I-TevI have provided a detailed picture of a member of the GIY-YIG family of endonucleases [60–62]. I-TevI and its close relative I-BmoI are endonucleases encoded within introns interrupting the thymidylate synthase genes of bacteriophage T4 and the soil bacterium *Bacillus mojavensis*, respectively. I-TevI and I-BmoI consist of an N-terminal globular catalytic domain, that is conserved with other family members, and a modular C-terminal DNA-binding domain connected by a flexible linker (Fig. 2a) [63–65]. GIY-YIG HEs cleave their homing site leaving 2 nt 3′ overhangs [14, 66, 67]. The N-terminal catalytic domain spans ~90 amino acids and contains five conserved regions [60]. The first GIY-YIG endonuclease visualized in complex with DNA is restriction enzyme R. Eco29k1, showing both tyrosines within the conserved motifs in the catalytic center [68]. For GIY-YIG restriction endonuclease Hpy1881, the GIY tyrosine was identified as a general base, with a conserved glutamate anchoring a single metal in the active site [69].

The C-terminal domains of both I-TevI and I-BmoI contribute the bulk of the DNA-binding energy. Indeed, this domain alone binds to the homing site with equal affinity to the full-length protein [14, 63]. Structural studies of the C-terminal domain of I-TevI in complex with its homing site has revealed that the DNA-binding domain itself consists of three smaller structured subdomains linked together by elongated unstructured regions (Fig. 2a). The DNA-binding domain wraps around the minor groove of its target DNA making few base-specific contacts [40]. Consistent with this minor groove binding is the demonstration that no single base-pair in the homing site is absolutely required for cleavage [36].

In addition to GIY-YIG motifs in some restriction enzymes and associated with the reverse transcriptase of Penelope retroelements [70], the motif also occurs in enzymes involved in DNA repair and maintenance of chromosomal genome stability. These include eukaryotic flap endonucleases, eukaryotic Slx1–Slx4 resolvase, and bacterial UvrC nucleotide excision repair protein (summarized in [61]). As another example of HE adaptation, I-TevI binds to an operator site overlapping its promoter and acts as an autorepressor [71]. The operator site is not efficiently cleaved, but is bound in a similar fashion as the homing site. Apparently this repression delays translation of I-TevI, facilitating splicing of the host intron [22].

3.3 HNH Endonucleases

Members of the HNH family have been found in group I introns in plastids and phages, group II introns, and free-standing ORFs of bacteria, phages, and a cyanobacterial intein [12]. The HNH motif consists of a stretch of approximately 30 amino acids that contain three highly conserved histidine and/or aspargine residues. Structural analysis indicates a role in metal binding for the first two conserved amino acids [72–74].

Many members of the HNH family violate one or more of the canonical properties for intron-encoded HEs. Unlike HEs from other families, many HNH endonucleases nick one strand of dsDNA 5′ to the intron insertion site [75, 76], although at least one HNH endonuclease, I-TevIII, has been shown to make a DSB by dimerization [77]. Another deviation from canonical HEs is demonstrated by the endonucleases I-HmuI and I-HmuII (Fig. 2a) encoded within homologous group I introns interrupting the DNA polymerase gene from the *B. subtilis* phages SPO1 and SP82, respectively. They are nicking endonucleases that cleave both intronless and intron-containing targets. These endonucleases prefer to cleave the DNA of the heterologous phage in vitro and

I-HmuII has been shown to exclude the SPO1 intron and flanking regions during mixed infection [75].

In group II intron retrohoming, the HNH nuclease is part of a multidomain intron-encoded protein [12] that forms a ribonucleoprotein particle (RNP) with its cognate intron. The group II intron portion of the RNP directs the complex to the appropriate target via base-pairing [6, 78]. The intron RNA reverse splices into the sense strand of the target while the HE nicks the template strand. This nick provides a 3′-OH which is used to prime RNA-dependent DNA synthesis by a separate reverse transcriptase domain of the HE.

The HNH motif has also been identified in nonspecific colicins nucleases [79, 80], resolvases [81] and type II restriction endonucleases [82]. Interestingly, the Cas9 endonuclease involved in the clustered, regularly interspaced, short palindromic repeats (CRISPR) system of bacterial adaptive immunity from *Streptococcus thermophilus* contains two catalytic nicking domains, one of which contains an HNH motif, and the other a RuvC-like motif. CRISPR endonucleases form RNP complexes with small bacterially-encoded RNAs that are homologous to short sequences derived from previously encountered phage or plasmids. Similar to group II intron HEs, Cas9 forms an RNP with the CRISPR RNA (crRNA), and is directed to target DNA by base pairing between the crRNA and target. A DSB results from two independent nicks in the target DNA, one by the HNH motif and the other by the RuvC-like motif [83, 84].

3.4 His-Cys Box Endonucleases

The His-Cys box endonucleases comprise a much smaller family of HEs, found in group I introns interrupting nuclear rRNA genes of lower eukaryotes [33, 85]. As the name implies, the signature of this family is a region that is rich in histidine and cysteine residues. The best studied of the His-Cys Box family is I-PpoI, a group I intron-encoded endonuclease from the slime mold *Physarum polycephalum*. The crystal structure of I-PpoI shows a dimer, where each monomer contains two histidine- and cysteine-rich sequences that coordinate separate zinc ions and function to stabilize the protein structure [86]. Like LHEs, I-PpoI binds its target DNA using antiparallel β-sheets. However, unlike the LHEs, I-PpoI induces a strong bend into the DNA target to bring the scissile phosphates into proximity with each active site (Fig. 2a).

Comparison of the structures of I-PpoI, the colicin E9 HNH endonuclease and a nonspecific nuclease from *Serratia* have identified a structural similarity at the active sites of all three. This similarity has suggested a reclassification of the two families into one known as ββα-Me which reflects the three secondary structural elements and the bound metal ion that define the motif [87].

3.5 PD-(D/E)xK and EDxHD Endonucleases

Restriction endonucleases share very little sequence conservation; however many contain a common catalytic fold from the PD-(D/E)xK super-family [88]. This motif appears to have been adapted by the group I intron-encoded homing endonuclease I-Ssp6803I from the cyanobacterium *Synechocystis* sp. 6803 [38, 89]. I-Ssp6803I is a small protein that functions as a tetramer to recognize a fairly large target site [38, 90, 91] (Fig. 2a). Like other HEs this recognition involves a paucity of protein–DNA contacts utilizing only one third of the possible hydrogen bonds [38]. This is in contrast to restriction endonucleases which use a high density of protein–DNA contacts to recognize small DNA target sites [38]. Presumably the under-saturation of contacts by I-Ssp6803I allows for cleavage of sequence variants as with other families of HEs. In addition to restriction endonucleases and HEs, the PD-(D/E)xK motif occurs in nucleases involved in DNA recombination, tRNA splicing, transposition, Holliday junction resolution, DNA repair, and Pol II termination [92].

Recently, a family of HEs, EDxHD, related to the PD-(D/E)xK was discovered by a bioinformatic analysis of the Global Ocean Sampling (GOS) environmental metagenomic sequence data [93, 94]. Like the GIY-YIG and HNH HE families, the EDxHD is modular. Although the overall fold in its catalytic domain is similar to the PD-(D/E)xK fold, the active site has diverged [95]. The crystal structure of I-Bth0305I, an EDxHD endonuclease encoded in a group I intron interrupting the *recA* gene of the *Bacillus thuringiensis* 0305r8–36 bacteriophage, supports bioinformatic evidence [96] that this family is homologous to very short patch repair (Vsr) endonucleases [19, 95]. The EDxHD HEs are also associated with split inteins encoded by a non-contiguous open reading frame, and several genes involved in DNA replication and repair [96]. Thus, involvement of enzymes related to HEs in nucleic transactions of the cell is a common feature among the different HE families.

4 Modular Semisynthetic DNA Cleavage Enzymes

In addition to HEs, modular cleavage enzymes are being assembled in the laboratory for genome engineering. Two modular types of semisynthetic site-specific cleavage enzymes are the zinc finger nucleases (ZFNs) [97] and the transcription activator-like effector (TALE) nucleases (TALENs) [98]. Both ZFNs and TALENs contain multiunit DNA-binding domains in which the specificity residues are fused to nonspecific DNA cleavage modules, typically from the type IIs restriction enzyme FokI (Fig. 2b). The DNA-binding domain of the ZFNs contain between three and five zinc fingers, which are fairly specific for stretches up to

~15 bp. Likewise, the modular TALE DNA-binding domain contains units of amino acid sequences each with specificity for a single nucleotide. A combination of these units forms a DNA-binding cartridge. The current most popular cleavage domain is again from FokI, which must dimerize. This requires a pair of DNA-binding cartridges each recognizing opposite strands to be fused at their C-termini to the FokI cleavage domain. The net result for both ZFNs and TALENs is a large tailored site-specific dimeric cleavage enzyme (Fig. 2b).

5 Reprogramming HEs

5.1 Birth of an Industry: Selection Systems

Almost two decades ago, when little was yet known of the basis for sequence specificity of the LHE enzymes, the ω endonuclease I-SceI was used to induce DSBR in mammalian chromosomes [99, 100]. It soon became clear that altered specificity variants of LHEs would have enormous application in both research and the biotech industry, and various selection systems were designed to isolate altered-specificity mutant enzymes. These include a bacterial blue-white colony screen based on β-galactosidase elimination [101], selection for growth based on control of a cell death protein [102], a bacterial two-hybrid selection system [103], and a yeast or mammalian white-blue screen based on repair of β-galactosidase [104]. More recently, other reporters such as URA3 and GFP have been used in the DNA repair assay, and yeast surface display has been employed as an effective way to isolate altered-specificity mutants [105]. The effectiveness of these selection systems and the perceived utility of retargeting HEs rapidly spawned several biotech companies, Cellectis Bioresearch in 1999, and both Precision Biosciences and Pregenen Genome Engineering in 2006.

5.2 HEs with Altered Specificity

As structures of members of all the HE families were solved, including monomeric and dimeric LHEs (Fig. 2a), rational design coupled with randomization and screening in a high-throughput format took hold, resulting in LHEs with altered specificity [106]. Simultaneously, highly sophisticated computational reprogramming resulted in redesigned LHE DNA binding and cleavage specificities [107]. RosettaDesign (RD) has been used to generate thousands of different mutants of the LHE I-CreI targeted towards 16 different base pair positions in the 22 bp I-CreI target site. Of these, over two-thirds had the intended new site specificity [108]. These results and those with the LHE I-MsoI demonstrate that specificity switches for multiple concerted base pair substitutions can be computationally designed, and that iteration between design and structure determination provides a route to large-scale specificity reprogramming [109].

Another approach to increasing the LHE specificity is by mining naturally occurring enzymes with different target sites that can be used as alternate platforms for reengineering [110–112]. A useful LAGLIDADG HE database and engineering server is LAHEDES http://homingendonuclease.net/.

5.3 Nicking Endonucleases

Another focus of attention has been enzymes that perform single-strand DNA cleavage rather than DSBs. Sequential cleavage among the GIY-YIG endonucleases [113, 114] and nicking among the HNH enzymes [115] were shown to result in gene conversion events. Later, LHEs were converted into nickases (reviewed by Chan et al. [116]). A nicking variant of I-AniI has been generated, that stimulates site-specific homologous recombination [117]. These nick-induced recombination events have two distinct and important advantages: they protect against error-prone nonhomologous end-joining reactions (Fig. 1b), and they reduce cellular toxicity [118].

5.4 Hybrid Endonucleases

Creation of hybrid endonucleases provides a different avenue to expanding HE specificity, such that the chimeric enzymes recognize corresponding hybrid target sites. This approach has been useful for both LHEs [119–122] and GIY-YIG enzymes [65] (Fig. 3). As new LHE nucleases are identified, closely related enzymes (e.g. from the OnuI family) can be easily combined to refine recognition specificity [123] or redesigned LHEs can undergo domain swapping (e.g. between engineered I-DmoI and I-CreI) to further expand specificities [124] (Fig. 3a). More adventuresome chimeras, true to the origin of the term in Greek mythology, have been constructed, as exemplified by a variant with a catalytically inactive LHE I-SceI fused to the restriction enzyme PvuII as the cleavage module [125] (Fig. 3b). Other restriction enzymes, particularly FokI, have been used as the cleavage domain with both ZF nucleases and TALENs (Fig. 2b). These ZFN and TALEN modular technologies are currently marketed by Addgene, Sangamo Biosciences, and Sigma-Aldrich.

The need for dimer formation by FokI and PvuII restricts cleavage to symmetrical target sequences (Figs. 2b and 3b). This limitation has been relieved by designing monomeric hybrid enzymes with the catalytic domain of GIY-YIG HE I-TevI fused to both ZF and inactive LHE scaffolds (Fig. 3c). Both the Tev-ZF and Tev-LHE monomers can induce site-specific DSBs and induce recombination in yeast [126].

Coupling designer endonucleases *in trans* with DNA end-processing enzymes is a recent strategy employed to drive productive homologous DNA repair pathways, in favor of unproductive nonhomologous end-joining [127]. This approach with TREX2, the 3′ repair exonuclease, has yielded improved targeted gene disruption in several different cell lines with ZFNs, TALENs, and an engineered I-CreI LHE.

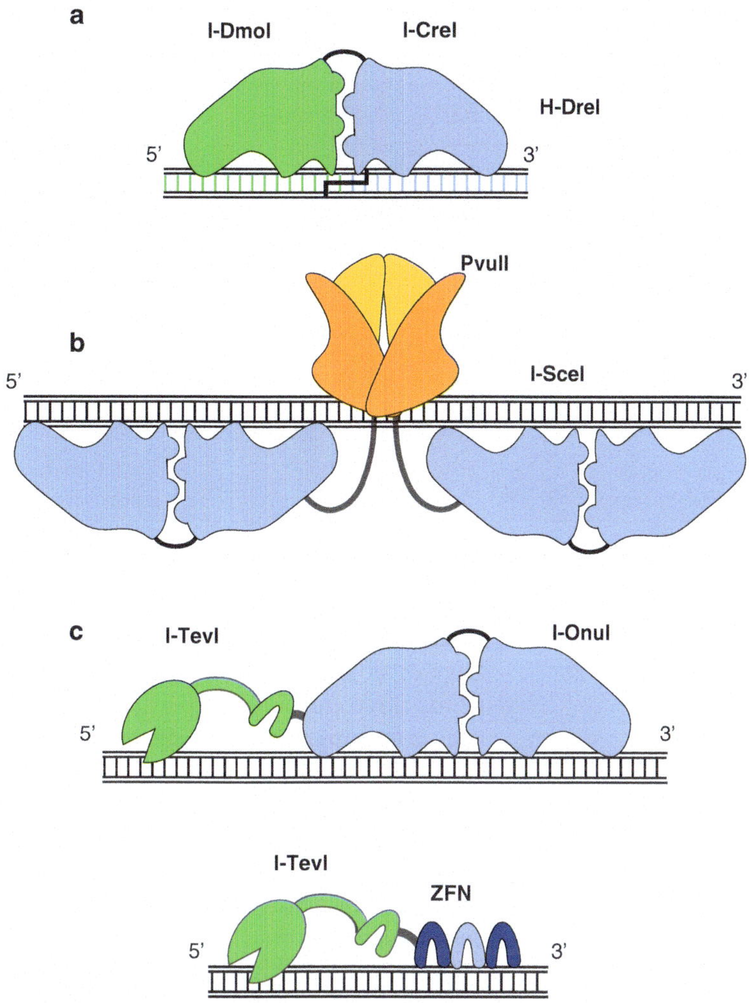

Fig. 3 Homing endonuclease fusions. (**a**) LAGLIDADG Fusion. An I-DmoI/I-CreI fusion is depicted cleaving a hybrid target site. (**b**) I-SceI-PvuII fusion is shown. (**c**) GIY-YIG fusions. The catalytic domain of I-TevI is fused with I-OnuI or zinc-finger units as DNA binding modules

6 Applications of HEs

Designer nucleases, including the HEs, ZFNs, and TALENs, are being used to modify the genomes of viruses, bacteria, yeast, plant, insect, and human cells, and they are revolutionizing genome engineering. Our focus here is on the HEs, which allow insertion, deletion, single-site mutation, and correction, in a highly site-specific and controlled fashion (Fig. 1b). A brief comparison of these different genome engineering tools will follow (Table 2).

Table 2
Comparison of protein-targeted genome engineering

	HE	ZFN	TALEN
DNA ends at DSB	3′ extension	5′ extension	5′ extension
Sequence availability	Limited by existing enzymes	Limited by triplet recognition of ZFs	Limited by first nucleotide (T)
Context effects	Common	Common	Uncommon
Specificity	High	Intermediate	High
Off-target effects	Low	Intermediate	Low
Enzyme engineering	Complex	Intermediate	Straightforward
Size compactness (modularity)	Small (monomeric)	Large (monomeric and dimeric)	Very large (dimeric)

6.1 HEs as Therapeutic Agents

Therapies built around HEs are still in their infancy (reviewed by Silva et al. and Marcaida et al. [128, 129]). Although the hurdles facing this technology are considerable, control over HE-engineered cells is far superior to random transgenics, as with viral vectors that can integrate indiscriminately causing off-target events. However, there is a trade-off of the relative safety of HEs; that is, efficiency on the scale of viral vectors will require not only versatile nuclease engineering for site specificity but also enhanced homologous recombination frequency. By using the aforementioned TREX2-coupling, a sevenfold increase in gene disruption of the endogenous HIV coreceptor *CCR5* was observed with an engineered I-CreI LHE over the uncoupled nuclease [127]. Targeted disruption of the *CCR5* with a ZFN is being tested in clinical trials as a therapeutic modality against HIV [130]. Additionally, LHE I-AniI is being used in a new therapeutic approach to cure cells of latent HIV infection by targeted mutagenesis of essential viral genes of the provirus [131].

Soon after the discovery of meganuclease-induced DSB repair, LHEs were used to stimulate recombination in mammalian cells [132]. This led to the notion of repairing defective genes, with potential application to patients with monogenic diseases, and particularly those that can be treated ex vivo. Among these are blood disorders, including thalasemas, porphyria, hemophilia, leukemia, skin ailments such as Xeroderma pigmentosum and melanoma, and immunodeficiencies [129]. Progress has already been made with engineered I-CreI targeted for correction of the XPC1 gene in cells from Xeroderma pigmentosum patients [133, 134]. Similar approaches have been used to tailor I-CreI for correction of mutation of the RAG1 gene in severe combined immunodeficiency disease (SCID) [135, 136]. The relative safety of gene correction for primary immunodeficiencies is an attractive feature of HE mediated gene therapy [137]. Similarly an I-CreI variant has been engineered to correct a dystrophin gene in Duchenne muscular dystrophy [138].

6.2 HEs in Insect Vector Control

An exciting potential application of HEGs is to effect insect vector control by promoting the spread of engineered insects through populations, and thereby curtailing spread of such diseases as malaria. Attempts at spreading deleterious mutations through vector populations using HE technology have therefore been initiated. The feasibility of the method has been demonstrated in both a *Drosophila melanogaster* model and in *Anopheles gambia* mosquitoes. In the fly model, using the LHE I-SceI it was shown that high rates of homing can be achieved within spermatogonia and in the female germline for successful deployment of a HEG-based gene drive strategy [139]. Likewise, it was demonstrated with the same HEG that a synthetic genetic element, consisting of mosquito regulatory regions, can increase transmission to progeny in transgenic *Anopheles* cage populations. Again, expression of I-SceI in the male germline induces high rates of site-specific gene conversion, resulting in the I-SceI gene acquisition that accounts for the observed genetic drive [140]. Similarly the I-PpoI gene in male mosquitoes was capable of introducing high levels of infertility in target populations in cage trials [141]. Through models of mosquito population genetics and malaria epidemiology combined with currently available HEG transmission data, it was concluded that HEG-based approaches could have a transformational effect on malaria control [142].

6.3 HEs in Agriculture

Recently, LHE I-CreI has been modified for agricultural applications in maize. The endonuclease gene was delivered to immature embryos to generate transgenic plants, with deletions and insertions detected at the HE cut site [143]. To improve prospects of nuclease-mediated improvement of plants, multigene plant transformation vectors have recently been constructed, with a cloning system based on ZFNs and HEs [144]. Viral vectors are also available for endonuclease delivery as a novel approach to plant engineering [145]. Thus HEs are being tailored for lofty applications in medicine, public health, and agronomy.

7 Comparison of Protein Targeting Technologies for Genome Engineering

TALENs (Fig. 2b) have been touted as "Genomic Cruise Missiles" as one of Science magazines breakthroughs of 2012. The striking structures of the TALENs explains their modular specificity [146, 147]. Nevertheless, TALENs like ZFNs and meganucleases must be engineered for each new DNA target, all with both targeting and cleavage constraints. The benefits and drawbacks of ZFNs, TALENs, and HEs for treatment of chronic viral infections such as HIV, hepatitis B, and Herpes Simplex viruses have been considered [148]. The difference among these three genome engineering tools is basically as follows (Table 2): First, the ZFNs and TALENs that exist as fusions to FokI generate DSBs with 5′ extensions, whereas HEs generate nicks or 3′ extensions. Second, since TALEN modules recognize single bases, they access broader sequence space than ZF modules which target nucleotide triplets, and as a result not all specific nucleotides are recognized. On the other hand, HEs are limited by the need to engineer naturally occurring enzymes, but these have an increasing cleavage repertoire as more enzymes are discovered. Third, these foregoing considerations result in higher cleavage specificity for the HEs and TALENs than for ZFNs. Fourth, context effects of ZFNs and HEs are more common than with TALENs [149]. Fifth, because of these specificity differences, off-target cleavage can be problematic with ZFNs and to a lesser extent with TALENs, whereas HEs tend to be most specific. Off-target effects not only can result in undesirable endpoints, but also can induce toxicity in the cell. Sixth, targeting the modular ZFNs and TALENs is technically more straightforward than redirecting the specificity of HEs, where the DNA-binding and cleavage determinants reside in the same molecule. Finally, this coexistence of specificity and cleavage functions of the HEs accounts for their compactness relative to the modular ZFNs and TALENs, a great advantage for application, particularly vectorization. For the above reasons, the pros and cons of the three types of protein endonucleases need to be considered for any particular application, and one needs to keep abreast of emerging subtleties, such as a recent report that associates different mutation profiles with ZFNs and TALENs [150].

8 RNA as a Player in Targeted Genome Editing

The challenge to protein endonuclease technology may come from readily programmable RNAs. Group II introns were the first RNAs to be used for gene targeting (reviewed by Lambowitz and Zimmerly and Cui and Davis [78, 151]). These introns themselves recognize their native DNA targets in an aforementioned retromobility reaction by an exon binding sequence (EBS) base-pairing with the DNA over ~14 bp. By reprogramming the intron EBS, highly site-specific retargeting occurs. Custom services for specific targetrons are available for use in several bacterial systems from both Sigma-Aldrich and Targetronics, although barriers remain in eukaryotes [78]. The targeting RNA must be in the form of a ribonucleoprotein, and entry into the nucleus remains the major obstacle. Nevertheless group II introns have the potential to not only integrate into DNA but also to make site-specific DSBs.

A new RNA-based approach for precise and efficient gene targeting that utilizes CRISPR/Cas9 derived from a bacterial immunity system [152, 153] is rapidly bursting on the scene [154–158]. The uniqueness of these RNA-guided endonucleases (RGENs) is based on designing guide RNAs that make the editing process highly flexible and facile, since the protein nuclease does not require engineering for retargeting. The bacterial CRISPR/Cas9 system has been successfully modified to accommodate mammalian and zebrafish transcription and translation requirements and kits are already being offered (by Addgene). Potential constraints of the CRISPR/Cas9 approach are the requirement for an adjacent proto-spacer sequence (an NGG triplet), genomic DNA accessibility due to chromatin and methylation states, and RNA secondary structure. However, a useful feature of the CRISPR/Cas9 engineering tool and one that distinguishes it from the other genome targeting technologies at this point is its utility for multiplexed editing. This versatility offers simultaneous change of several sites within a single genome.

Ironically, in both these cases of RNA-based targeting, the protein clippers are members of the HNH HE family. The technology is coming full circle, and developing at an extraordinary pace: >40 years since the genetic effects of ω were observed, 10 years since ZFNs were developed, 2 years since TALENs were offered to the world, and months since RGENs were discovered and spread through the literature, research, and corporate laboratories alike.

9 Summary

The HE field has indeed come a long way over 40 years, from the discovery of a robust recombination event that accounted for non-Mendelian inheritance in yeast mitochondria, through a detailed understanding of the structure and function of the endonuclease family responsible for the event, followed by clever manipulation of DNA cleavage enzymes used to edit genomes in highly targeted fashion. Tailoring endonuclease specificity has found broad application that started with research in genome engineering of bacterial, fungal, and mammalian cells and is now being used in the fields of agriculture and human health. Endonucleases are not only being groomed as anti-viral agents, but also for gene therapy of monogenic diseases, while the prospect of controlling malaria mosquitoes in sub-Saharan Africa, using the very enzyme that led to the discovery of HEs, becomes ever more promising.

Acknowledgements

We thank Matt Stanger for preparing the figures, Barry Stoddard for providing the images for Fig. 1a, and Rebecca McCarthy for preparing the manuscript. Research in the Belfort lab is supported by NIH grants GM39422 and GM44844.

References

1. Netter P, Petrochilo E, Slonimski PP, Bolotin-Fukuhara M, Coen D, Deutsch J, Dujon B (1974) Mitochondrial genetics. VII. Allelism and mapping studies of ribosomal mutants resistant to chloramphenicol, erythromycin and spiramycin in *S. cerevisiae*. Genetics 78:1063–1100
2. Bolotin M, Coen D, Deutsch J, Dujon B, Netter P, Petrochilo E, Slonimski PP (1971) La recombinaison des mitochondries chez Saccharomyces cerevisiae. Bull Inst Pasteur 69:215–239
3. Bos JL, Heyting C, Borst P, Arnberg AC, van Bruggen EFJ (1978) An insert in the single gene for the large ribosomal RNA in yeast mitochondrial DNA. Nature 275:336–338
4. Faye G, Dennebouy N, Kujawa C, Jacq C (1979) Inserted sequence in the mitochondrial 23S ribosomal RNA gene of the yeast Saccharomyces cerevisiae. Mol Gen Genet 168:101–109
5. Jacquier A, Dujon B (1985) An intron-encoded protein is active in a gene conversion process that spreads an intron into a mitochondrial gene. Cell 41:383–394
6. Lambowitz AM, Belfort M (1993) Introns as mobile genetic elements. Annu Rev Biochem 62:587–622
7. Szostak JW, Orr-Weaver TL, Rothstein RJ, Stahl FW (1983) The double-strand-break repair model for recombination. Cell 33:25–35
8. Belfort M (1990) Phage T4 introns: self-splicing and mobility. Annu Rev Genet 24:363–385
9. Dalgaard JZ, Garrett RA, Belfort M (1993) A site-specific endonuclease encoded by a typical archaeal intron. Proc Natl Acad Sci USA 90:5414–5417
10. Kane PM, Yamashiro CT, Wolczyk DF, Neff N, Goebl M, Stevens TH (1990) Protein splicing converts the yeast TFP1 gene product to the 69-kD subunit of the vacuolar H(+)-adenosine triphosphatase. Science 250:651–657
11. Hirata R, Oshumi Y, Nakano A, Kawasaki H, Suzuki K, Anraku Y (1990) Molecular structure of a gene, VMA1, encoding the catalytic subunit of H+-translocating adenosine triphosphatase from vacuolar membranes of Saccharomyces cerevisiae. J Biol Chem 265:6726–6733
12. Dalgaard JZ, Klar A, Moser MJ, Holley WR, Chatterjee A, Mian IS (1997) Statistical modeling and analysis of the LAGLIDADG family of site-specific endonucleases and identification of an intein that encodes a site-specific endonuclease of the H-N-H family. Nucleic Acids Res 25:4626–4638
13. Sharma M, Ellis RL, Hinton DM (1992) Identification of a family of bacteriophage T4 genes encoding proteins similar to those present in group I introns of fungi and phage. Proc Natl Acad Sci USA 89:6658–6662
14. Edgell DR, Shub DA (2001) Related homing endonucleases I-BmoI and I-TevI use different strategies to cleave homologous recognition sites. Proc Natl Acad Sci USA 98:7898–7903
15. Mota EM, Collins RA (1988) Independent evolution of structural and coding regions in a *Neurospora* mitochondrial intron. Nature 332:654–656
16. Quirk SM, Bell-Pedersen D, Belfort M (1989) Intron mobility in the T-Even phages: high frequency inheritance of group I introns promoted by intron open reading frames. Cell 56:455–465
17. Bonocora RP, Shub DA (2009) A likely pathway for formation of mobile group I introns. Curr Biol 19:223–228
18. Macreadie IG, Scott RM, Zinn AR, Butow RA (1985) Transposition of an intron in yeast mitochondria requires a protein encoded by that intron. Cell 41:395–402
19. Taylor GK, Stoddard BL (2012) Structural, functional and evolutionary relationships between homing endonucleases and proteins from their host organisms. Nucleic Acids Res 40:5189–5200

20. Haber JE, Wolfe KH (2005) Function and evolution of HO and VDE endonucleases in fungi. In: Belfort M, Derbyshire V, Stoddard BL, Wood DW (eds) Homing endonucleases and inteins. Springer, Berlin, pp 161–175
21. Pietrokovski S (1994) Conserved sequence features of inteins (protein introns) and their use in identifying new inteins and related proteins. Protein Sci 3:2340–2350
22. Gibb EA, Edgell DR (2010) Better late than never: delayed translation of intron-encoded endonuclease I-TevI is required for efficient splicing of its host group I intron. Mol Microbiol 78:35–46
23. Stoddard BL, Belfort M (2010) Social networking between mobile introns and their host genes. Mol Microbiol 78:1–4
24. Goddard MR, Burt A (1999) Recurrent invasion and extinction of a selfish gene. Proc Natl Acad Sci USA 96:13880–13885
25. Gimble FS (2000) Invasion of a multitude of genetic niches by mobile endonuclease genes. FEMS Microbiol Lett 185:99–107
26. Gogarten JP, Hilario E (2006) Inteins, introns, and homing endonucleases: recent revelations about the life cycle of parasitic genetic elements. BMC Evol Biol 6:94
27. Liu Q, Belle A, Shub DA, Belfort M, Edgell DR (2003) SegG endonuclease promotes marker exclusion and mediates co-conversion from a distant cleavage site. J Mol Biol 334:13–23
28. Zeng Q, Bonocora RP, Shub DA (2009) A free-standing homing endonuclease targets an intron insertion site in the psbA gene of cyanophages. Curr Biol 19:218–222
29. Bonocora RP, Zeng Q, Abel EV, Shub DA (2011) A homing endonuclease and the 50-nt ribosomal bypass sequence of phage T4 constitute a mobile DNA cassette. Proc Natl Acad Sci USA 108:16351–16356
30. Loizos N, Tillier ERM, Belfort M (1994) Evolution of mobile group I introns: recognition of intron sequences by an intron-encoded endonuclease. Proc Natl Acad Sci USA 91:11983–11987
31. Edgell DR, Stanger MJ, Belfort M (2004) Coincidence of cleavage sites of intron endonuclease I-TevI and critical sequences of the host thymidylate synthase gene. J Mol Biol 343:1231–1241
32. Koufopanou V, Goddard MR, Burt A (2002) Adaptation for horizontal transfer in a homing endonuclease. Mol Biol Evol 19:239–246
33. Stoddard BL (2011) Homing endonucleases: from microbial genetic invaders to reagents for targeted DNA modification. Structure 19:7–15
34. Paques F, Duchateau P (2007) Meganucleases and DNA double-strand break-induced recombination: perspectives for gene therapy. Curr Gene Ther 7:49–66
35. Colleaux L, D'Auriol L, Galibert F, Dujon B (1988) Recognition and cleavage site of the intron-encoded *omega* transposase. Proc Natl Acad Sci USA 85:6022–6026
36. Bryk M, Quirk SM, Mueller JE, Loizos N, Lawrence C, Belfort M (1993) The *td* intron endonuclease makes extensive sequence tolerant contacts across the minor groove of its DNA target. EMBO J 12:2141–2149
37. Pingoud A, Jeltsch A (2001) Structure and function of type II restriction endonucleases. Nucleic Acids Res 29:3705–3727
38. Zhao L, Bonocora RP, Shub DA, Stoddard BL (2007) The restriction fold turns to the dark side: a bacterial homing endonuclease with a PD-(D/E)-XK motif. EMBO J 26:2432–2442
39. Jurica MS, Monnat RJ Jr, Stoddard BL (1998) DNA recognition and cleavage by the LAGLIDADG homing endonuclease I-*Cre*I. Mol Cell 2:469–476
40. Van Roey P, Waddling CA, Fox KM, Belfort M, Derbyshire V (2001) Intertwined structure of the DNA-binding domain of intron endonuclease I-TevI with its substrate. EMBO J 20:3631–3637
41. Silva G, Dalgaard JZ, Belfort M, Van Roey P (1999) Crystal structure of the thermostable archaeal intron-encoded endonuclease I-*Dmo*I. J Mol Biol 286:1123–1136
42. Duan X, Gimble FS, Quiocho FA (1997) Crystal structure of PI-SceI, a homing endonuclease with protein splicing activity. Cell 89:555–564
43. Ichiyanagi K, Ishino Y, Ariyoshi M, Komori K, Morikawa K (2000) Crystal structure of an archaeal intron-encoded homing endonuclease PI-PfuI. J Mol Biol 300:889–901
44. Heath PJ, Stephens KM, Monnat RJ Jr, Stoddard BL (1997) The structure of I-*Cre*I, a group I intron-encoded homing endonuclease. Nat Struct Biol 4:468–476
45. Gimble FS, Stephens BW (1995) Substitutions in conserved dodecapeptide motifs that uncouple the DNA binding and DNA cleavage activities of PI-*Sce*I endonuclease. J Biol Chem 270:5849–5856

46. Schottler S, Wende W, Pingoud V, Pingoud A (2000) Identification of Asp218 and Asp326 as the principal Mg2+ binding ligands of the homing endonuclease PI-SceI. Biochemistry 39:15895–15900
47. Chevalier BS, Monnat RJ Jr, Stoddard BL (2001) The homing endonuclease I-CreI uses three metals, one of which is shared between the two active sites. Nat Struct Biol 8:312–316
48. Chevalier B, Sussman D, Otis C, Noel AJ, Turmel M, Lemieux C, Stephens K, Monnat RJ Jr, Stoddard BL (2004) Metal-dependent DNA cleavage mechanism of the I-CreI LAGLIDADG homing endonuclease. Biochemistry 43:14015–14026
49. Klar AJ (2010) The yeast mating-type switching mechanism: a memoir. Genetics 186:443–449
50. Bolduc JM, Spiegel PC, Chatterjee P, Brady KL, Downing ME, Caprara MG, Waring RB, Stoddard BL (2003) Structural and biochemical analyses of DNA and RNA binding by a bifunctional homing endonuclease and group I intron splicing factor. Genes Dev 17:2875–2888
51. Ho Y, Kim SJ, Waring RB (1997) A protein encoded by a group I intron in *Aspergillus nidulans* directly assists RNA splicing and is a DNA endonuclease. Proc Natl Acad Sci USA 94:8994–8999
52. Wenzlau JM, Saldanha RJ, Butow RA, Perlman PS (1989) A latent intron-encoded maturase is also an endonuclease needed for intron mobility. Cell 56:421–430
53. Belfort M (2003) Two for the price of one: a bifunctional intron-encoded DNA endonuclease-RNA maturase. Genes Dev 17:2860–2863
54. Szczepanek T, Lazowska J (1996) Replacement of two non-adjacent amino acids in the *S. cerevisiae* bi2 intron-encoded RNA maturase is sufficient to gain a homing-endonuclease activity. EMBO J 15:3758–3767
55. Chatterjee P, Brady KL, Solem A, Ho Y, Caprara MG (2003) Functionally distinct nucleic acid binding sites for a group I intron encoded RNA maturase/DNA homing endonuclease. J Mol Biol 329:239–251
56. Geese WJ, Kwon YK, Wen X, Waring RB (2003) *In vitro* analysis of the relationship between endonuclease and maturase activities in the bi-functional group I intron-encoded protein, I-AniI. Eur J Biochem 270:1543–1554
57. Knizewski L, Ginalski K (2007) Bacterial DUF199/COG1481 proteins including sporulation regulator WhiA are distant homologs of LAGLIDADG homing endonucleases that retained only DNA binding. Cell Cycle 6:1666–1670
58. Kaiser BK, Clifton MC, Shen BW, Stoddard BL (2009) The structure of a bacterial DUF199/WhiA protein: domestication of an invasive endonuclease. Structure 17:1368–1376
59. Kaiser BK, Stoddard BL (2011) DNA recognition and transcriptional regulation by the WhiA sporulation factor. Sci Rep 1:156
60. Kowalski JC, Belfort M, Stapleton MA, Holpert M, Dansereau JT, Pietrokovski S, Baxter SM, Derbyshire V (1999) Configuration of the catalytic GIY-YIG domain of intron endonuclease I-*Tev*I: coincidence of computational and molecular findings. Nucleic Acids Res 27:2115–2125
61. Dunin-Horkawicz S, Feder M, Bujnicki JM (2006) Phylogenomic analysis of the GIY-YIG nuclease superfamily. BMC Genomics 7:98
62. Van Roey P, Meehan L, Kowalski J, Belfort M, Derbyshire V (2002) Catalytic domain structure and hypothesis for function of GIY-YIG intron endonuclease I-TevI. Nat Struct Biol 9:806–811
63. Derbyshire V, Kowalski JC, Dansereau JT, Hauer CR, Belfort M (1997) Two-domain structure of the *td* intron-encoded endonuclease I-TevI correlates with the two-domain configuration of the homing site. J Mol Biol 265:494–506
64. Liu QQ, Dansereau JT, Puttamadappa SS, Shekhtman A, Derbyshire V, Belfort M (2008) Role of the interdomain linker in distance determination for remote cleavage by homing endonuclease I-TevI. J Mol Biol 379:1094–1106
65. Liu Q-Q, Derbyshire V, Belfort M, Edgell DR (2006) Distance determination by GIY-YIG intron endonucleases: discrimination between repression and cleavage functions. Nucleic Acids Res 34:1755–1764
66. Bell-Pedersen D, Quirk SM, Bryk M, Belfort M (1991) I-TevI, the endonuclease encoded by the mobile *td* intron, recognizes binding and cleavage domains on its DNA target. Proc Natl Acad Sci USA 88:7719–7723
67. Belle A, Landthaler M, Shub DA (2002) Intronless homing: site-specific endonuclease SegF of bacteriophage T4 mediates localized marker exclusion analogous to homing endonucleases of group I introns. Genes Dev 16:351–362

68. Mak AN, Lambert AR, Stoddard BL (2010) Folding, DNA recognition, and function of GIY-YIG endonucleases: crystal structures of R.Eco29kI. Structure 18:1321–1331
69. Sokolowska M, Czapinska H, Bochtler M (2011) Hpy188I-DNA pre- and post-cleavage complexes–snapshots of the GIY-YIG nuclease mediated catalysis. Nucleic Acids Res 39:1554–1564
70. Pyatkov KI, Arkhipova IR, Malkova NV, Finnegan DJ, Evgen'ev MB (2004) Reverse transcriptase and endonuclease activities encoded by Penelope-like retroelements. Proc Natl Acad Sci USA 101:14719–14724
71. Edgell DR, Derbyshire V, Van Roey P, LaBonne S, Stanger MJ, Li Z, Boyd TM, Shub DA, Belfort M (2004) Intron-encoded homing endonuclease I-TevI also functions as a transcriptional autorepressor. Nat Struct Mol Biol 11:936–944
72. Ko T-P, Liao C-C, Ku W-Y, Chak K-F, Yuan HS (1999) The crystal structure of the DNase domain of colicin E7 in complex with its inhibitor Im7 protein. Structure 7:91–102
73. Kleanthous C, Kuhlmann UC, Pommer AJ, Ferguson N, Radford SE, Moore GR, James R, Hemmings AM (1999) Structural and mechanistic basis of immunity toward endonuclease colicins. Nat Struct Biol 6: 243–252
74. Friedhoff P, Franke I, Meiss G, Wende W, Krause KL, Pingoud A (1999) A similar active site for non-specific and specific endonucleases. Nat Struct Biol 6:112–113
75. Goodrich-Blair H, Shub DA (1996) Beyond homing: competition between intron endonucleases confers a selective advantage on flanking genetic markers. Cell 84:211–221
76. Landthaler M, Begley U, Lau NC, Shub DA (2002) Two self-splicing group I intron in the ribonucleotide reductase large subunit of Staphylococcus aureus phage Twort. Nucleic Acids Res 30:1935–1943
77. Robbins JB, Stapleton M, Smith D, Dansereau JT, Derbyshire V, Belfort M (2007) Homing endonuclease I-TevIII: dimerization as a means to a double strand break. Nucleic Acids Res 35:1589–1600
78. Lambowitz AM, Zimmerly S (2011) Group II introns: mobile ribozymes that invade DNA. Cold Spring Harb Perspect Biol 3:a003616
79. Gorbalenya AE (1994) Self-splicing group I and group II introns encode homologous (putative) DNA endonucleases of a new family. Protein Sci 3:1117–1120
80. Shub DA, Goodrich-Blair H, Eddy SR (1994) Amino acid sequence motif of group I intron endonucleases is conserved in open reading frames of group II introns. Trends Biochem Sci 19:402–404
81. Raaijmakers H, Toro I, Birkenbihl R, Kemper B, Suck D (2001) Conformational flexibility in T4 endonuclease VII revealed by crystallography: implications for substrate binding and cleavage. J Mol Biol 308:311–323
82. Bujnicki JM, Radlinska M, Rychlewski L (2001) Polyphyletic evolution of type II restriction enzymes revisited: two independent sources of second-hand folds revealed. Trends Biochem Sci 26:9–11
83. Wiedenheft B, Sternberg SH, Doudna JA (2012) RNA-guided genetic silencing systems in bacteria and archaea. Nature 482:331–338
84. Gasiunas G, Barrangou R, Horvath P, Siksnys V (2012) Cas9-crRNA ribonucleoprotein complex mediates specific DNA cleavage for adaptive immunity in bacteria. Proc Natl Acad Sci USA 109:E2579–E2586
85. Johansen S, Embley TM, Willassen NP (1993) A family of nuclear homing endonucleases. Nucleic Acids Res 21:4405
86. Flick KE, Jurica MS, Monnat RJ Jr, Stoddard BL (1998) DNA binding and cleavage by the nuclear intron-encoded homing endonuclease I-*PpoI*. Nature 394:96–101
87. Kuhlmann UC, Moore GR, James R, Kleanthous C, Hemmings AM (1999) Structural parsimony in endonuclease active sites: should the number of homing endonuclease families be redefined? FEBS Lett 463:1–2
88. Bujnicki JM (2003) Crystallographic and bioinformatic studies on restriction endonucleases: inference of evolutionary relationships in the "midnight zone" of homology. Curr Protein Pept Sci 4:327–337
89. Orlowski J, Boniecki M, Bujnicki JM (2007) I-Ssp6803I: the first homing endonuclease from the PD-(D/E)XK superfamily exhibits an unusual mode of DNA recognition. Bioinformatics 23:527–530
90. Bonocora RP, Shub DA (2001) A novel group I intron-encoded endonuclease specific for the anticodon region of tRNA(fMet) genes. Mol Microbiol 39:1299–1306
91. Zhao L, Pellenz S, Stoddard BL (2009) Activity and specificity of the bacterial PD-(D/E)XK homing endonuclease I-Ssp6803I. J Mol Biol 385:1498–1510
92. Steczkiewicz K, Muszewska A, Knizewski L, Rychlewski L, Ginalski K (2012) Sequence,

structure and functional diversity of PD-(D/E)XK phosphodiesterase superfamily. Nucleic Acids Res 40:7016–7045

93. Rusch DB, Halpern AL, Sutton G, Heidelberg KB, Williamson S, Yooseph S, Wu D, Eisen JA, Hoffman JM, Remington K et al (2007) The Sorcerer II Global Ocean Sampling expedition: northwest Atlantic through eastern tropical Pacific. PLoS Biol 5:e77

94. Yooseph S, Sutton G, Rusch DB, Halpern AL, Williamson SJ, Remington K, Eisen JA, Heidelberg KB, Manning G, Li W et al (2007) The Sorcerer II Global Ocean sampling expedition: expanding the universe of protein families. PLoS Biol 5:e16

95. Taylor GK, Heiter DF, Pietrokovski S, Stoddard BL (2011) Activity, specificity and structure of I-Bth0305I: a representative of a new homing endonuclease family. Nucleic Acids Res 39:9705–9719

96. Dassa B, London N, Stoddard BL, Schueler-Furman O, Pietrokovski S (2009) Fractured genes: a novel genomic arrangement involving new split inteins and a new homing endonuclease family. Nucleic Acids Res 37:2560–2573

97. Porteus MH, Carroll D (2005) Gene targeting using zinc finger nucleases. Nat Biotechnol 23:967–973

98. Mussolino C, Morbitzer R, Lutge F, Dannemann N, Lahaye T, Cathomen T (2011) A novel TALE nuclease scaffold enables high genome editing activity in combination with low toxicity. Nucleic Acids Res 39:9283–9293

99. Choulika A, Perrin A, Dujon B, Nicolas JF (1995) Induction of homologous recombination in mammalian chromosomes by using the I-*SceI* system of *Saccharomyces cerevisiae*. Mol Cell Biol 15:1968–1973

100. Rouet P, Smih F, Jasin M (1994) Introduction of double-strand breaks into the genome of mouse cells by expression of a rare-cutting endonuclease. Mol Cell Biol 14:8096–8106

101. Seligman LM, Chisholm KM, Chevalier BS, Chadsey MS, Edwards ST, Savage JH, Veillet al (2002) Mutations altering the cleavage specificity of a homing endonuclease. Nucleic Acids Res 30:3870–3879

102. Gruen M, Chang K, Serbanescu I, Liu DR (2002) An *in vivo* selection system for homing endonuclease activity. Nucleic Acids Res 30:e29

103. Gimble FS, Moure CM, Posey KL (2003) Assessing the plasticity of DNA target site recognition of the PI-SceI homing endonuclease using a bacterial two-hybrid selection system. J Mol Biol 334:993–1008

104. Chames P, Epinat JC, Guillier S, Patin A, Lacroix E, Paques F (2005) In vivo selection of engineered homing endonucleases using double-strand break induced homologous recombination. Nucleic Acids Res 33:e178

105. Jarjour J, West-Foyle H, Certo MT, Hubert CG, Doyle L, Getz MM, Stoddard BL, Scharenberg AM (2009) High-resolution profiling of homing endonuclease binding and catalytic specificity using yeast surface display. Nucleic Acids Res 37:6871–6880

106. Arnould S, Chames P, Perez C, Lacroix E, Duclert A, Epinat JC, Stricher F, Petit AS, Patin A, Guillier S et al (2006) Engineering of large numbers of highly specific homing endonucleases that induce recombination on novel DNA targets. J Mol Biol 355:443–458

107. Ashworth J, Havranek JJ, Duarte CM, Sussman D, Monnat R Jr, Stoddard BL, Baker D (2006) Computational redesign of endonuclease DNA binding and cleavage specificity. Nature 441:656–659

108. Ulge UY, Baker DA, Monnat RJ Jr (2011) Comprehensive computational design of mCreI homing endonuclease cleavage specificity for genome engineering. Nucleic Acids Res 39:4330–4339

109. Ashworth J, Taylor GK, Havranek JJ, Quadri SA, Stoddard BL, Baker D (2010) Computational reprogramming of homing endonuclease specificity at multiple adjacent base pairs. Nucleic Acids Res 38:5601–5608

110. Takeuchi R, Lambert AR, Mak AN, Jacoby K, Dickson RJ, Gloor GB, Scharenberg AM, Edgell DR, Stoddard BL (2011) Tapping natural reservoirs of homing endonucleases for targeted gene modification. Proc Natl Acad Sci USA 108:13077–13082

111. Jacoby K, Metzger M, Shen BW, Certo MT, Jarjour J, Stoddard BL, Scharenberg AM (2012) Expanding LAGLIDADG endonuclease scaffold diversity by rapidly surveying evolutionary sequence space. Nucleic Acids Res 40:4954–4964

112. Li H, Ulge UY, Hovde BT, Doyle LA, Monnat RJ Jr (2012) Comprehensive homing endonuclease target site specificity profiling reveals evolutionary constraints and enables genome engineering applications. Nucleic Acids Res 40:2587–2598

113. Mueller JE, Smith D, Bryk M, Belfort M (1995) Intron-encoded endonuclease I-TevI

binds as a monomer to effect sequential cleavage via conformational changes in the *td* homing site. EMBO J 14:5724–5735

114. Carter JM, Friedrich NC, Kleinstiver B, Edgell DR (2007) Strand-specific contacts and divalent metal ion regulate double-strand break formation by the GIY-YIG homing endonuclease I-BmoI. J Mol Biol 374:306–321

115. Landthaler M, Shub DA (2003) The nicking homing endonuclease I-BasI is encoded by a group I intron in the DNA polymerase gene of the Bacillus thuringiensis phage Bastille. Nucleic Acids Res 31:3071–3077

116. Chan SH, Stoddard BL, Xu SY (2011) Natural and engineered nicking endonucleases–from cleavage mechanism to engineering of strand-specificity. Nucleic Acids Res 39:1–18

117. McConnell Smith A, Takeuchi R, Pellenz S, Davis L, Maizels N, Monnat RJ Jr, Stoddard BL (2009) Generation of a nicking enzyme that stimulates site-specific gene conversion from the I-AniI LAGLIDADG homing endonuclease. Proc Natl Acad Sci USA 106:5099–5104

118. Metzger MJ, McConnell-Smith A, Stoddard BL, Miller AD (2011) Single-strand nicks induce homologous recombination with less toxicity than double-strand breaks using an AAV vector template. Nucleic Acids Res 39:926–935

119. Chevalier BS, Kortemme T, Chadsey MS, Baker D, Monnat RJ, Stoddard BL (2002) Design, activity, and structure of a highly specific artificial endonuclease. Mol Cell 10:895–905

120. Epinat JC, Arnould S, Chames P, Rochaix P, Desfontaines D, Puzin C, Patin A, Zanghellini A, Paques F, Lacroix E (2003) A novel engineered meganuclease induces homologous recombination in yeast and mammalian cells. Nucleic Acids Res 31:2952–2962

121. Silva GH, Belfort M (2004) Analysis of the LAGLIDADG interface of the monomeric homing endonuclease I-DmoI. Nucleic Acids Res 32:3156–3168

122. Silva GH, Belfort M, Wende W, Pingoud A (2006) From monomeric to homodimeric endonucleases and back: engineering novel specificity of LAGLIDADG enzymes. J Mol Biol 361:744–754

123. Baxter S, Lambert AR, Kuhar R, Jarjour J, Kulshina N, Parmeggiani F, Danaher P, Gano J, Baker D, Stoddard BL et al (2012) Engineering domain fusion chimeras from I-OnuI family LAGLIDADG homing endonucleases. Nucleic Acids Res 40:7985–8000

124. Grizot S, Epinat JC, Thomas S, Duclert A, Rolland S, Paques F, Duchateau P (2010) Generation of redesigned homing endonucleases comprising DNA-binding domains derived from two different scaffolds. Nucleic Acids Res 38:2006–2018

125. Fonfara I, Curth U, Pingoud A, Wende W (2012) Creating highly specific nucleases by fusion of active restriction endonucleases and catalytically inactive homing endonucleases. Nucleic Acids Res 40:847–860

126. Kleinstiver BP, Wolfs JM, Kolaczyk T, Roberts AK, Hu SX, Edgell DR (2012) Monomeric site-specific nucleases for genome editing. Proc Natl Acad Sci USA 109:8061–8066

127. Certo MT, Gwiazda KS, Kuhar R, Sather B, Curinga G, Mandt T, Brault M, Lambert AR, Baxter SK, Jacoby K et al (2012) Coupling endonucleases with DNA end-processing enzymes to drive gene disruption. Nat Methods 9:973–975

128. Silva G, Poirot L, Galetto R, Smith J, Montoya G, Duchateau P, Paques F (2011) Meganucleases and other tools for targeted genome engineering: perspectives and challenges for gene therapy. Curr Gene Ther 11:11–27

129. Marcaida MJ, Munoz IG, Blanco FJ, Prieto J, Montoya G (2010) Homing endonucleases: from basics to therapeutic applications. Cell Mol Life Sci 67:727–748

130. Cannon P, June C (2011) Chemokine receptor 5 knockout strategies. Curr Opin HIV AIDS 6:74–79

131. Aubert M, Ryu BY, Banks L, Rawlings DJ, Scharenberg AM, Jerome KR (2011) Successful targeting and disruption of an integrated reporter lentivirus using the engineered homing endonuclease Y2 I-AniI. PLoS One 6:e16825

132. Jasin M (1996) Genetic manipulation of genomes with rare-cutting endonucleases. Trends Genet 12:224–228

133. Arnould S, Perez C, Cabaniols JP, Smith J, Gouble A, Grizot S, Epinat JC, Duclert A, Duchateau P, Paques F (2007) Engineered I-CreI derivatives cleaving sequences from the human XPC gene can induce highly efficient gene correction in mammalian cells. J Mol Biol 371:49–65

134. Redondo P, Prieto J, Munoz IG, Alibes A, Stricher F, Serrano L, Cabaniols JP, Daboussi F, Arnould S, Perez C et al (2008) Molecular basis of xeroderma pigmentosum group C

DNA recognition by engineered meganucleases. Nature 456:107–111
135. Grizot S, Smith J, Daboussi F, Prieto J, Redondo P, Merino N, Villate M, Thomas S, Lemaire L, Montoya G et al (2009) Efficient targeting of a SCID gene by an engineered single-chain homing endonuclease. Nucleic Acids Res 37:5405–5419
136. Munoz IG, Prieto J, Subramanian S, Coloma J, Redondo P, Villate M, Merino N, Marenchino M, D'Abramo M, Gervasio FL et al (2011) Molecular basis of engineered meganuclease targeting of the endogenous human RAG1 locus. Nucleic Acids Res 39:729–743
137. Pessach IM, Notarangelo LD (2011) Gene therapy for primary immunodeficiencies: looking ahead, toward gene correction. J Allergy Clin Immunol 127:1344–1350
138. Chapdelaine P, Pichavant C, Rousseau J, Paques F, Tremblay JP (2010) Meganucleases can restore the reading frame of a mutated dystrophin. Gene Ther 17:846–858
139. Chan YS, Naujoks DA, Huen DS, Russell S (2011) Insect population control by homing endonuclease-based gene drive: an evaluation in Drosophila melanogaster. Genetics 188:33–44
140. Windbichler N, Menichelli M, Papathanos PA, Thyme SB, Li H, Ulge UY, Hovde BT, Baker D, Monnat RJ Jr, Burt A et al (2011) A synthetic homing endonuclease-based gene drive system in the human malaria mosquito. Nature 473:212–215
141. Klein TA, Windbichler N, Deredec A, Burt A, Benedict MQ (2012) Infertility resulting from transgenic I-PpoI male Anopheles gambiae in large cage trials. Pathog Glob Health 106:20–31
142. Deredec A, Godfray HC, Burt A (2011) Requirements for effective malaria control with homing endonuclease genes. Proc Natl Acad Sci USA 108:E874–E880
143. Gao H, Smith J, Yang M, Jones S, Djukanovic V, Nicholson MG, West A, Bidney D, Falco SC, Jantz D et al (2010) Heritable targeted mutagenesis in maize using a designed endonuclease. Plant J 61:176–187
144. Zeevi V, Liang Z, Arieli U, Tzfira T (2012) Zinc finger nuclease and homing endonuclease-mediated assembly of multigene plant transformation vectors. Plant Physiol 158:132–144
145. Vainstein A, Marton I, Zuker A, Danziger M, Tzfira T (2011) Permanent genome modifications in plant cells by transient viral vectors. Trends Biotechnol 29:363–369
146. Mak AN, Bradley P, Cernadas RA, Bogdanove AJ, Stoddard BL (2012) The crystal structure of TAL effector PthXo1 bound to its DNA target. Science 335:716–719
147. Deng D, Yan C, Pan X, Mahfouz M, Wang J, Zhu JK, Shi Y, Yan N (2012) Structural basis for sequence-specific recognition of DNA by TAL effectors. Science 335:720–723
148. Schiffer JT, Aubert M, Weber ND, Mintzer E, Stone D, Jerome KR (2012) Targeted DNA mutagenesis for the cure of chronic viral infections. J Virol 86:8920–8936
149. Meckler JF, Bhakta MS, Kim MS, Ovadia R, Habrian CH, Zykovich A, Yu A, Lockwood SH, Morbitzer R, Elsaesser J et al (2013) Quantitative analysis of TALE-DNA interactions suggests polarity effects. Nucleic Acids Res 41:4118–4128
150. Kim Y, Kweon J, Kim JS (2013) TALENs and ZFNs are associated with different mutation signatures. Nat Methods 10:185
151. Cui X, Davis G (2007) Mobile group II intron targeting: applications in prokaryotes and perspectives in eukaryotes. Front Biosci 12:4972–4985
152. Burgess DJ (2013) Technology: a CRISPR genome-editing tool. Nat Rev Genet 14:80
153. Jinek M, Chylinski K, Fonfara I, Hauer M, Doudna JA, Charpentier E (2012) A programmable dual-RNA-guided DNA endonuclease in adaptive bacterial immunity. Science 337:816–821
154. Mali P, Yang L, Esvelt KM, Aach J, Guell M, Dicarlo JE, Norville JE, Church GM (2013) RNA-guided human genome engineering via Cas9. Science 339:823–826
155. Cong L, Ran FA, Cox D, Lin S, Barretto R, Habib N, Hsu PD, Wu X, Jiang W, Marraffini LA et al (2013) Multiplex genome engineering using CRISPR/Cas systems. Science 339:819–823
156. Cho SW, Kim S, Kim JM, Kim JS (2013) Targeted genome engineering in human cells with the Cas9 RNA-guided endonuclease. Nat Biotechnol 31:230–232
157. Hwang WY, Fu Y, Reyon D, Maeder ML, Tsai SQ, Sander JD, Peterson RT, Yeh JR, Joung JK (2013) Efficient genome editing in zebrafish using a CRISPR-Cas system. Nat Biotechnol 31:227–229
158. Jiang W, Bikard D, Cox D, Zhang F, Marraffini LA (2013) RNA-guided editing of bacterial genomes using CRISPR-Cas systems. Nat Biotechnol 31:223–239

Chapter 2

Bioinformatic Identification of Homing Endonucleases and Their Target Sites

Eyal Privman

Abstract

Homing endonuclease genes (HEGs) are a large, phylogenetically diverse superfamily of enzymes with high specificity for especially long target sites. The public genomic sequence databases contain thousands of HEGs. This is a large and diverse arsenal of potential genome editing tools. To make use of this natural resource, one needs to identify candidate HEGs. Due to their special relationship with a host gene, it is also possible to predict their cognate target sequences. Here I describe the HomeBase algorithm that was developed to this end. A detailed description of the computational pipeline is provided with emphasis on technical and methodological caveats of the approach.

Key words Homing endonucleases, Homology search, Target-site prediction, HomeBase

1 Introduction

In this chapter, I describe a computational approach that identifies novel Homing endonuclease genes (HEGs) in nucleotide sequence databases, infers their native target sequence, and then generalizes this single target to a predicted range of possible targets. This approach was first used to construct the HomeBase collection [1], which is searchable using the HomeBase web server at http://homebase-search.tau.ac.il/. The first part of Subsection 3 describes the usage of the web server to search for HEGs in the existing HomeBase collection that are predicted to have targets within a given query DNA sequence. The second part describes running the HomeBase pipeline on a local computer in order to search for novel HEGs in a nucleotide sequence database.

The HomeBase pipeline involves first the inference of the native target sequence. This stage relies on the observation that the two halves of the target sequence are found in the exons/exteins flanking the intron/intein in which the HEG resides (Fig. 1). In the second stage HomeBase infers the range of nonnative targets that may be cleaved by the nuclease. This stage relies on the

David R. Edgell (ed.), *Homing Endonucleases: Methods and Protocols*, Methods in Molecular Biology, vol. 1123, DOI 10.1007/978-1-62703-968-0_2, © Springer Science+Business Media New York 2014

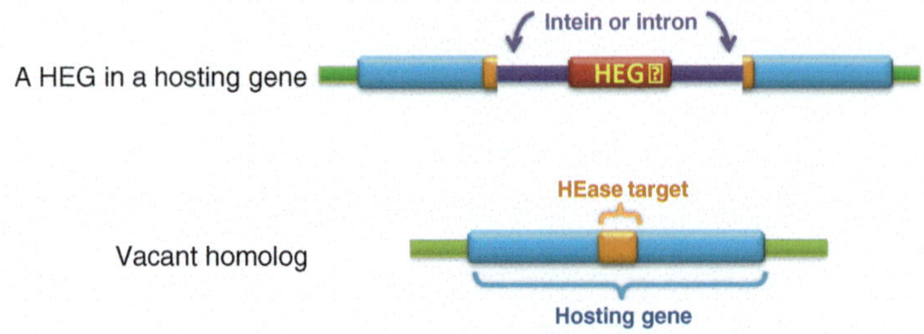

Fig. 1 A HEG residing in an intron/intein of a hosting gene compared to a homolog of the hosting gene that does not contain the intron/intein

observation that the long target sequences allow considerable plasticity: some nucleotides can be substituted while cleavage efficiency is retained [2]. For HEGs that reside in introns or inteins of a protein-coding host gene, the target sequence is typically coding for a conserved amino acid translation, yet synonymous (silent) mutations are still frequent. Therefore, HEGs evolved tolerance to variation in these synonymous sites [3, 4]. This observation is useful for the prediction of the range of targets that a nuclease can recognize beyond the native target sequence found in its host. In HomeBase, the range is defined by the translated amino acid sequence of the target site.

2 Materials

The computational protocol described below requires the following:

1. A query set consisting of the protein sequences of known HEGs. For example, HEG-containing intron sequences may be obtained from the Group I Intron Sequence and Structure Database [5] (http://www.rna.whu.edu.cn/gissd) and HEG-containing inteins from the intein database INBASE [6] (http://www.neb.com/neb/inteins.html). The manual curation of these databases ensures that these protein sequences do not include exonic or exteinic parts. This is essential for our purpose because otherwise the BLAST search will retrieve many homologs of the hosting gene instead of homologs of the HEGs. The query sequences will include only the HEG sequence for the case of introns. For the case of inteins we will use the full intein sequence, which includes both the nuclease domain and the protein-splicing domains.
2. The nucleotide database that will be searched for HEGs needs to be available in plain FASTA format, and also formatted as a BLAST-searchable database.
3. A local installation of a BLAST implementation such as the NCBI BLAST package (ftp://ftp.ncbi.nlm.nih.gov/blast/

executables/blast+/LATEST/). Special-purpose software to parse and analyze the BLAST results, such as the Perl scripts described below, which were used in the original production of the HomeBase database. These scripts are available for download from http://homebase-search.tau.ac.il/. A BLAST parser implementation such as the BioPerl Bio::SearchIO module is also recommended [7] (http://www.bioperl.org/).

4. Depending on the size of the database to be searched, considerable computing resources may be necessary. As an example, the original HomeBase collection was constructed in 2009 by searching a total of 36 Gbp genomic and metagenomic sequences, and the run-time of the pipeline was roughly 5 weeks on a 40-core cluster. A key factor is the number of queries in the BLAST-2 stage (see below). In that case we had about 10,000 queries, which were eventually narrowed down to less than 1,000 in the final output.

3 Methods

3.1 Using the HomeBase Web Server to Search the Existing HomeBase Collection of Putative HEGs

The results of the original HomeBase pipeline that was run in 2009 are online and searchable using the HomeBase web server at the address: http://homebase-search.tau.ac.il/. The web server allows searching for HEGs in the existing HomeBase collection that are predicted to cut within a query DNA sequence provided by the user.

1. Enter the query sequence by either copy-paste into the text box or by choosing to upload a file. You may enter the sequence either as a plain DNA sequence or as a FASTA formatted file.
2. Click "submit" and wait for the search to complete, which normally takes up to a few minutes for an input sequence of a few thousands of bases. A results page will be shown with a table in the following format:

HEN name	Cleavage site position	Blast pairwise alignment	Match score
ref|XM_001554561.1|_TRUNC_1_4210_INS_774_3208	24058	EKASGFEESM EKASGFEESM EKASGFEESM	9/10
gb|AY836254.1|_TRUNC_1_1788_INS_177_1744	24058	EKASGFEESM EK+SGFEESM EKSSGFEESM	8/9
gb|AACY023780064.1|_TRUNC_1_1533_INS_431_1389	20035	AGLPAQPYSMS AG P QP+S+S AGFPCQPFSIS	6/14
gb|AY863213.1|_TRUNC_14408_21647_INS_742_1726	12808	LHQGKTNNPLTV LHQ +NNPL + LHQNGSNNPLGI	7/12

HEN name: The ID for the nuclease in the HomeBase collection, which includes the NCBI Nucleotide ID in which this gene was identified by the HomeBase pipeline. Clicking the link gives the coding sequence for this gene, as defined by the HomeBase pipeline.

Cleavage site position: The position of the predicted target site in the query sequence.

Blast pairwise alignment: An alignment of translated sequences of the native target sequence (from the HomeBase database) and the predicted target sequence (in the query sequence).

Match score: The number of identical amino acid residues out of the total alignment length.

3.2 Searching for Novel HEGs in a Sequence Database Using a Local Implementation of the HomeBase Pipeline

Two consecutive rounds of BLAST searches will be run: The first identifies candidate novel HEGs (BLAST-1). These sequences are used as queries in the second search (BLAST-2) to identify *vacant homologs*: for a HEG-hosting gene that contains a HEG inside one of its introns/inteins, a vacant homolog is a homolog of the hosting gene that does not contain the intron/intein (Fig. 1). Subsequently, the alignments to the vacant homologs are used to infer the target sequences.

3.2.1 Search for HEGs (BLAST-1)

1. The set of known HEGs is used as queries in a translated BLAST (tblastn) search against the nucleotide database to find novel HEGs. Using the NCBI BLAST package:
   (a) Create a BLAST formatted database from a FASTA formatted file of the DNA sequences that will be searched for novel HEGs:

   ```
   makeblastdb -in database.fasta -dbtype nucl
   ```

   (b) Run translated BLAST (tblastn) of the known HEG protein sequences against the database:

   ```
   blastn -query known_hegs.fa -db database.
   fasta -out blast1.out
   ```

   (c) Retain all hits of E-value < 10, which is the default in NBCI BLAST (*see* **Note 1**).
2. For long hit sequences it is necessary to divide the hit sequence to disjoint loci by clustering the HSPs from all queries on the same hit sequence (*see* **Note 2**). Overlapping HSPs are clustered, as well as neighboring HSPs less than 2,000 bases apart. Additional 1,000 bp flanking sequences are included on either side. An implementation of this procedure is available in the script getBlastSeqs.pl (see lines 81–96).

3.2.2 Defining Target Sites Based on Vacant Homologs (BLAST-2)

1. Each hit sequence from the BLAST-1 results (as defined by the above clustering procedure) is used as a query in the second BLAST search (BLAST-2) to identify vacant homologs (*see* **Note 3**). A translated BLAST (tblastx) search is performed

to allow any of the three possible translations of the strand that was aligned to the known HEG in BLAST-1. The BLAST database should be the same as for BLAST-1:

```
tblastx -query blast1_hits.fa -db database.
fasta -out blast2.out -strand plus
```

2. Filter the BLAST-2 hits using two Perl scripts: findVacantAllele.pl that parses the BLAST-2 output and prints summary information for candidate hits; and findTargetSite.pl that reads this information, applies the more complicated filters and predicts the target sequence of the putative HEG:
   (a) The alignment to the hit sequence must fit the expected homology between a full and a vacant homolog, as shown in Fig. 1: the homology is in the two exonic sequences, which flank the intron/intein that contains the HEG in the query sequence, and are adjacent in the hit sequence (the vacant homolog). Each of these two homology regions will result in a separate HSP. The two HSPs should be approximately adjacent in the hit sequence and separated by a large insertion in the query sequence. We also require that they be on the same strand and the same reading frame in the hit (findVacantHomolog.pl lines 73–114).
   (b) The insertion is classified as an intein if both BLAST-2 HSPs are in the same reading frame. Otherwise, the insertion is classified as an intron.
   (c) Similarly, the BLAST-1 HSP inside the insertion (which represents region coding for the homing endonuclease) must be in the same frame of the BLAST-2 HSPs to be classified as an intein (*see* **Note 4**). Where the insertion contains several BLAST-1 HSPs in different frames a majority vote is taken (findTargetSite.pl lines 464–501).
   (d) If all three reading frames are consistent, we translate the insertion sequence in this frame and check for stop codons, which are not expected in an intein. If stop codons are found the insertion is classified as an intron (findTargetSite.pl lines 503–516).
   (e) Discard insertions classified as introns if they are <450 bases long, and insertions classified as inteins if they are <930 bases long (*see* **Note 5**) (findVacantHomolog.pl lines 103–114, findTargetSite.pl lines 913–920).
   (f) Check that some BLAST-1 HSP (an alignment to a known HEG) lies inside the insertion in the query sequence, which represents the intron/intein. Do not allow the BLAST-1 HSP(s) to overlap with more than 15 bases of either of the BLAST-2 HSPs (*see* **Note 6**) (findVacantHomolog.pl lines 116–148, findTargetSite.pl lines 899–911).

(g) Correct overlap or gap between the two HSPs (*see* **Note 7**): To correct an overlap some bases must be removed from either or both HSPs (findTargetSite.pl lines 395–430). To correct for a gap some bases must be added (findTargetSite.pl lines 647–887). To decide on the correct number of bases that should be removed/added to each HSP, iterate over all possibilities for the position of the insertion of the intron/intein in the hit sequence, and choose the position that gives the highest similarity between the query and the hit in the extended/shortened HSPs. The rationale behind this is that true homology is expected to produce higher similarity than chance similarity. Formally, let L be the overlap length and S be the splice site position relative to the 5′ start of the overlap region, and so $0 \leq S \leq L$. We seek S that maximizes the sum of similarity in the two HSPs after resolving the overlap region. Similarity is measured between the translated amino acid sequences using the same BLOSUM62 matrix that is used by translated BLAST. For introns, the insertion site should be allowed to reside in the middle of a codon. That is, one nucleotide is added/removed to one HSP and two nucleotides to the other HSP. This should not be allowed for sequences identified as inteins.

(h) At least three out of the five amino acid positions at the end of each HSP after overlap/gap correction (the end of the exon) are required to be identical or similar between the query and hit sequences. Similarity is defined as in BLAST: a positive score in the BLOSUM62 matrix. This criterion ensures that the homology is sufficient to resolve the splice sites reliably (findTargetSite.pl lines 432–462).

(i) For introns, search the insertion for the largest open reading frame (ORF) that overlaps one or more BLAST-1 hit in the same frame. This ORF is the putative coding sequence of the HEG. We require it to be at least 85 codons long, to accommodate the shortest known HEGs (findTargetSite.pl lines 518–645).

(j) Filter noncoding host genes (*see* **Note 8**): Translate 50 codons (where available) on either side of the intron in the reading frame of the BLAST-2 HSPs. If a stop codon was found the host gene is classified as a noncoding RNA. For classification of partial sequences as a protein-coding host, we require that 50 codons will be available for at least one of the exons (findTargetSite.pl lines 889–897).

(k) The final output of this procedure is the target sequence flanking the inferred splice sites in the query (the HEG-containing sequence), and the coding sequence of the HEG (in introns) or the HEG-containing intein. As the predicted target sequence we report the first seven codons after the splice sites of each exon, giving a total of 14 codons or 42 bases (*see* **Note 9**) (findTargetSite.pl lines 968–990).

(l) For many BLAST-2 hit sequences, several overlapping pairs of HSPs satisfy the above requirements. Therefore, it is necessary to choose the splice site positions that have the highest support, from more HSP pairs (findTargetSite.pl lines 1017–1051).

3. One BLAST-2 query sequence may contain several HEGs. After identifying the first HEG-containing intron/intein in a given BLAST-2 query, repeat the above procedure using only BLAST-1 HSPs that do not overlap this intron/intein. Only BLAST-2 HSP pairs that contain these BLAST-1 HSPs are considered for additional introns/inteins. We repeat this process until no BLAST-1 HSPs remain for the given BLAST-2 query (Loop starting in findTargetSite.pl line 155).

3.2.3 Quality Assurance and Control

The above protocol inevitably outputs some proportion of false-positive results. Therefore, it is necessary to estimate this proportion and adjust the protocol if it is higher than acceptable. False positives can be classified into two classes: (1) sequences that do not contain a HEG and (2) sequences that contain a HEG but whose predicted target sequence is wrong. To assess the accuracy of prediction putative HEGs can be sampled randomly and manually inspected for all stages of the automatic pipeline, including: confidence in the BLAST-1 homology and conservation of known HEG motifs; confidence and accuracy of the BLAST-2 alignment to the vacant homology, especially of the exon/extein boundaries after correction of gaps/overlaps. Where possible the prediction can be compared to existing annotation of intron position in the host genes. One can also check for the intron/intein splice sites consensus sequences (e.g., inteins sequences typically start with a C and end with HN). If too many results appear to be false positives, then some of the criteria and thresholds in the relevant stages of the protocol may be adjusted.

3.2.4 Searching for Possible Target Sites in a Genome of Interest

Predicted HEGs and their target sequences are potentially useful to target specific sites in a genome of interest, as a tool for genetic manipulation. Potential targets can be identified using a translated BLAST (tblastn) search to find a match between the translations of the predicted target sequences and all possible six-frame translations of the genome. This approach relies on the observation that target specificity is defined mainly by the non-synonymous sites of the target sequence [1]. Since the actual target sequence is a subsequence of the 14 codons of the predicted target, then candidate hits should be preferred to have no mismatches in the central region of the target sequence.

4 Notes

1. Each BLAST-1 hit sequence often has several high scoring pairs (HSPs): pairwise alignments of a segment of the query to a segment of the hit. Furthermore, several queries often match the same hit and yield overlapping HSPs. Especially for hits in whole chromosome sequences, which may contain several different HEGs, the number of HSPs can be large.
2. These subsequences of the hit sequences are extracted to serve as the query sequences for the second BLAST search (BLAST-2). This procedure may result in a cluster of several HEG-containing intron/inteins in the same host gene.
3. In this search we are looking for a homolog of the hosting gene in which the HEG is embedded, so we will be interested in alignments involving the sequences flanking the BLAST-1 hit. These coding sequences may be in a different reading frame than the HEG, but have to be on the same strand.
4. By chance alone we expect one out of nine introns to satisfy the requirement for the three reading frames to be consistent (the two exons and the HEG). These introns will be mistaken for inteins according to this criterion.
5. We require that insertions classified as introns are at least 450 bases long, because this is the length of the shortest known HEG-containing introns. For inteins we require 930 bases, which is the length of the shortest HEG-containing inteins. This allows filtering out hits to homologs that contain mini-inteins (inteins lacking a HEG). Such homologs appear to have a large deletion compared to the homolog with the full (HEG-containing) intein. However, these deletions would be shorter than a deletion of the whole intein. Thus, they will not pass the 930 bases cutoff.
6. A large overlap is not expected because BLAST-1 queries do not include any exonic sequences, and so the BLAST-2 HSPs (which are supposed to be the exons) should not bear homology to the known HEG. This requirement is necessary to filter out hits to repetitive sequences as well as homologs containing a mini-intein that result in pairs of HSPs with a similar structure resembling a vacant homolog.
7. Ideally, the two HSPs (the exon sequences) should be exactly adjacent in the hit sequence (the vacant homolog that is missing the insertion of the intron/intein). However, in practice, they often overlap because BLAST extends the alignment from the true homology of the exons, to chance similarities in the intron/intein. In other cases, the HSPs may have a small gap between them (in the hit sequence) if the homology is distant enough so that some substitutions have obscured

the similarity in the positions adjacent to the splice sites. To correct an overlap some bases must be removed from either or both HSPs. To correct for a gap some bases must be added.

8. Many HEG-containing introns reside in noncoding RNA genes, especially rRNA. To identify protein-coding host genes we search for stop codons in the exons surrounding the intron. For noncoding sequences of 150 bases, we expect <10 % probability of not containing stop codons, and thus being erroneously classified as coding. Note that this step is required for the use of the translated target sequence as the range of putative targets for the novel HEG. However, other applications that are not based on this approach may not require this filter. For example, if the goal is only to identify novel HEGs (prediction of the target sequence is not required) then this step can be removed.
9. This is enough to encompass the target sequence of any known HEG, but for many the actual target would be a shorter subsequence.

References

1. Barzel A, Privman E, Peeri M, Naor A, Shachar E, Burstein D, Lazary R, Gophna U, Pupko T, Kupiec M (2011) Native homing endonucleases can target conserved genes in humans and in animal models. Nucleic Acids Res 39: 6646–6659
2. Gimble FS, Wang J (1996) Substrate recognition and induced DNA distortion by the PI-SceI endonuclease, an enzyme generated by protein splicing. J Mol Biol 263:163–180
3. Kurokawa S, Bessho Y, Higashijima K, Shirouzu M, Yokoyama S, Watanabe KI, Ohama T (2005) Adaptation of intronic homing endonuclease for successful horizontal transmission. FEBS J 272:2487–2496
4. Scalley-Kim M, McConnell-Smith A, Stoddard BL (2007) Coevolution of a homing endonuclease and its host target sequence. J Mol Biol 372:1305–1319
5. Zhou Y, Lu C, Wu QJ, Wang Y, Sun ZT, Deng JC, Zhang Y (2008) GISSD: group I intron sequence and structure database. Nucleic Acids Res 36:D31–D37
6. Perler FB (2002) InBase: the Intein Database. Nucleic Acids Res 30:383–384
7. Stajich JE, Block D, Boulez K, Brenner SE, Chervitz SA, Dagdigian C, Fuellen G, Gilbert JG, Korf I, Lapp H et al (2002) The Bioperl toolkit: Perl modules for the life sciences. Genome Res 12:1611–1618

Chapter 3

PCR-Based Bioprospecting for Homing Endonucleases in Fungal Mitochondrial rRNA Genes

Mohamed Hafez, Tuhin Kumar Guha, Chen Shen, Jyothi Sethuraman, and Georg Hausner

Abstract

Fungal mitochondrial genomes act as "reservoirs" for homing endonucleases. These enzymes with their DNA site-specific cleavage activities are attractive tools for genome editing and gene therapy applications. Bioprospecting and characterization of naturally occurring homing endonucleases offers an alternative to synthesizing artificial endonucleases. Here, we describe methods for PCR-based screening of fungal mitochondrial rRNA genes for homing endonuclease encoding sequences, and we also provide protocols for the purification and biochemical characterization of putative native homing endonucleases.

Key words Homing endonucleases, Data mining, Biotechnology, Protein purification, Endonuclease assays

1 Introduction

Homing endonucleases (HEs) are DNA site-specific cutting enzymes that are encoded by homing endonuclease genes (HEGs). HEGs can be freestanding genes, or they can be encoded within archaeal introns, and they are frequently embedded within self-splicing elements such as group I, group II introns, and inteins [1–6]. HEs promote mobility of the elements that encode them as HEs are site-specific endonucleases that introduce a double-strand break, or a single-stranded nick, at specific sites within cognate alleles that lack HEG insertions. This process is referred to as "homing" and involves homologous recombination [4]. HEs have applications in biotechnology as they can be applied to promote DNA modification or genome editing such as site-directed mutagenesis or gene repair [7, 8].

HEGs and their host introns are quite invasive and significantly contribute towards the size of fungal mitochondrial (mt) DNA genomes [9]. In particular, the mtDNA rDNA genes appear to be a reservoir for potentially mobile introns and HEs [10–12].

David R. Edgell (ed.), *Homing Endonucleases: Methods and Protocols*, Methods in Molecular Biology, vol. 1123, DOI 10.1007/978-1-62703-968-0_3, © Springer Science+Business Media New York 2014

Based on conserved amino-acid motifs, HEs are classified into different families, of which the LAGLIDADG and GIY-YIG families of HEs are most frequently encountered among fungal mitochondrial group I introns [13].

So far, most applications are based on a very limited number of well-characterized native homing endonucleases (I-SceI, I-CreI, I-DmoI, I-AniI, and I-OnuI; [14, 15]), and therefore, there is a need to bioprospect for more native HEs [11, 12] so that a wider choice of target sites can be used as substrates for HEs [16]. Bioprospecting for native HEGs and the possibility of using these HEs as scaffolds for engineering a variety of novel chimeric HEs with new sequence specificities will make this group of rare cutting molecular scissors a set of valuable tools in biotechnology [17–19].

2 Materials

2.1 DNA Extraction and mtDNA rDNA Gene Amplification

1. 2 % Malt Extract Agar medium (MEA): 20 g/L Malt extract supplemented with 1 g/L yeast extract (YE) and 20 g/L bacteriological agar.
2. Peptone, glucose, yeast extract (PYG) liquid medium: 1 g/L peptone, 1 g/L yeast extract, and 3 g/L glucose).
3. 15 mL polypropylene Falcon conical tubes.
4. Cetyl trimethyl ammonium bromide (CTAB) nucleic acids extraction buffer: 150 mM NaCl, 50 mM EDTA, 10 mM Tris–HCl, 1 % CTAB (w/v), 1 M NaCl, pH 7.4.
5. Sodium dodecylsulfate (SDS) or sodium lauryl sulfate (SLS) stock solution: 20 % (w/v) in H_2O.
6. Glass beads (0.5-mm).
7. 70 % Ethanol.
8. 95 % Ethanol.
9. Solution of chloroform–isoamyl alcohol (24:1, v/v).
10. DNA storage buffer: 1× Tris-EDTA (TE) buffer (10 mM Tris–HCl, pH 7.6, 1 mM $Na_2EDTA \cdot 2H_2O$).
11. PCR reaction mixture (total volume 50 μL) ingredients (μL/reaction): 10× Taq DNA polymerase buffer [5]; 50 mM $MgCl_2$ (0.5); 2.5 mM dNTP [4]; 40 pmol each forward and reverse primer (0.5+0.5); H_2O (38.25); genomic DNA template (1 μL~10–100 ng); and Taq DNA polymerase (0.25; ~2.5 units).
12. Tris-borate EDTA buffer: 1× TBE buffer (89 mM Tris-borate, 10 mM EDTA, pH 8.0).
13. PCR product Cleanup System.
14. BigDye®Terminator system v 3.1 (Life Technologies/Applied Biosystems).

15. TOPO cloning kit (Invitrogen/Life technologies).
16. Agarose gel loading buffer (6×): 3 mL glycerol (30 %), 25 mg bromophenol blue (0.25 %) dH_2O to 10 mL.

2.2 Protein Overexpression and Purification

1. Luria-Bertani Broth (LB) media: For 1 L of LB mix the following reagents in a 2 L glass container and stir thoroughly 10 g Tryptone, 5 g Yeast extract, 5 g NaCl, 1 L MilliQ water, add 200 μL of 5 N NaOH and autoclave.
2. 2×-YT medium: Measure ~900 mL of distilled H_2O, 16 g Tryptone, 10 g Yeast Extract, 5 g NaCl, adjust pH to 7.0 with 5 N NaOH, adjust to 1 L and autoclave.
3. Super Optimal broth with Catabolite repression (SOC): 2 % Tryptone, 0.5 % Yeast Extract, 10 mM NaCl, 2.5 mM KCl, 10 mM $MgCl_2$, 10 mM $MgSO_4$, 20 mM glucose.
4. Salt solution for TOPO kit: 1.2 M NaCl, 0.06 M $MgCl_2$.
5. Champion™ pET200 directional TOPO® expression cloning system (Invitrogen, Life Technologies).
6. Cell Lysis (CL) buffer: 50 mM Tris–HCl, pH 8.0, 1 mM EDTA.
7. Wash Buffer 1: 50 mM Tris–HCl, pH 8.0, 10 mM HEPES.
8. Wash Buffer 2 (for HiTrap column): 50 mM Tris–HCl, pH 8.0.
9. Buffer D1: 40 mM HEPES, pH 7.5, 50 mM NaCl, 3 mM β-mercaptoethanol.
10. Protein storage buffer: 40 mM HEPES, pH 7.5, 50 mM NaCl, 1 mM dithiothreitol (DTT), 30 % (w/v) glycerol.
11. Endonuclease reaction buffer: Reaction buffer #3 (Invitrogen/Life technologies): 50 mM Tris–HCl, pH 8.0, 10 mM $MgCl_2$, 100 mM NaCl supplemented with 1 mM DTT.
12. Wizard® Plus Minipreps DNA purification kit (Promega, Madison).

3 Methods

3.1 Fungal Growth and DNA Extraction

Many fungi are fairly easy to cultivate using standard microbiological techniques, and for the nonexpert they can be obtained from various culture collections (*see* **Note 1**).

1. For routine culturing fungi are maintained in petri plates containing 2 % MEA.
2. From 5- to 12-day old cultures, remove small agar plugs with growth and inoculate 250 mL Erlenmeyer flasks containing 50 mL PYG liquid medium to generate biomass for DNA extraction. For fungi that produce large masses of spores, generate

spore suspensions by adding 5–10 mL of water to the agar plate and gently shaking the plates to dislodge the spores. Inoculate the PYG media with the spore suspension.

3. Incubate liquid cultures at 20–25 °C for 3–7 days (temperature and days of incubation will vary among different fungi) to generate biomass for DNA extractions.
4. Extract fungal whole cell DNA by filtering the cultures through a Whatman # 1 filter paper.
5. Add 100–200 mg harvested mycelium to a sterile 15 mL Falcon polypropylene conical tube, and add 3 mL of lysis buffer plus 4 g of acid-washed and baked-dry 0.5-mm glass beads.
6. Vortex the mixtures for 2–3 min, and add an additional 3 mL of lysis buffer to each tube.
7. Add ~660 μL of 20 % SDS or SLS solution to a final concentration of 1 % along with NaCl (1 M final concentration) and CTAB (1 % final concentration). Mix tubes gently and incubate for a minimum of 1 h [or overnight (O/N)] at 55 °C.
8. Extract cell debris, glass beads, denatured proteins, and lipids in 7 mL of chloroform–isoamyl alcohol (24:1, v/v) and pellet by centrifugation at 2,500 rpm in a table top centrifuge for 20 min at room temperature. Transfer the top aqueous layer to a 15 mL Falcon conical tube and precipitate the DNA by adding 2.5 volumes of ice-cold 95 % ethanol.
9. Store tubes for about 3 h (or O/N) at −20 °C and recover the nucleic acids by centrifugation (1,500 × *g* for 30 min). Wash resulting pellets with 1 mL of 70 % ethanol to remove excess salts. Air-dry the pellets and resuspend in 300 μL of 1× TE buffer. Approximately 50–100 ng of DNA can be recovered from each strain.
10. *See* refs. 20, 21 for more information about whole cell DNA extraction.

3.2 PCR Primer Design

Primer design for gaining access to the mtDNA *rns* and *rnl* genes will depend on the fungi being investigated; here, we can only suggest some primers sequences (Figs. 1 and 2) that work for members of the Ophiostomatales and related taxa. However, we present an "intron landscape" for both rRNA genes to provide an overview as to possible locations that could be investigated for the presence of HE/intron insertions. There are many fungal mtDNA sequences available in public databases that allow for designing PCR primers and for data mining for possible HE sequences (such as NCBI, http://www.ncbi.nlm.nih.gov/genomes/).

3.3 PCR Amplification (rns Gene)

1. The *rns* gene can be fully or partially amplified with the primer pairs rns-5′/rns-3′ or mtsr-1/mtsr-2, respectively; also *see* Fig. 1 for additional primer sequences that can bind highly conserved regions in the *rns* gene of ascomycetous fungi.

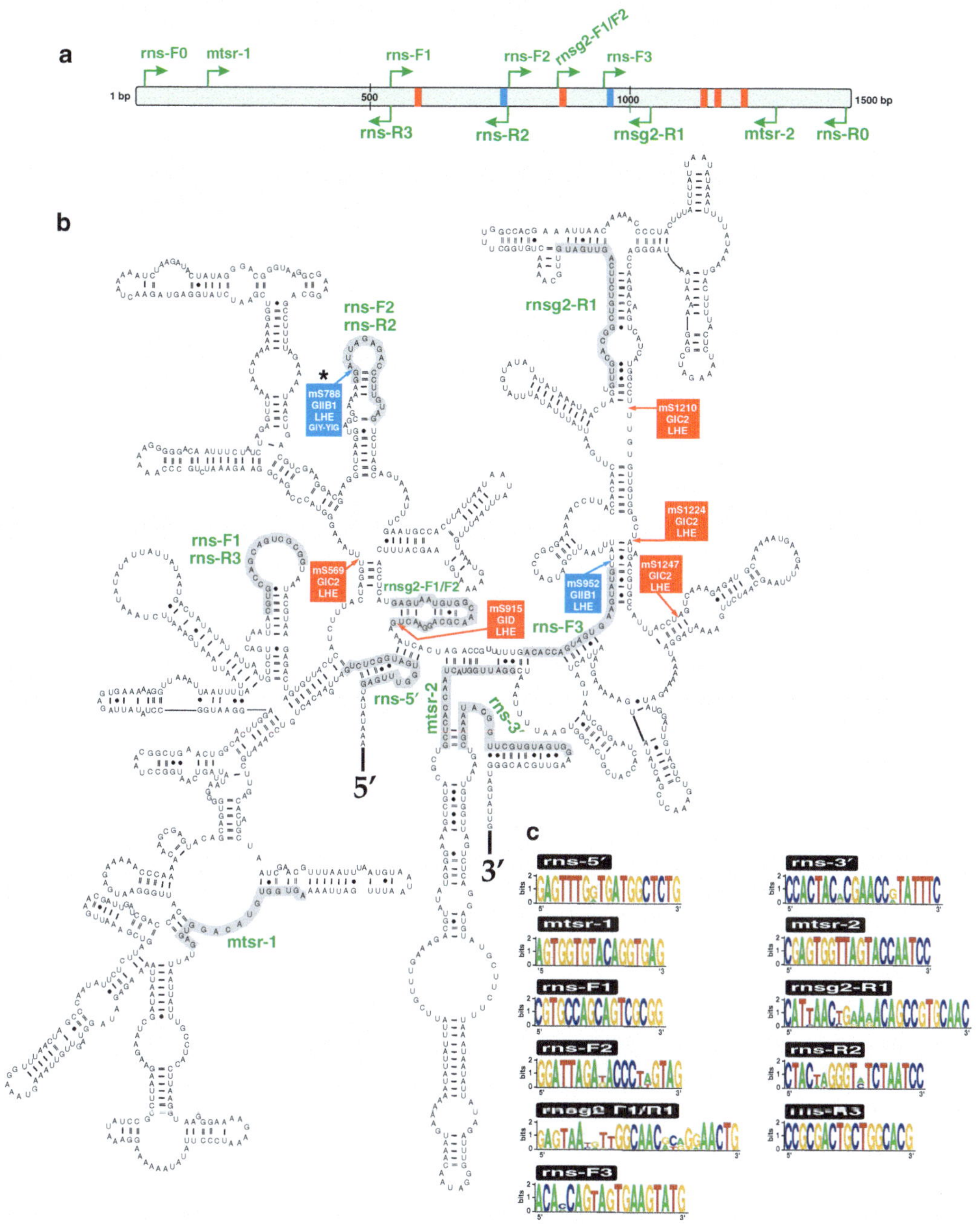

Fig. 1 (**a**) Schematic representation of the *rns* gene in ascomycetous fungi indicating some primer-binding sites. (**b**) Nucleotide sequence and secondary structure model for the *O. novo-ulmi* subsp. *americana* mitochondrial *rns* RNA (GenBank accession number HQ292074) showing the four structural domains indicated by Roman numbers (I–IV). Indicated on this secondary structure are some reported locations of group I (GI) and group II (GII) introns for ascomycetous fungi; intron naming is with reference to the *E. coli* SSU rRNA sequence (*see* ref. 36). Intron encoded ORFs (LHE = LAGLIDAG HEs or GIY type HEs) are also indicated. Primer-binding sites are highlighted in gray. (**c**) Nucleotide sequence logos are shown for the exon-specific primers that bind to highly conserved sequences in the *rns* gene of ascomycetous fungi. The sequence logos were generated by the online program WebLogo version 2.8.2 [37]

a

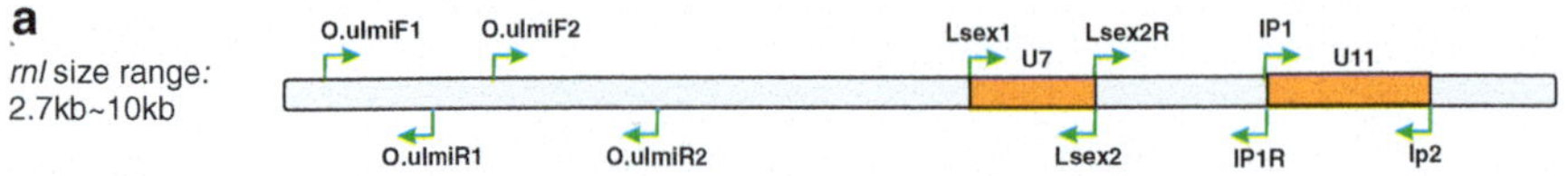

b

Lsex1	GCTAGTAGAGAATACGAAGGC	O.ulmiF1	GGGGAACCTTCCTCAAAGAC
Lsex2	GACCGCATTTAACGGCCAAGG	O.ulmiR1	CTTGTTCTGGGCTGTTTCCC
IP1	GGAAAAGCTACGCTAGGG	O.ulmiF2	GTTGCAAAGATATCCGATG
IP2	CTTGCGCAAATTAGCC	O.ulmiR2	CACCATAAGCTTTAGGCG
Lsex2R	CCTTGGCCGTTAAATGCGGTC		
IP1R	CCGTAGCGTAGCTTTTCC		

c

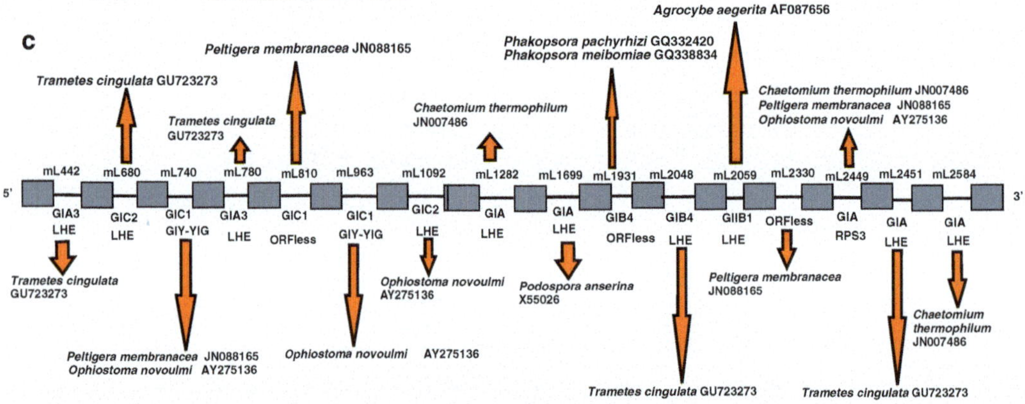

Fig. 2 (**a**) Schematic representation of the *rnl* gene in ascomycetous fungi showing the relative location of some potential PCR primers. The *rnl* universally conserved U7 and U11 regions have been noted to be prone to intron insertions [11]. Size range for the *rnl* gene is based on the *Sporothrix schenckii* (without the mL2449 intron; AB568599) and *Agrocybe aegerita rnl* (including all introns) genes. (**b**) List of primers that might be suitable for amplifying segments of the *rnl* gene in ascomycetous fungi; specifically members of the Ophiostomatales [11]. (**c**) The *rnl* intron landscape for ascomycetous and selected basidiomycetous fungi, indicating some of the reported positions occupied by introns (intron type and ORF category are also indicated). Introns are named based on insertion sites with respect to the *E. coli* rDNA sequences [36]

2. Carry out PCR amplifications in a reaction mix of 50 μL (*see* Subheading 2.1, **item 10**).

3. The PCR conditions for the rns-5′/rns-3′ and mtsr-1/mtsr-2 primer pairs can be as follows: an initial denaturation step at 94 °C for 1 min followed by 25–30 cycles of denaturation (94 °C for 1 min), annealing (55 °C for 30 s), and extension (70 °C for 1 min/1 kbp) with a final extension step at 70 °C for 10 min.

3.4 Gel Electrophoresis

1. Preparation of a 1 % Agarose gel: Add 1 g ultrapure agarose (Life technologies) to 100 mL of 1× TBE buffer then mix and melt agarose in microwave oven. Once the agarose has completely dissolved, allow to cool down (~55–60 °C) and pour into an assembled gel casting tray with positioned comb. Allow the gel to solidify at room temperature and carefully remove the comb and place the gel into electrophoresis box containing 1× TBE buffer.

2. Mix each DNA sample with the agarose gel loading buffer and load samples into the wells of the gel. Electrophorese at 80–120 V until the tracking dye migrates to the positive electrode end of the gel. The DNA fragments are sized with a DNA ladder (such as 1Kb plus DNA ladder by Invitrogen/Life technologies).
3. Stain nucleic acids by soaking gel in 1× TBE buffer supplemented with 0.5 μg/mL ethidium bromide (EtBr) and expose the stained gel with ultraviolet light.

3.5 PCR Product Purification

1. Purify the PCR products using a PCR purification kit according to manufacturer's instructions.

3.6 DNA Sequencing

1. Purified DNA fragments are cloned into pCR4 TOPO plasmids (Life Technologies/Invitrogen) for sequencing to improve the sequence quality; however, PCR fragments can be directly sequenced using appropriate primers.
2. Initially, for cloned PCR products, use plasmid specific primers as supplied by the TOPO cloning kit: M13 Forward, M13 Reverse, T7 (forward), and T3 (reverse) to obtain sequences; thereafter primers are designed as needed (also *see* Figs. 1 and 2 for *rns* and potential *rnl* primers) to complete all sequences in both directions.
3. Sequence reactions are prepared in our lab by using the BigDye® Terminator v3.1 Cycle Sequencing Kit (Applied Biosystems) following the manufacturer's instructions.
4. Denatured sequencing products are resolved on a 3130 genetic analyzer (Applied Biosystems) or any other suitable automated sequencing platform.

3.7 Sequence Analysis and Comparative Sequence Analysis and Data Mining for HEs

It can be useful to extract HE-like sequences from NCBI databases; an alternate approach of "prospecting" for homing endonucleases. Plus it is useful to compare HE-like sequences obtained by data mining or by one's own sequencing efforts with other HEs. Such comparative sequence analysis might provide clues as to the state/condition of the HE, i.e., are conserved sequence motifs present, or are there signs of "ORF erosion"? [22]. In addition, it has been recognized that there are several clades of HE-like elements that sometimes share similar properties. Knowing the phylogenetic position of the HE may allow one to use existing literature on related HEs to formulate approaches for characterization and possibly reengineering the HE element to target alternative DNA sites [16, 23].

1. Assemble individual sequences manually into contigs using the GeneDoc program v2.7.000 [24].
2. Identify potential ORFs within the sequences with the ORF finder program (http://www.ncbi.nlm.nih.gov/gorf/gorf.html) (setting 4: genetic code for mtDNA of molds).

3. Use the online resource BLASTp [25] to retrieve sequences that are related to the putative HE-like ORFs obtained from your own sequencing efforts.
4. Align nucleotide sequences with Clustal-X [26] and edit the alignments manually with the aid of the GeneDoc program.
5. Generate amino-acid sequence alignments with the online PRALINE multiple sequence alignment program [27] and then refined alignments with GeneDoc.
6. For phylogenetic analyses, only those segments of the alignment where all sequences can be aligned unambiguously are to be retained. Phylogenetic estimates can be generated by programs contained within the PHYLIP package [28] and the MrBayes program v3.1 [29]. In PHYLIP (version 3.69c), phylogenetic trees are obtained by analyzing alignments with either PROTPARS (protein parsimony algorithm) or DNAPARS programs for amino-acid or nucleotide sequences, respectively. In combination with bootstrap analysis (SEQBOOT) and CONSENSE, majority rule consensus trees can be generated. Phylogenetic estimates can also be generated within PHYLIP using the NEIGHBOR program using distance matrices generated by PROTDIST (setting: Dayhoff PAM250 substitution matrix) or DNADIST (K84 setting).
7. The MrBayes program can be used for Bayesian analysis and the parameters for amino-acid alignments can be as follows: mixed model. The Bayesian inference of phylogenies is initiated from a random starting tree, and four chains are run simultaneously typically for 1–5 million generations; the trees are usually sampled every 1,000 generations. The first 25 % of trees generated are discarded ("burn-in"), and the remaining trees are used to compute the posterior probability values.
8. In addition, for the novice there are "user" friendly online (http://www.phylogeny.fr/) [30] and downloadable phylogenetic analysis programs available (such as MEAG 5 [31]).

3.8 HE Protein Overexpression in E. coli

The activity of HEs has to be verified by performing endonuclease assays. Various procedures have been described previously [32]; herein we describe a protocol that can be performed by someone who has some basic familiarity with molecular biology protocols.

3.9 Testing for Endonuclease Activity: Cloning Strategies

Various cloning and expression systems can be used to construct an expression plasmid but herein, the protocol followed is based on using the pET200 Directional TOPO® cloning system (Invitrogen/ Life Technologies), for expression of a recombinant protein in *E. coli* with an N-terminal tag containing the Xpress epitope and 6× His tag. *See* **Note 2** for alternative vector systems.

3.10 Codon Optimization and Gene Synthesis

A codon-optimized version of the HE gene sequence should be synthesized to account for differences between the fungal mitochondrial and bacterial genetic code and codon biases (*see* **Note 3**).

3.11 Designing PCR Primers for pET TOPO Subcloning of HE ORFs

1. Typically gene synthesis projects return the gene of interest inserted within a "generic vector" not suitable for protein overexpression. In order for a gene sequences to be subcloned in frame into the pET200 TOPO expression vector, one needs to amplify the insert with a modified forward primer that has a CACC sequence added to its 5′ end this allows for the directional cloning of the gene into the pET200 TOPO vector. Given below is the DNA sequence of the N-terminus of a theoretical ORF and the proposed sequence of the forward PCR primer:

DNA sequence **5′-ATG GAT TTA TTT AAA ...**

Proposed Forward PCR primer **5′-<u>C ACC</u> ATG GAT TTA TTT AAA ...**

2. Design the reverse PCR primer to include the stop codon (*see* **Note 4**).

3.12 Producing Blunt-End PCR Products

1. Blunt ended PCR products are generated with a thermostable proofreading DNA polymerase such as Pfusion (New England Biolabs).
2. Optimize the PCR conditions in order to produce a single, discrete PCR product of the correct size (*see* **Note 5**).
3. Perform agarose gel electrophoresis to verify the quality and quantity of the PCR product. Concentrations can be measured using a spectrophotometer or NanoDrop device.

3.13 Performing TOPO Ligation/Cloning Reaction

For optimal results, make sure to use a 2:1 molar ratio of PCR product to TOPO vector in the cloning reaction.

1. Set up the TOPO cloning reaction by mixing 0.5–4 μL of fresh PCR product, 1 μL of salt solution (TOPO kit), 1 μL of TOPO vector and add sterile water to achieve a final volume of 6 μL.
2. Optional: One can also set up a control ligation mixture by mixing 1 μL of salt solution, 1 μL of TOPO vector plus sterile water to a final volume of 6 μL.
3. Mix reaction mixtures gently and incubate for 5–30 min at room temperature (22–23 °C) (*see* **Note 6**).
4. Place the reactions on ice and proceed to transform the chemical competent *E. coli* DH5α cells.

3.14 Chemical Transformation Protocol

Chemical competent cells (strain DH5α) can be prepared in-house [33] or can be purchased from various suppliers. Transforming electrocompetent cells requires modification of the concentration of the salt solution. To prevent arcing during electroporation, reduce the salt concentration to 50 mM NaCl and 2.5 mM $MgCl_2$.

1. Add 3 μL of the TOPO ligation mixture into a vial containing 100 μL of chemically competent *E. coli* DH5α cells and mix gently. Avoid pipetting up and down.
2. Optional: Repeat **step 1** for the control experiment (Subheading 3.13, **step 2**).
3. Incubate both the vials on ice for 5–30 min (*see* **Note 7**).
4. Heat-shock the cells for 1 min at exactly 42 °C without shaking.
5. Transfer the vials onto ice and keep for 2 min.
6. Add 300 μL of pre-warmed SOC medium at room temperature to both vials (in case **step 2** is included).
7. Tightly cap the tubes and shake horizontally (200 rpm) at 37 °C for 1 h.

Spread 100–150 μL (may have to set up a series of plates with different amounts to get optimal spread of colonies) of the mixture on a warm LB agar plate containing the appropriate antibiotic(s) (i.e., 100 μg/mL Kanamycin) and incubate at 37 °C.

3.15 Analyzing Positive Clones

1. Expect to have no colonies in the control plate (*see* Subheading 3.13, **step 2**) and plenty in the experimental plate which is indicative of successful ligation and transformation.
2. Pick at least ten colonies and streak them onto LB plates with the appropriate antibiotics (100 μg/mL Kanamycin). Incubate the plate overnight at 37 °C.
3. From the above LB agar plate, use cells of potential recombinant clones to inoculate 5 mL LB cultures that are kept at 37 °C with agitation for 14–18 h.
4. 1–2 mL of the LB culture is collected for extracting plasmid DNAs. Plasmid DNAs are recovered by various methods [33]; however, one can also perform colony PCR (from **step 2** above) screening [34] to confirm positive clones.
5. Perform restriction enzyme digestion to confirm the presence of the correct construct. Ideally one should use a restriction enzyme or a combination of enzymes that cuts once in the vector and once in the insert.
6. Visualize restriction digests by agarose gel electrophoresis (Refer to Subheading 3.4).

3.16 Preparing the Cells (Transformants) for Long-Term Storage

1. Once a correct clone has been identified, re-streak the original colony on a LB plate (containing the appropriate antibiotic) to obtain a single colony.
2. Isolate a single colony and inoculate a 5 mL LB tube containing the appropriate antibiotic. Grow until the culture reaches stationary phase (usually overnight).

3. Mix 0.85 mL of the culture with 0.15 mL of 50 % sterile glycerol and transfer to a cryovial and store at −80 °C.
4. As an additional backup always store an aliquot of purified plasmid DNA at −20 °C.

3.17 Sequencing the Plasmid Construct

Sequence the plasmid insert in order to confirm the orientation and to ensure that the ORF is in frame with the vector that provides the start codon and the N-terminal His-tag. To perform sequencing, use universal primers like T7 forward and T7 reverse. We also recommend using gene specific forward and reverse primers (Refer to Subheading 3.6).

3.18 Overexpression of the Fungal Protein in E. coli

E. coli BL21 Star (DE3) is specifically designed for the overexpression of genes regulated by the T7 promoter. However, *do not* use this strain for the propagation and maintenance of plasmids as this strain has leaky T7 RNA polymerase expression, which might lead to instability and eventual loss of the plasmid (*see* **Note 8**). Each Directional TOPO kit provides a positive control vector where the *β-galactosidase* is directionally cloned into the pET TOPO vector. Perform the control experiment using the positive control vector using the same protocol outlined below.

3.19 Transforming BL21 Star (DE3) Cells

1. Thaw BL21 Star (DE3) competent cells (~100 μL) on ice for 10 min.
2. Add 5–10 ng (1–5 μL) recombinant plasmid DNA into the vial and stir gently with a pipette tip.
3. Incubate on ice for 30 min.
4. Heat-shock the cells for 1 min at 42 °C without shaking and transform as outlined in Subheading 3.14, **steps 4–8**.
5. Analyze the positive clones as outlined in Subheading 3.15.
6. Prepare glycerol stocks of the positive clones as outlined in Subheading 3.16.

3.20 Small-Scale Expression of the Protein

1. Inoculate small flasks (50 mL of LB media containing 100 μg/mL kanamycin supplemented with 0.25 % w/v glucose) with 500 μL of O/N culture of *E. coli* (BL21 Star (DE3) containing the recombinant plasmid construct). Inoculate another small flask with just *E. coli* BL21 Star (DE3) containing only the control plasmid (plasmid containing no insert).
2. Grow the cultures (with agitation) at 37 °C till O.D. reaches 0.6 and then induce with 0.2 mM IPTG (low) and 1 mM IPTG (high) to the respective flasks and shift flasks to various temperatures. Several trials may be required to optimize the concentration of IPTG (range: 0.1–1 mM) and temperature (range: 15–37 °C) for proper induction.
3. Incubate the flasks at various temperatures for 6 h or overnight.

4. Centrifuge the cells at 7,000 rpm for 10 min in a 4 °C high-speed centrifuge.
5. Discard supernatant and resuspend cell pellet in 2 mL of cell lysis buffer.
6. Sonicate thoroughly to lyse the cells. Keep vial on ice during the entire period.
7. Centrifuge at 12,000 rpm for 15 min at 4 °C with a high-speed centrifuge and collect the crude protein extract in microcentrifuge tubes. Keep on ice.
8. Determine the concentration of the crude protein mixture by A_{260}/A_{280} ratio using a spectrophotometer.
9. Analyze the samples by SDS-PAGE using about 8 μg of each of the protein extracts plus the control sample(s).
10. Determine the molecular weight of the protein of interest by submitting the sequence of the protein (in fasta format) to online programs such as http://www.bioinformatics.org/sms/prot_mw.html.
11. Check the SDS-PAGE protein gel for overexpression of the protein of interest by scanning for a band in the appropriate size range that is absent in the control lane. One can also perform a western blot to further confirm the presence of the desired protein [33]. Once specific parameters have been determined for the overexpression, one can proceed to the large-scale overexpression of the protein.

3.21 Large-Scale Overexpression of the Protein

1. Inoculate 5 mL LB media (supplemented with 100 μg/mL kanamycin and 0.25 % w/v glucose) with a small amount of the glycerol culture for *E. coli* BL21 star transformed with pET200-recombinant construct and incubate overnight (ON) at 37 °C in a rotatory incubator.
2. Inoculate 5 mL of O/N culture into 1 L of 2×—YT medium (supplemented with 100 μg/mL of kanamycin and with 0.25 % w/v glucose).
3. Grow the culture at 37 °C with agitation and induce when the OD600 reaches ~0.6–0.8 with the desired concentration of IPTG and grow further at the predetermined conditions for overexpression.
4. Harvest the cells by centrifugation at 5,000 rpm for 15 min and freeze pellet at −80 °C.

3.22 Purification of the Protein

1. Thaw the cell pellet in a warm water bath and resuspend in 10 mL of CL buffer per 1 g wet weight of the cell. Stir the suspension for 30 min at 4 °C in order to make it homogeneous.
2. Lyse the cells using a French press two times and centrifuge lysate at 18,000 rpm for 30 min at 4 °C to pellet cell debris.

3. Add the clear lysate to 3 mL of Ni-NTA resin (Qiagen, Toronto) and incubate at 4 °C with shaking for 30–60 min.
4. Load the crude-extract onto a Ni-NTA super flow column (Qiagen, Toronto).
5. Carry out the following series of washings with wash 1: 30 mL the wash buffer (WB) supplemented with 20 mM of imidazole; wash 2: 30 mL of WB buffer with 30 mM of imidazole; and wash 3: 30 mL of WB buffer with 40 mM of imidazole. Collect and save 1 mL of each wash.
6. Elute the protein in Elution buffer (wash buffer containing either 125 mM or 250 mM imidazole, pH adjusted to 8 with NaOH). Collect the eluting samples in 1 fraction (1 mL) with 125 mM imidazole and then in two fractions (1 mL/fraction) with 250 mM imidazole.
7. Remove excess imidazole by dialysis in buffer D1 using a Slide-A-Lyzer dialysis cassette (Millipore, Billerica, USA) with a desired molecular weight (MW) cutoff.
8. Check the concentration of the protein with a spectrophotometer and analyze the fractions by performing SDS-PAGE. If there are background proteins, try to adjust the concentration of imidazole in the wash buffer.
9. A second purification step can be carried out using a 1 mL HiTrap™ heparin HP column (GE Healthcare, Europe).
10. Equilibrate the column with 5 column volumes of wash buffer (without any addition of salt).
11. Take ~20 μg (concentration already determined in **step 5**) of the sample in a syringe and pass it through the column. Collect the flow-through.
12. Wash with two column volumes of wash buffer over a range of 200 mM to 1.5 M NaCl. (Increase the NaCl concentration by 100 mM in each step). Collect each of the fractions into 1.5 mL microcentrifuge tubes.
13. Analyze the fractions by performing SDS-PAGE.
14. Pool the desired fractions to a final volume of 9 mL in a protein storage buffer and concentrate using Amicon Ultracel centrifugal filters (Millipore, Billerica, MA) with a predetermined MW cutoff and centrifuge at 4,000 × *g* at 4 °C until the sample is concentrated in a final volume of 500 μL. Keep the protein at −80 °C. Check the concentration of the protein before freezing.

3.23 Biochemical Characterization: In Vitro Endonuclease Cleavage Assay

1. Construct a substrate plasmid by inserting a DNA segment that contains the target site for the HE, such as an allele that does not contain the HE (and/or associated intron) sequence. Also generate a control plasmid by inserting a DNA fragment

that lacks the HE target site such as the allele with the HEG (and/or intron) insertion. The latter plasmid should not be cleaved by the HE as the cleavage site is disrupted by the HE/intron sequence. One could also obtain substrates and controls by using PCR products of alleles that lack the HE/intron insertion and alleles that contain the HE/intron; however, some HEs appear to prefer plasmid DNAs as substrates.

2. Transform the plasmids (substrate and control) into *E. coli* DH5α and then purify the constructs from ~2 mL LB O/N cultures with any suitable plasmid purification kit.
3. Combine: 15 μL of substrate plasmid (25 μg/mL), 5 μL Invitrogen Buffer React #3 supplemented with 1 mM DTT, 5 μL putative HE protein (~50 μg/mL) and 25 μL H_2O. In addition, the linearized substrate plasmid can be tested as a substrate for endonuclease activity.
4. Set up an identical reaction as in **step 3** but with the control plasmid which contains an insert that comprises the HE/intron containing allele.
5. Incubate the cleavage reactions at 37 °C and 10 μL aliquots are taken at the following time intervals 0, 30 and 60 min; stop the reactions by adding 2 μL of 200 mM EDTA (pH 8.0) and 1 μL of proteinase K (1 mg/mL) to each 10 μL aliquots followed by incubation for 30 min at 37 °C.
6. Resolve the cleavage reaction products on a 1 % agarose gel; in addition, samples representing an untreated version of the substrate and the control plasmid(s) should be resolved on this gel along with a suitable molecular weight marker.

3.24 Cleavage Site Mapping

1. Treat substrate plasmid with HE under optimal conditions (as outlined above).
2. Resolve the cleaved substrate plasmid on a 1 % agarose gel and recover the band from the gel with any suitable PCR product gel cleanup/extraction system.
3. Treat the linearized substrate plasmid with T4 DNA polymerase under conditions that generate blunt ends [35]; reaction mixture contains 40 μL linearized plasmid (25 μg/mL), 2 μL T4 DNA polymerase (5 units/μL), 20 μL 5× T4 DNA polymerase buffer, 20 μL dNTP mixture (0.5 mM) and the total volume is adjusted to 100 μL with sterile distilled water.
4. Incubate the reaction mixture at room temperature (~24 °C) for 20 min and place on ice for 5 min and terminate the reaction by heating at 70 °C for 10 min.
5. Purify the linearized DNA and treat with 2 μL of T4 DNA Ligase (1 unit/μL) in the presence of 10 μL 5× Ligase buffer in a total volume of 40 μL. Incubate the ligation reaction at room temperature for 2 h to generate the desired religated plasmid.

6. Dilute the ligation reaction fivefold and use 10 μL of this dilution to transform chemical competent *E. coli* DH5α cells.
7. Purify the plasmid from the transformed overnight cultures with a suitable plasmid purification kit (such as Wizard® Plus Minipreps DNA purification kit, Promega) and sequence using the BigDye® Terminator v3.1 Cycle Sequencing Kit (Applied Biosystems) following the manufacturer's instructions.
8. Denature the sequencing products and resolve the sequencing products on a suitable automated sequencing platform.
9. Compare the chromatogram with the sequencing reaction for the original uncleaved substrate plasmid using the same primers for both types of constructs.
10. Nucleotides missing in the HE and T4 DNA polymerase treated substrate plasmid when compared to the original untreated substrate sequence define the nucleotides removed by T4 DNA polymerase. The staggered ends (3′ overhangs) are generated by the HE at the cleavage site on the substrate plasmid. This approach works for LAGLIDADG type HEs that typically generate four nucleotide 3′ overhangs at the cleavage site [11, 13].

4 Notes

1. Culture collections: UAMH = University of Alberta Microfungus Collection & Herbarium, Devonian Botanic Garden, Edmonton, AB, Canada, T6G 2E1; CBS = Centraal Bureau voor Schimmelcultures, Utrecht, The Netherlands; ATCC = American Type Culture Collection, P.O. Box 1549, Manassas, VA 20108, USA). Also keep in mind that suitable import permits may have to be obtained for ordering and importing microorganisms.
2. An alternative to directional TOPO cloning may include cloning in pET expression systems (pET28b+) as these also have the N-terminal 6× His-tag. If the desired protein is insoluble (evident from the first few trials), one can switch to fusion protein expression systems such as Glutathione S-Transferase (GST) tagging or the maltose binding protein (MBP) using the pMAL expression and purification system.
3. Codon optimization: Several online programs assist in codon optimization, e.g., http://www.encorbio.com/protocols/Codon.htm, http://genomes.urv.es/OPTIMIZER/. Several commercial outfits will perform codon optimization and gene synthesis such as GenScript (http://www.genscript.com/), GeneArt (Life Technologies), Gene Oracle (Sigma-Aldrich), etc.
4. The reverse PCR primer cannot be complementary to the CACC sequence at the 5′ end of the forward primers.

5. If a single discreet band is not obtained during PCR, increase the annealing temperature of the reaction to optimize the PCR reaction or gel purification of the desired DNA band might be necessary.
6. If the PCR products are large (>3 kb), increasing the ligation time may yield more colonies. The TOPO ligation reaction incubation time can be varied from 30 s to 30 min.
7. You may store the TOPO ligation reaction at −20 °C overnight; however, it is recommended not to store it for more than 24 h.
8. For potentially toxic gene(s) or a gene that encodes an unstable mRNA, or when the overexpressed protein is prone to degradation by the host cell, then the BL21 Star™ (DE3) pLysS strain (Life Technologies/Invitrogen) can be used.

Acknowledgments

This work is supported by a Discovery grant from the Natural Sciences and Engineering Research Council of Canada (NSERC) to G.H. M.H. is supported by the Egyptian Ministry of Higher Education and Scientific Research; and T.K.G. and C.S. would like to acknowledge funding support from the Faculty of Science Graduate Award program (University of Manitoba).

References

1. Michel F, Ferat JL (1995) Structure and activities of group II introns. Annu Rev Biochem 64:435–461
2. Toor N, Hausner G, Zimmerly S (2001) Coevolution of the group II intron RNA structure with its intron-encoded reverse transcriptase. RNA 7:1142–1152
3. Gogarten JP, Senejani AG, Zhaxybayeva O, Olendzenski L, Hilario E (2002) Inteins: structure, function, and evolution. Annu Rev Microbiol 56:263–287
4. Belfort M, Derbyshire V, Parker MM, Cousineau B, Lambowitz AM (2002) Mobile introns: pathways and proteins. In: Craig NL, Craigie R, Gellert M, Lambowitz AM (eds) Mobile DNA II. ASM Press, Washington, DC, pp 761–783
5. Mullineux ST, Costa M, Bassi GS, Michel F, Hausner G (2010) A group II intron encodes a functional LAGLIDADG homing endonuclease and self-splices under moderate temperature and ionic conditions. RNA 16:1818–1831
6. Barzel A, Privman E, Peeri M, Naor A, Shachar E, Burstein D, Lazary R, Gophna U, Pupko T, Kupiec M (2011) Native homing endonucleases can target conserved genes in humans and in animal models. Nucleic Acids Res 39:6646–6659
7. Stoddard BL (2011) Homing endonucleases: from microbial genetic invaders to reagents for targeted DNA modification. Structure 19:7–15
8. Hafez M, Hausner G (2012) Homing endonucleases: DNA scissors on a mission. Genome 55:553–569
9. Hausner G (2003) Fungal mitochondrial genomes, introns and plasmids. In: Arora DK, Khachatourians GG (eds) Applied mycology and biotechnology, vol III, Fungal genomics. Elsevier Science, New York, pp 101–131
10. Haugen P, Bhattacharya D (2004) The spread of LAGLIDADG homing endonuclease genes in rDNA. Nucleic Acids Res 32:2049–2057
11. Sethuraman J, Majer A, Friedrich NC, Edgell DR, Hausner G (2009) Genes-within-genes: multiple LAGLIDADG homing endonucleases

target the ribosomal protein S3 gene encoded within a rnl group I intron of *Ophiostoma* and related taxa. Mol Biol Evol 26:2299–2315

12. Hafez M, Hausner G (2011) The highly variable mitochondrial small-subunit ribosomal RNA gene of *Ophiostoma minus*. Fungal Biol 115:1122–1137
13. Stoddard BL (2006) Homing endonuclease structure and function. Q Rev Biophys 38: 49–95
14. Jacoby K, Metzger M, Shen BW, Certo MT, Jarjour J, Stoddard BL, Scharenberg AM (2012) Expanding LAGLIDADG endonuclease scaffold diversity by rapidly surveying evolutionary sequence space. Nucleic Acids Res 40:4954–4964
15. Prieto J, Molina R, Montoya G (2012) Molecular scissors for in situ cellular repair. Crit Rev Biochem Mol Biol 47:207–221
16. Takeuchi R, Lambert AR, Mak AN, Jacoby K, Dickson RJ, Gloor GB, Scharenberg AM, Edgell DR, Stoddard BL (2011) Tapping natural reservoirs of homing endonucleases for targeted gene modification. Proc Natl Acad Sci U S A 108:13077–13082
17. Stoddard BL, Scharenberg AM, Monnat RJ Jr (2008) Advances in engineering homing endonucleases for gene targeting: ten years after structures. In: Bertolotti R, Ozawa K (eds) Progress in gene therapy 3: autologous and cancer stem cell gene therapy. World Scientific Press, Hackensack, NJ, pp 135–167
18. Marcaida MJ, Muñoz IG, Blanco FJ, Prieto J, Montoya G (2010) Homing endonucleases: from basics to therapeutic applications. Cell Mol Life Sci 67:727–748
19. Taylor GK, Stoddard BL (2012) Structural, functional and evolutionary relationships between homing endonucleases and proteins from their host organisms. Nucleic Acids Res 40:5189–5200
20. Kim WK, Mauthe W, Hausner G, Klassen GR (1990) Isolation of high molecular weight DNA and double-stranded RNAs from fungi. Can J Bot 68:1898–1902
21. Hausner G, Reid J, Klassen GR (1992) Do galeate-ascospore members of the Cephaloascaceae, Endomycetaceae and Ophiostomataceae share a common phylogeny? Mycologia 84:870–881
22. Goddard MR, Burt A (1999) Recurrent invasion and extinction of a selfish gene. Proc Natl Acad Sci U S A 96:13880–13885
23. Taylor GK, Petrucci LH, Lambert AR, Baxter SK, Jarjour J, Stoddard BL (2012) LAHEDES: the LAGLIDADG homing endonuclease database and engineering server. Nucleic Acids Res 40:W110–W116
24. Nicholas KB, Nicholas HB, Deerfield DW II (1997) GeneDoc: analysis and visualization of genetic variation. EMB News 4:14
25. Altschul SF, Madden TL, Schaffer AA, Zhang J, Zhang Z, Miller W, Lipman DJ (1997) Gapped BLAST and PSI-BLAST: a new generation of protein database search programs. Nucleic Acids Res 25:3389–3402
26. Thompson JD, Gibson TJ, Plewniak F, Jeanmougin F, Higgins DG (1997) The Clustal-X windows interface: flexible strategies for multiple sequence alignment aided by quality analysis tools. Nucleic Acids Res 25: 4876–4882
27. Simossis VA, Heringa J (2005) PRALINE: a multiple sequence alignment toolbox that integrates homology-extended and structure information. Nucleic Acids Res 33:W289–W294
28. Felsenstein FJ (2006) PHYLIP (Phylogeny Inference Package). Version 3.6a. Distributed by the author, Department of Genetics, University of Washington, Seattle, WA
29. Ronquist F, Huelsenbeck JP (2003) MRBAYES 3: Bayesian phylogenetic inference under mixed models. Bioinformatics 19: 1572–1574
30. Dereeper A, Guignon V, Blanc G, Audic S, Buffet S, Chevenet F, Dufayard J-F, Guindon S, Lefort V, Lescot M, Claverie J-M, Gascuel O (2008) Phylogeny.fr: robust phylogenetic analysis for the non-specialist. Nucleic Acids Res 36:W465–W469
31. Tamura K, Peterson D, Peterson N, Stecher G, Nei M, Kumar S (2011) MEGA5: molecular evolutionary genetics analysis using maximum likelihood, evolutionary distance, and maximum parsimony methods. Mol Biol Evol 28: 2731–2739
32. Kowalski JC, Derbyshire V (2002) Characterization of homing endonucleases. Methods 28:365–373
33. Sambrook J, Russell DW (2001) Molecular cloning: a laboratory manual, 3rd edn. Cold Spring Harbor Laboratory Press, New York
34. Dafa'alla TH, Hobom G, Zahner H (2000) Direct colony identification by PCR-miniprep. Mol Biol Today 1:65–66
35. Bae H, Kim KP, Song JM, Kim JH, Yang JS, Kwon ST (2009) Characterization of intein homing endonuclease encoded in the DNA polymerase gene of *Thermococcus marinus*. FEMS Microbiol Lett 297:180–188
36. Johansen S, Haugen P (2001) A new nomenclature of group I introns in ribosomal DNA. RNA 7:935–936
37. Crooks GE, Hon G, Chandonia JM, Brenner SE (2004) WebLogo: a sequence logo generator. Genome Res 14:1188–1190

Chapter 4

Mapping Homing Endonuclease Cleavage Sites Using In Vitro Generated Protein

Richard P. Bonocora and Marlene Belfort

Abstract

Mapping the precise position of endonucleolytic cleavage sites is a fundamental experimental technique used to describe the function of a homing endonuclease. However, these proteins are often recalcitrant to cloning and over-expression in biological systems because of toxicity induced by spurious DNA cleavage events. In this chapter we outline the steps to successfully express a homing endonuclease in vitro and use this product in nucleotide-resolution cleavage assays.

Key words Selfish DNA, Intron, Intein, Mobile DNA, Mobile genetic element

1 Introduction

Homing endonucleases (HEs) are site-specific DNA endonucleases that promote the horizontal transfer of the gene encoding them (i.e., homing endonuclease gene or HEG) and flanking DNA. These HEGs are considered selfish/parasitic elements and can be found as free-standing genes or embedded within intervening sequences such as introns and inteins. Typical HEs recognize relatively large sequences (14–40 bp) as compared to restriction endonucleases (4–8 bp). These recognition sequences usually contain a distinguishing characteristic that enables the HE differentiate between a genome containing the HEG and one that lacks the sequence [1–3]. The associated intervening sequence of intron- and intein-encoded HEGs lies within the HE recognition sequence thereby disrupting it and preventing self-cleavage. However, the same uninterrupted sequence in another genome is sensitive to the HE. A HE-induced DNA break initiates a gene conversion event whereby the HEG is copied into the new location (intron/intein homing). The process for free-standing HEGs is similar, but the mode of protection can vary; the recognition site typically contains

David R. Edgell (ed.), *Homing Endonucleases: Methods and Protocols*, Methods in Molecular Biology, vol. 1123, DOI 10.1007/978-1-62703-968-0_4, © Springer Science+Business Media New York 2014

sequence changes that prevent its cleavage (intronless homing) [4], but it has also been observed that an intron, which does not encode the HEG [5] or other genetic element positioned at this location, can prevent cleavage (collaborative homing) [6].

One of the key pieces of information describing the function of an HE is the exact position where the nuclease cuts DNA. The location of the cleavage site (CS) is important for biotechnological application and can also be informative as to how the genome is protected from cleavage. For example, is the CS near an intervening sequence, or are there sequence differences between sensitive and resistant sites? Cleavage information also provides insight into the enzymatic properties of the nuclease, such as whether both strands are cut and the nature of the extensions if there is a double-strand break. The experimental design to map CSs is conceptually straightforward; one simply mixes the endonuclease with potential DNA targets and looks for changes in mobility of the target DNA by gel electrophoresis. The exact cleavage point is mapped by comparing cleavage product migration to a DNA sequencing ladder.

The toxicity associated with HEs is a less tractable problem. Efforts to alleviate the toxicity of certain proteins by tuning their over-expression in vivo have been a long-standing course of investigation [7–9]. Various strategies to resolve this problem have been employed, including the use of tightly controlled araP_{BAD} [10] and rhaP_{BAD} [11] inducible promoters, antisense promoters [12, 13], the generation of new hybrid promoters [14], introduction of RNA polymerase (RNAP) by bacteriophage infection [15], and control of plasmid copy number [16, 17]. One way to overcome this hurdle is to produce the endonuclease in vitro using commercially available cell-free in vitro transcription/translation systems. This method is fast and simple, and since no living cells are involved, toxicity is not an issue [18, 19]. Cell-free protein synthesis has been used to produce several HEGs including I-TevI [20, 21], I-TevII [22], SegF [4], Hef [23], SegA [24], I-DmoI [25], F-CphI [5], MobA [6], MobE [23], I-TslI [3], and I-Ssp6803I [26]. Such studies employing in vitro-synthesized HEs have contributed greatly to our understanding of the structure–function and evolution of these remarkable proteins.

Coupling in vitro-synthesized HEs with radiolabeled DNA targets in endonuclease cleavage assays can provide a vast amount of information about the enzyme. Although such data are generally qualitative in nature, results are produced quickly by simple standard protocols and can be extremely valuable in determining a path that the research will take. In this chapter, we focus on producing a homing endonuclease in vitro using a cell-free system and mapping its cleavage site at nucleotide resolution with single-strand, end-radiolabeled DNA targets.

2 Materials

2.1 Kinasing Reagents

1. T4 polynucleotide kinase (10 Units/mL).
2. 10× Kinase buffer A (500 mM Tris–HCl pH 7.6, 100 mM $MgCl_2$, 50 mM DTT, 1 mM spermidine).
3. γ-^{32}P-ATP (3,000 Ci/mmol; 10 uCi/μL).
4. 5 % Trichloroacetic acid.
5. Glass microfiber filter (Whatman) 2.3 cm grade GF/B.
6. Scintillation cocktail.

2.2 Polymerase Chain Reaction Reagents

1. Taq DNA polymerase (5 Units/mL).
2. 10× PCR buffer (100 mM Tris–HCl pH 8.8, 500 mM KCl, 0.8 % v/v, Nonidet P40).
3. dNTP Mix (2 mM each dATP, dCTP, dGTP, and dTTP).
4. 25 mM $MgCl_2$.

2.3 Sequencing Reagents

1. 2 N NaOH.
2. 5 mM EDTA.
3. Sequenase v2.0 (Affymetrix/USB).
4. 5× Sequenase buffer (200 mM Tris–HCl, pH 7.5, 100 mM $MgCl_2$, 250 mM NaCl).
5. Modified labeling mix (7.5 μM each dNTP).
6. ddATP, ddCTP, ddGTP, and ddTTP mixes (Each mix contains: 80 μM each dNTP, 8 μM specific ddNTP, 50 mM NaCl).
7. 0.1 M DTT.
8. Sequencing loading dye (95 % formamide, 20 mM EDTA, 0.05 % bromophenol blue, 0.05 % xylene cyanol FF).

2.4 In Vitro Transcription/Translation Reagents

1. TNT T7 Quick Coupled Transcription/Translation System (Promega) *see* **Note 1** for discussion of different cell-free protein expression systems.
2. 1 mM methionine.
3. ^{35}S-methionine (>1,000 Ci/mmol).
4. Nuclease-free water.

2.5 Endonuclease Assay Reagents

1. 10× ECA buffer (0.5 M Tris–HCl pH 7.5, 0.5 M NaCl).
2. 0.25 mg/mL Poly (dI-dC) (Sigma).
3. 0.1 M $MgCl_2$.
4. Phenol, saturated, pH 6.6/7.9.

2.6 Denaturing Polyacrylamide Gel Electrophoresis (PAGE) Reagents

1. 40 % (19:1 acrylamide–bis acrylamide) polyacrylamide (*see* **Note 2**).
2. 10× Tris-borate EDTA (TBE) buffer (890 mM Tris, 890 mM Boric Acid, 20 mM EDTA, pH 8.3).
3. Ultrapure urea.
4. 10 % ammonium per sulfate (APS) stored at 4 °C.
5. *N,N,N,N′*-tetramethyl-ethylenediamine (TEMED) stored at 4 °C.

2.7 Molecular Biology

1. Plasmid miniprep kit.
2. PCR purification kit.

2.8 Equipment

1. Thermal cycler.
2. Spectrophotometer.
3. High-voltage power supply.
4. Phosphor imager.
5. Model S2 Sequencing Gel Electrophoresis Apparatus (or equivalent) (*see* **Note 3**).
6. Glass plates (30 cm × 40 cm) for Gel Electrophoresis Apparatus.
7. 0.4 mm spacers and combs to fit Gel Electrophoresis Apparatus.
8. Whatman filter paper (30 cm × 40 cm).
9. Gel dryer.
10. Vacuum apparatus to fit glass microfiber filters for scintillation counting.
11. Scintillation counter.

3 Methods

The overall strategy is straightforward. A primer is designed to incorporate a T7 promoter upstream of a HEG (Fig. 1a, *see* **Note 4** for discussion of optimal parameters) for PCR amplification. This PCR product is mixed with a coupled in vitro T7 RNA polymerase transcription/translation system and the T7 promoter directs transcription of the PCR amplified HEG. The RNA is then translated by ribosomes in the cell-free system producing the homing endonuclease (Fig. 1b). This in vitro-synthesized HE is mixed directly with the PCR-generated putative target DNA that has been radiolabeled on the 5′ end of a single strand (Fig. 1c). Double-strand cleavage by the HE results in a total of four DNA strands, only one of which is physically connected to the radioactive atom (Fig. 1d). The products are then separated by denaturing gel electrophoresis alongside a DNA sequencing ladder containing DNA molecules with the identical 5′ radiolabeled end as the substrate DNA (Fig. 2a).

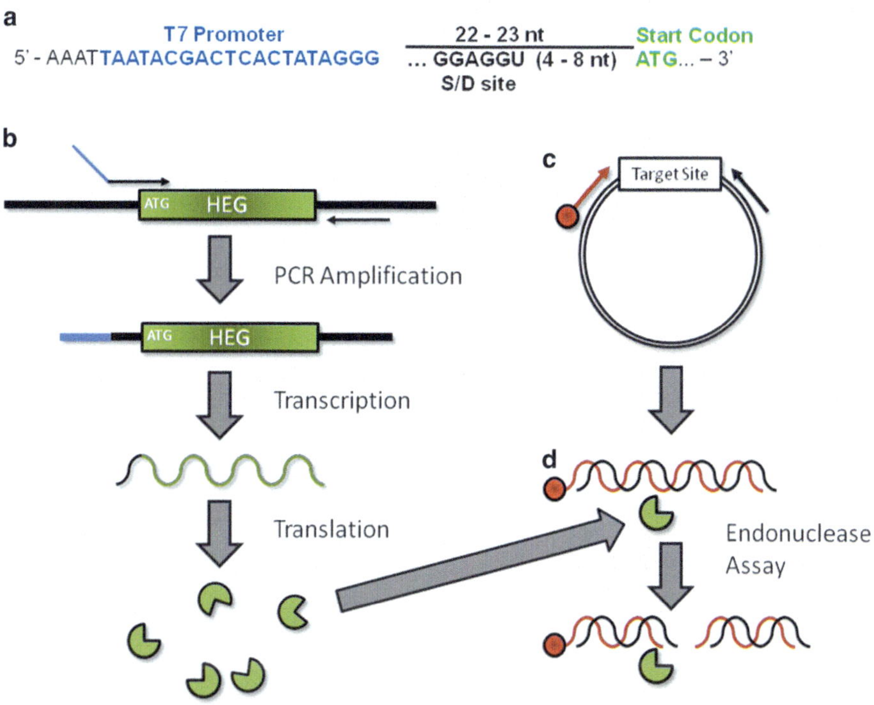

Fig. 1 Schematic of in vitro homing endonuclease expression and endonuclease assay. (**a**) Upstream primer design for amplification of a HEG. T7 promoter (*blue*) should be placed ~22–23 bp upstream of the HEG initiation codon (*green*). Sequence at the 5′ end in black is thought to stabilize RNAP–promoter interaction. If using a prokaryotic system a ribosome binding site/Shine–Delgarno sequence (S-D) should be incorporated. (**b**) In vitro homing endonuclease production. PCR amplification results in the incorporation of a T7 promoter (*blue*) upstream of the HEG. This product is used to direct protein synthesis in the coupled in vitro transcription/translation reaction. (**c**) Target site DNA is amplified from a plasmid with a 5′-end labeled (*red*) and unlabeled primer resulting in a duplex DNA molecule labeled on one strand. The same labeled primer and plasmid DNA is used to generate a DNA sequencing ladder. The resulting ladder and the PCR product are labeled at the exact same position. (**d**) Endonuclease Assay. The HE (from **b**) is mixed with precursor dsDNA labeled on the 5′-end. Cleavage of both strands results in two dsDNA products. Of the resulting four individual strands, only one is covalently linked to the label and therefore visible by phosphor imaging

3.1 Generation of Singly 5′-End Labeled Target DNA Substrate

1. Kinase primer by mixing 9.5 μL H_2O, 1.5 μL 10× buffer A, 1.0 μL 6 μM oligo, 2.0 μL γ-^{32}P-ATP, and 1.0 μL T4 PNK.
2. Incubate at 37 °C for 30 min.
3. Place reactions on ice or store at −20 °C.
4. PCR amplify the target DNA by adding 56.0 μL H_2O, 10.0 μL 10× PCR buffer, 10.0 μL dNTP mix, 6.0 μL 25 mM $MgCl_2$, 1.0 μL 6 μM unlabeled paired oligonucleotide, 1.0 μL template DNA (1:100 dilution of a miniprep), and 1.0 μL Taq DNA polymerase to the labeled primer.
5. Incubate for the following cycle: 95 °C for 10 min (hot start), followed by 25 cycles of 95 °C for 30 s, 50 °C for 30 s, 72 °C for 30 s. Do a final extension at 72 °C for 5 min (*see* **Note 5**).

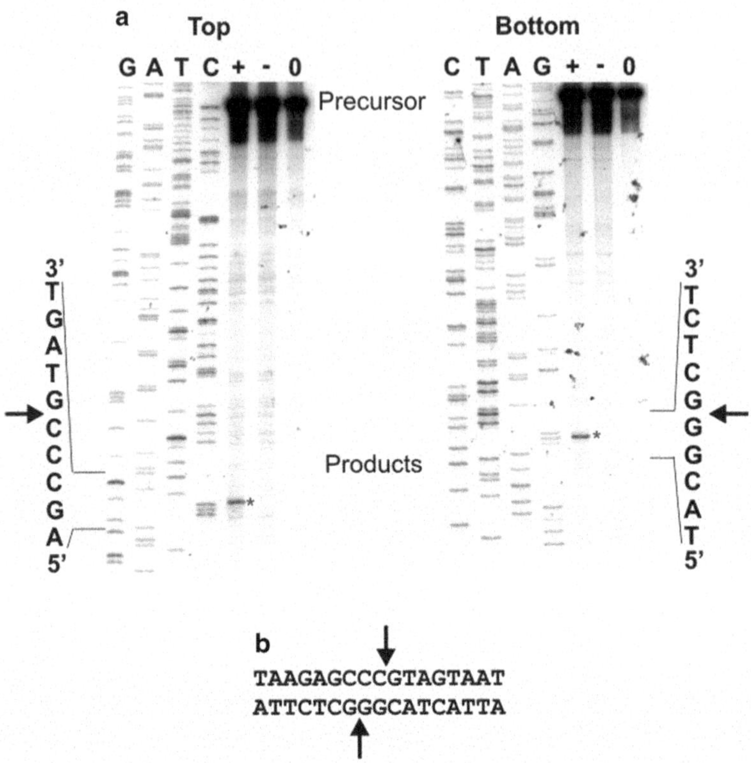

Fig. 2 Cleavage site mapping using in vitro-synthesized protein. (**a**) Precursor/substrate DNA was amplified, and each strand was individually 5′-end-labeled with ^{32}P in different PCRs. Sequencing ladders were generated using the corresponding labeled primer. The ddNTP used in the sequencing reactions is indicated above each lane. The sequence of the target DNA immediately flanking the cut site (represented by an *arrowhead*) is indicated next to each autoradiogram image. Precursor/substrate DNA was incubated with in vitro-synthesized protein (+), a mock unprogrammed in vitro synthesis reaction (–) or DNA only (0) and the reactions were separated on a denaturing polyacrylamide gel by electrophoresis. The regions where the precursors and products (*asterisk*) migrate are indicated. (**b**) Summary of the cleavage reaction. The DNA sequence for a portion of the region is shown below the images. The cut sites on each strand are indicated by *arrows* and result in 2 nt 3′ extensions

6. The PCR product should be purified with a commercially available PCR cleanup kit and eluted in 30–50 μL H_2O.
7. Reactions should be placed on ice or stored at −20 °C.
8. Check incorporation of label by placing a glass microfiber filter onto the vacuum apparatus and wet filter with 5 % TCA.
9. Add 1 μL of purified PCR product and wash for 1 min with 5 % TCA followed by 95 % ethanol to help dry the filter.
10. Place filter in scintillation vial with scintillation cocktail and read on ^{32}P channel. This allows calculation of the amount of radiolabel incorporated into the DNA.

11. Measure total label by placing 1 μL of purified PCR product on glass filter and add directly (no washing) to scintillation vial with cocktail (*see* **Note 6**).

3.2 Generation of DNA Sequencing Ladder

1. Kinase primer by mixing 1.5 μL H_2O, 0.5 μL 10× Buffer A, 1.0 μL 2 μM oligo, 2.0 μL γ-^{32}P-ATP, and 0.2 μL T4 PNK.
2. Incubate at 37 °C for 30 min.
3. Place reactions on ice or store at −20 °C.
4. Prepare plasmid DNA from 1.5 to 3.0 mL culture using a commercially available miniprep kit.
5. Elute DNA in 100 μL H_2O.
6. Denature plasmid DNA for sequencing by mixing 15.0 μL plasmid DNA, 1.75 μL 2 N NaOH, and 0.7 μL 5 mM EDTA.
7. Incubate at 37 °C for 30 min.
8. Precipitate by adding 1.7 μL 3 M sodium acetate and 50.0 μL 100 % ethanol and vortex.
9. Incubate for at least 30 min at −80 °C or > 2 h at −20 °C.
10. Centrifuge at high speed for 10 min.
11. Remove the supernatant.
12. Dry and resuspend the pellet in 5.4 μL H_2O.
13. Anneal primer to plasmid DNA by mixing 5.4 μL denatured plasmid DNA, 2.6 μL kinased primer, and 2.0 μL 5× Sequenase buffer.
14. Heat to >90 °C and cool slowly to 37 °C in a heat block with heat turned off.
15. Spin briefly and place on ice.
16. Aliquot 2.5 μL of each ddNTP mix into four separate tubes labeled G, A, T, and C that have been pre-warmed to 37 °C.
17. For the extension and termination reactions add 1.0 μL 0.1 M DTT, 2.0 μL modified labeling mix and 0.5 μL H_2O to denatured DNA mix.
18. Begin extension reaction by adding 2.5 μL of a 1:8 dilution of Sequenase (in the provided dilution buffer).
19. Incubate for 4 min at 37 °C.
20. Aliquot 3.5 μL of this extension mix to the pre-aliquotted ddNTP termination mix.
21. Incubate at 37 °C for 4 min.
22. Add 5 μL sequencing loading dye.
23. Place reactions on ice or store at −20 °C.

3.3 Amplification of HEG for In Vitro Expression

1. PCR amplify the HEG by mixing 68.0 μL H_2O, 10.0 μL 10× PCR buffer, 10.0 μL dNTP mix, 6.0 μL 25 mM $MgCl_2$, 2.0 μL 20 μM upstream (T7 promoter-containing) primer (*see* **Note 4**), 2.0 μL 20 μM downstream primer, 1.0 μL template DNA, and 1.0 μL Taq DNA polymerase.
2. Incubate for the following cycle 95 °C for 10 min (hot start), followed by 25 cycles of: 95 °C for 30 s, 50 °C for 30 s, 72 °C for 1 min (*see* **Notes 5** and **7**).
3. Purify the PCR product with a commercially available PCR cleanup kit and elute in 30–50 μL H_2O.
4. Quantify DNA on a spectrophotometer.

3.4 Generation of In Vitro-Synthesized Homing Endonuclease

1. To produce the homing endonuclease mix 40 μL TNT T7 Quick Master Mix, 1 μL 1 mM methionine, 2.5–5 μL PCR-generated DNA template (~100–800 ng), and bring the total volume up to 50 μL with H_2O (*see* **Note 8**).
2. In parallel produce a mock in vitro synthesis control (*see* **Note 9**) by mixing 40 μL TNT T7 Quick Master Mix, 1 μL 1 mM methionine, and 9 μL H_2O.
3. Incubate the reactions at 30 °C for 60–90 min.
4. The protein products are ready to use and should be kept on ice or frozen at −70 °C in aliquots to prevent numerous freeze–thaw cycles for storage.

3.5 Endonucleolytic Cleavage Assay (EC Assay)

1. Thaw all reagents at 25 °C and place on ice before mixing.
2. Gently Mix 2 μL 10× ECA buffer, 2 μL 0.25 mg/mL Poly (dI-dC), 2 μL 0.1 M $MgCl_2$, ~10^5 cpm target DNA, 2 μL in vitro-synthesized protein, and H_2O to 20 μL on ice.
3. Briefly centrifuge to move all liquid to the bottom of the tube.
4. Incubate reactions at 30 °C for 30 min (*see* **Note 10**).
5. Stop reactions on ice.

3.6 Phenol Extraction

1. Add an equal volume (20 μL) of phenol.
2. Vortex and centrifuge for 2 min.
3. Transfer aqueous phase to a fresh 1.7 mL tube.
4. Add 5 μL sequencing loading buffer to each reaction.

3.7 Denaturing Polyacrylamide Gel

1. To make an 8 % polyacrylamide/7 M urea denaturing gel mix 38 mL H_2O, 20 mL 40 % polyacrylamide, 10 mL 10× TBE buffer, and 36.7 g urea in a 250 mL beaker with a stir bar (*see* **Note 11**).
2. Gently heat and stir the mixture to dissolve the urea (*see* **Note 12**). While the solution is mixing proceed to **step 3**.

3. Clean glass plates, spacers, and combs with H_2O and 95 % ethanol. Assemble glass plates and spacers. Tape the bottom and sides of the gel to prevent leaking.
4. Add 0.5 mL 10 % APS and 20 μL TEMED to the dissolved gel solution, mix well, and slowly pour the gel solution into one corner of the opening between the two plates with this "sandwich" held at a ~45° angle, until the gel is filled.
5. Lay the gel flat, place the comb between the glass plates, and clamp each side of the gel with three clamps per side. Allow the gel to polymerize completely (typically at least an hour).
6. Before running the gel, remove the comb and tape carefully. Place the gel into the electrophoresis apparatus and fill the upper and lower tanks with 1× TBE, ensuring that the glass plates are submerged in buffer.

3.8 Electrophoresis

1. Pre-run the gel at 60 W for at least 30 min (*see* **Note 13**).
2. Heat the samples at 95 °C for 5 min, place on ice for 1–2 min, centrifuge for 30 s, and place the tubes back on ice.
3. Prior to loading, rinse the wells with buffer using a syringe, being careful not to damage the wells.
4. Load 1–10 μL into each well (*see* **Note 14**).
5. Run gel for 2.5 h at a constant 60 W (*see* **Note 11**).
6. When the electrophoresis is complete, carefully disassemble the gel sandwich by prying the glass plates apart with a spatula. If the gel remains attached to only one plate, press a large filter paper on top of the gel, flip so that the paper is on the bottom, and transfer the gel to the paper. If the gel remains attached to both plates, gently vibrate the glass plate to help the gel fall onto one plate.
7. Place on gel dryer and cover with plastic wrap. Dry under vacuum with heat until completely dry (~30–60 min) (*see* **Note 15**).
8. Expose to film or phosphor imaging screen overnight (*see* **Note 16**).

4 Notes

1. Cell-free protein synthesis systems are mainly derived from *E. coli*, rabbit reticulocyte lysates and wheat germ extracts. The protocol described here uses a rabbit reticulocyte cell-free system, but can easily be adapted for other cell-free protein synthesis systems. In fact, cell-free systems derived from wheat germ and *E. coli* have also been successfully used in our hands [20, 21, 26]. The manufacturers provide detailed protocols for using their systems and typically no modification is necessary.

The *E. coli* system generally provides the highest yield, but our personal experience suggests that this system can be contaminated by DNA exonucleases and is not the best choice when studying homing endonucleases. We have not experienced any nonspecific nuclease problems with either of the eukaryotic systems. Additionally, eukaryotic systems have better RNA unwinding activities and should be used when translational start signals are sequestered in stem-loop structures.

2. Acrylamide is a neurotoxin. Use protective equipment when handling.
3. Our preferred electrophoresis apparatus is a Model S2 Sequencing Gel Electrophoresis Apparatus that fits 30 cm × 40 cm gel sequencing plates. This apparatus is equipped with an aluminum plate that disperses heat generated during the electrophoresis and minimizes gel "smiling." It also has a convenient built-in drain for the top buffer chamber.
4. The upstream primer needs to incorporate a T7 promoter at a sufficient distance upstream of the translational start of the HEG (~22 bp Fig. 1a). The sequence between the T7 promoter and translational start can anneal to the template though this is not necessary. If using a prokaryotic cell-free extract, production of the HE will require the inclusion of a ribosome binding or Shine–Delgarno (S-D) site (5′-GGAGGU-3′) between 4 and 8 nt upstream of the translational start [27]. A naturally occurring site can be used, but if no site exists upstream of the HEG, one can be incorporated into the upstream primer.
5. The annealing temperature, extension time and number of cycles should be determined empirically for each set of oligonucleotides and template. As the optimal PCR conditions will be dependent on the particular template and primers, the conditions should first be established using unlabeled primers. In our experience, for cleavage site mapping target DNA PCR products between 0.2 and 0.4 kb work best and therefore only a short extension time is necessary.
6. The yield of labeled DNA ranges from 10^3 to 10^5 cpm/μL depending on the individual oligonucleotide used.
7. To ensure that the HEG is wild type, the template should be genomic DNA since clones of HEGs are often mutated. However, if a stable, wild-type clone of the HEG exists that is not conducive to over-expression that would work also.
8. The Master Mix should be thawed, distributed into 20 μL or 40 μL aliquots and frozen at −70 °C. When handling the Master Mix, it should be placed on ice when not used in incubation. Also, protein synthesis can be monitored by the incorporation of radioactive methionine. Simply substitute 2 μL ^{35}S-methionine for the unlabeled methionine.

The technical manual for this system is quite detailed and no modifications are necessary.

9. A mock synthesis reaction, where the cell-free system lacks a PCR product and is therefore not programmed to produce an HEG should be performed in parallel. This allows non-HE derived cleavages of the target to be "subtracted" from the result.

10. Endonuclease cleavage assay conditions including: buffer, pH, monovalent ion and divalent ion, temperature, and time should be optimized for each protein, although the conditions set out here are a good starting point. We use a lower temperature (30 °C vs. 37 °C) to minimize the action of nonspecific nucleases. Also, each set of reactions should include the HEG programmed in vitro synthesis reaction, a mock synthesis reaction and DNA only control (Fig. 2a).

11. The percentage gel and duration of run required for optimal resolution will vary depending on fragment lengths and will need to be determined empirically. This protocol is based on product fragments between 0.1 and 0.2 kb.

12. Make sure that the solution is not too warm or the gel may polymerize too quickly.

13. Wattage, pre-run and run conditions will vary depending on apparatus, gel thickness, and buffer composition. This has been optimized for 1× TBE gels cast with 0.4 mm spacers on a Model S2 apparatus. Thicker gels need to be pre-run and run at a lower wattage or voltage.

14. The ratio of sequencing reaction to cleavage assay reaction will have to be determined by trial and error and depends on the quality of each of the reactions.

15. It is imperative that the gel is completely dry; otherwise it may crack when removed from the vacuum.

16. For ease of interpretation, the reaction containing the in vitro-synthesized HEG, and therefore, the cleavage product to be mapped should be loaded immediately adjacent to the sequencing ladder. Ideally it should be loaded next to the ddNTP lane that corresponds to the last base contained in the cleavage product (Fig. 2a). This allows for the most accurate mapping of the cleavage site. To interpret the data, one simply has to look for the ddNTP band that aligns with the cleavage product. This corresponds to the 3′-most base contained within the cleavage product and the DNA strand is cut immediately 3′ to this base (Fig. 2a, b). Sometimes, the sequencing band that lines up with the cleavage product remains ambiguous. Since the sequencing and cleavage reactions are contained in different buffers and are not typically precipitated, slight migration differences can occur. If this is a problem, both sets of reactions could be phenol extracted, precipitated and resuspended in loading dye.

Alternatively, the cleavage reactions and the sequencing reaction can be mixed before loading. In this instance, the product should produce an extra band in each of the sequencing reactions except for the ddNTP that has a band that aligns with the product (*see* Fig. 4.3a in [3] for an example).

Acknowledgements

We would like to thank Caren J. Stark, Matthew Stanger, Dorie Smith, and Carol Lyn Piazza for critical reading of the manuscript. Research in the Belfort Lab is supported by NIH grants GM39422 and GM44844.

References

1. Lambowitz AM, Belfort M (1993) Introns as mobile genetic elements. Annu Rev Biochem 62:587–622
2. Stoddard BL (2011) Homing endonucleases: from microbial genetic invaders to reagents for targeted DNA modification. Structure 19(1): 7–15
3. Bonocora RP, Shub DA (2009) A likely pathway for formation of mobile group I introns. Curr Biol 19(3):223–228
4. Belle A, Landthaler M, Shub DA (2002) Intronless homing: site-specific endonuclease SegF of bacteriophage T4 mediates localized marker exclusion analogous to homing endonucleases of group I introns. Genes Dev 16(3):351–362
5. Zeng Q, Bonocora RP, Shub DA (2009) A free-standing homing endonuclease targets an intron insertion site in the psbA gene of cyanophages. Curr Biol 19(3):218–222
6. Bonocora RP et al (2011) A homing endonuclease and the 50-nt ribosomal bypass sequence of phage T4 constitute a mobile DNA cassette. Proc Natl Acad Sci U S A 108(39): 16351–16356
7. Studier FW (1991) Use of bacteriophage T7 lysozyme to improve an inducible T7 expression system. J Mol Biol 219(1):37–44
8. Moffatt BA, Studier FW (1987) T7 lysozyme inhibits transcription by T7 RNA polymerase. Cell 49(2):221–227
9. Studier FW, Moffatt BA (1986) Use of bacteriophage T7 RNA polymerase to direct selective high-level expression of cloned genes. J Mol Biol 189(1):113–130
10. Guzman LM et al (1995) Tight regulation, modulation, and high-level expression by vectors containing the arabinose PBAD promoter. J Bacteriol 177(14):4121–4130
11. Giacalone MJ et al (2006) Toxic protein expression in *Escherichia coli* using a rhamnose-based tightly regulated and tunable promoter system. Biotechniques 40(3):355–364
12. Worrall AF, Connolly BA (1990) The chemical synthesis of a gene coding for bovine pancreatic DNase I and its cloning and expression in *Escherichia coli*. J Biol Chem 265(35): 21889–21895
13. O'Connor CD, Timmis KN (1987) Highly repressible expression system for cloning genes that specify potentially toxic proteins. J Bacteriol 169(10):4457–4462
14. Lutz R, Bujard H (1997) Independent and tight regulation of transcriptional units in *Escherichia coli* via the LacR/O, the TetR/O and AraC/I1-I2 regulatory elements. Nucleic Acids Res 25(6):1203–1210
15. Miao F, Drake SK, Kompala DS (1993) Characterization of gene expression in recombinant *Escherichia coli* cells infected with phage lambda. Biotechnol Prog 9(2):153–159
16. Bowers LM et al (2004) Bacterial expression system with tightly regulated gene expression and plasmid copy number. Gene 340(1): 11–18
17. Saida F et al (2006) Expression of highly toxic genes in *E. coli*: special strategies and genetic tools. Curr Protein Pept Sci 7(1):47–56
18. Katzen F, Chang G, Kudlicki W (2005) The past, present and future of cell-free protein synthesis. Trends Biotechnol 23(3):150–156
19. Katzen F, Peterson TC, Kudlicki W (2009) Membrane protein expression: no cells required. Trends Biotechnol 27(8):455–460

20. Bell-Pedersen D et al (1989) A site-specific endonuclease and co-conversion of flanking exons associated with the mobile td intron of phage T4. Gene 82(1):119–126
21. Bell-Pedersen D et al (1991) I-TevI, the endonuclease encoded by the mobile td intron, recognizes binding and cleavage domains on its DNA target. Proc Natl Acad Sci U S A 88(17):7719–7723
22. Bell-Pedersen D et al (1990) Intron mobility in phage T4 is dependent upon a distinctive class of endonucleases and independent of DNA sequences encoding the intron core: mechanistic and evolutionary implications. Nucleic Acids Res 18(13):3763–3770
23. Sandegren L, Nord D, Sjoberg BM (2005) SegH and Hef: two novel homing endonucleases whose genes replace the mobC and mobE genes in several T4-related phages. Nucleic Acids Res 33(19):6203–6213
24. Sharma M, Ellis RL, Hinton DM (1992) Identification of a family of bacteriophage T4 genes encoding proteins similar to those present in group I introns of fungi and phage. Proc Natl Acad Sci U S A 89(14):6658–6662
25. Dalgaard JZ, Garrett RA, Belfort M (1993) A site-specific endonuclease encoded by a typical archaeal intron. Proc Natl Acad Sci U S A 90(12):5414–5417
26. Bonocora RP, Shub DA (2001) A novel group I intron-encoded endonuclease specific for the anticodon region of tRNA(fMet) genes. Mol Microbiol 39(5):1299–1306
27. Shine J, Delgarno L (1975) Determinant of cistron specificity in bacterial ribosomes. Nature 254:34–38

Chapter 5

Mapping Free-Standing Homing Endonuclease Promoters Using 5′RLM-RACE

Ewan A. Gibb

Abstract

5′RLM-RACE is a PCR-based technique used to map the 5′ termini of transcripts in both eukaryotic and prokaryotic organisms. Free-standing homing endonuclease promoters often lack recognizable promoters making predicting the transcriptional start site challenging. Furthermore, homing endonucleases are often expressed at very low levels making transcript mapping a challenge. Here, I present a 5′RLM-RACE protocol with special considerations for the expected abundance of homing endonucleases and for their potential to be subjected to RNA processing events.

Key words Homing endonuclease, 5′RLM-RACE, Promoter mapping

1 Introduction

The regulation of homing endonuclease expression is poorly understood. This is especially true when considering the free-standing homing endonucleases, where the endonuclease open reading frame is inserted between host genes, effectively integrating into an existing host operon [1–5]. Some of these endonucleases may have ambiguous—or entirely lack—promoters, leaving the precise transcript initiation site difficult to predict. Moreover, transcripts encoding homing endonuclease open reading frames can be subject to multiple layers of regulation, including RNA processing by various cellular RNases [6–9]. These regulatory features may limit homing endonuclease expression to very low levels. Collectively, these features make experimental mapping of free-standing endonucleases challenging.

A number of experimental approaches are suitable for mapping homing endonuclease promoters, including primer extension [10], RNase protection assays (RPA) [11], and RNA ligase-mediated rapid amplification of cDNA ends (RLM-RACE) [12]. In the case of the latter, there are several advantages: (1) RLM-RACE does not require radioactivity or specialized gel apparatus, (2) RLM-RACE is

David R. Edgell (ed.), *Homing Endonucleases: Methods and Protocols*, Methods in Molecular Biology, vol. 1123, DOI 10.1007/978-1-62703-968-0_5, © Springer Science+Business Media New York 2014

suitable for mapping variant initiation sites and/or RNA processing events while simultaneously mapping the transcript starts, and (3) as RLM-RACE is a PCR based approach, it is especially useful in mapping the 5′ termini of homing endonucleases which are expressed at low levels. Here, I describe a 5′RLM-RACE protocol adopted to free-standing endonucleases.

2 Materials

Prepare all solutions with DEPC-treated (0.1 %) or nuclease-free ultrapure water (prepared by purifying deionized water to attain a sensitivity of 18 MΩ cm at 25 °C) and analytical grade reagents. Diligently follow RNA-free handling protocols to avoid contamination with RNases and RNA degradation. Total RNA should be of the highest possible quality (*see* **Note 1**) and suspended in nuclease-free ultrapure water.

2.1 Dephosphorylation Reaction

1. High-quality total RNA.
2. Tobacco Acid Pyrophosphatase Reaction Buffer (10×): 50 mM sodium acetate (NaOAc; pH 6.0), 1 % β-mercaptoethanol, 10 mM ethylenediaminetetraacetic acid (EDTA), 0.1 % Triton X-100. Store in aliquots at −20 °C.
3. Water bath or heating block set to 37 °C.
4. Commercial or DEPC treated nuclease-free H_2O.
5. Tobacco Acid Pyrophosphatase (TAP).
6. Sterile, RNase-free 1.5 ml Eppendorf tubes.
7. Ethanol (95–100 %).
8. Sodium acetate (3 M, pH 5.2).

2.2 RNA Adaptor Ligation Reaction

1. Dephosphorylated RNA (from Subheading 3.1).
2. RNA Ligase Buffer (10×): 500 nM Tris–HCl, pH 7.5 at 25 °C, 100 mM $MgCl_2$, 10 mM ATP, 10 mM Dithiothreitol (DTT; *see* **Note 2**). Store in aliquots at −20 °C.
3. RNAadaptor:5′GCUGAUGGCGAUGAAUGAACACUGCG UUUGCUGGCUUUGAUGAAA.
4. T4 RNA ligase.
5. Commercial silica-based RNA purification columns. We used RNeasy columns from Qiagen.
6. Commercial or DEPC treated nuclease-free H_2O.
7. Water bath or heating block set to 37 °C.
8. Sterile, RNase-free 1.5 ml Eppendorf tubes.

2.3 Reverse Transcription Reaction

1. RT buffer (10×): 500 mM Tris–HCl, pH 8.3 at 25 °C, 750 mM KCl, 30 mM $MgCl_2$, 100 mM DTT. Store in aliquots at −20 °C.
2. Homing endonuclease specific primer (20 μM).
3. Thermocycler.
4. Commercial or DEPC treated nuclease-free H_2O.
5. Sterile, RNase-free 100 μl PCR tubes.
6. dNTPs (10 mM); mix 10 μl each stock (100 mM) dATP, dGTP, dCTP, and dTTP with 360 μl nuclease-free H_2O.
7. Reverse transcriptase.
8. RNase inhibitor (optional).
9. Pre-chilled cooling block or ice.
10. Column purified, adaptor-ligated RNA (from Subheading 3.2).

2.4 PCR1: Outer PCR

1. PCR Reaction Buffer (10×): 100 mM Tris–HCl, pH 8.3 at 25 °C, 500 mM KCl, 15 mM $MgCl_2$. Store in aliquots at −20 °C.
2. dNTPs (10 mM); prepared as Subheading 2.3, **item 6**.
3. Homing endonuclease specific reverse primer (10 μM).
4. Outside adaptor primer (10 μM): 5′ GCTGATGGCGATGAAT GAACACTG.
5. Taq DNA polymerase.
6. MilliQ H_2O, the same nuclease-free water as Subheadings 2.1 and 2.2 can be used, but it is not necessary.
7. Thermocycler.
8. Sterile 100 μl PCR tubes.
9. RT-PCR reaction as prepared in Subheading 3.3.

2.5 PCR2: Nested PCR

1. PCR Reaction Buffer (10×): prepared as Subheading 2.4, **item 1**.
2. dNTPs (10 mM); prepared as Subheading 2.3, **item 6**.
3. Homing endonuclease specific reverse primer (10 μM).
4. Inside adaptor primer: 5′ ACACTGCGTTTGCTGGCTTT GATG.
5. Taq DNA polymerase.
6. MilliQ H_2O, the same nuclease-free water as Subheadings 2.1 and 2.2 can be used, but it is not necessary.
7. Thermocycler.
8. Sterile 100 μl PCR tubes.
9. PCR1 reaction as prepared in Subheading 2.4.

2.6 Gel Purification

1. Ethanol (95–100 %).
2. Sodium acetate (3 M, pH 5.2).
3. Electrophoresis grade agarose.

4. TAE buffer (10×): 48.4 g of Tris–HCl, 3.7 g of EDTA, and 11.4 ml of glacial acetic acid (17.4 M) brought to 1 l with deionized water. Dilute to 1× for TAE running buffer.
5. Commercial gel extraction kit. We used the QIAquick gel extraction kit from Qiagen.
6. 6× DNA loading dye: 0.25 % bromophenol blue, 0.25 % xylene cyanol FF, 30 % glycerol in water.

2.7 Cloning and Sequencing

1. Gel-purified 5′RLM-RACE products from Subheading 3.5.
2. T4 DNA ligase buffer (10×): 500 mM Tris–HCl, pH 7.5 at 25 °C, 100 mM $MgCl_2$, 10 mM ATP, 100 mM Dithiothreitol.
3. T4 DNA ligase.
4. Sterile deionized H_2O.
5. Water bath set to 16 °C.
6. Cloning vector for TA cloning. We used the pCR® 2.1 vector from Invitrogen.
7. Competent *E. coli* cells.
8. Water bath set to 42 °C.
9. SOC media: 2 g tryptone, 0.5 g yeast extract, 0.2 ml of 5 M NaCl, 0.25 ml of 1 M KCl, 1 ml of 1 M $MgCl_2$, 1 ml of 1 M $MgSO_4$, and 2 ml of 1 M glucose brought to 100 ml with deionized H_2O. Sterilize by autoclaving.
10. LB plates with appropriate antibiotic.
11. Incubator set to 37 °C.

3 Methods

Decontaminate all surfaces and pipettes using commercially available RNAse decontamination solutions prior to starting the experiments (we used RNaseZap, Ambion). All reactions should be assembled on ice to minimize RNA degradation. While the RNA can be frozen at each intermediate step, it is not recommended to do so, as multiple freeze–thaw cycles have been shown to promote RNA degradation. Instead, it would be best to complete up to Subheading 3.3 (cDNA conversion) prior to freezing the samples. Starting material should be 10 μg of high-quality total RNA as assessed by denaturing gel electrophoresis or by an Agilent 2100 bioanalyzer.

3.1 Removing Triphosphates from Unprocessed Transcripts Using TAP

1. Keeping samples on ice add the following components to a sterile 1.5 ml Eppendorf tube: 10 μg total RNA in H_2O, 2 μl TAP (10×) buffer and 25 units TAP or H_2O for control reaction (*see* **Note 3**). Bring to 20 μl with nuclease-free H_2O.

2. Mix by pipetting or gentle vortexing, spin briefly, and incubate at 37 °C for 1 h, then place on ice.
3. Next, precipitate the RNA with 2 μl 3 M NaOAc and 60 μl of 95–100 % ethanol. Set on ice for 30 min, and centrifuge at max speed for 30 min.
4. Remove the ethanol, air-dry briefly to remove ethanol traces, and resuspend the RNA pellet in 30 μl nuclease-free H_2O.
5. Place reaction on ice. Inactivating the enzyme is not required.

3.2 Ligating Adaptor RNAs to Decapped Substrates

1. In sterile 1.5 ml Eppendorf tubes, assemble the following components on ice: 30 μl dephosphorylated RNA (from Subheading 3.1), 4 μl 10× T4 RNA ligase buffer, 1 μl RNA adaptor (1.5 μg/μl), and 40 units T4 RNA ligase (*see* **Notes 4** and **5**). Bring to a total volume of 40 μl with nuclease-free H_2O.
2. Mix by pipetting or gentle vortexing, spin briefly, and incubate at 37 °C for 1 h.
3. Purify the adaptor-ligated total RNA using Qiagen RNeasy columns according to the manufacturer's instructions (*see* **Note 6**).
4. Elute in 40 μl nuclease-free water.
5. Place reaction on ice. Inactivating the enzyme is not required.

3.3 Converting Adaptor-Ligated RNA to cDNA

1. Keeping samples on ice add the following components to a sterile 100 μl PCR tube: 1 μg adaptor-ligated total RNA in 10 μl H_2O and 1 μl 20 μM homing endonuclease-specific primer (*see* **Note 7**). Mix and incubate in a pre-warmed 70 °C thermocycler or water bath for 5 min.
2. Immediately snap cool on ice or pre-chilled cooling block.
3. Next, add the following components to the reaction: 5 μl dNTPs (10 mM), 2 μl 10× RT buffer, 0.5 μl RNase inhibitor, and 1 μl reverse transcriptase (200 units).
4. Carefully mix by pipetting or gentle vortexing, spin briefly, and cycle reactions at 25 °C for 5 min, 37 °C for 1 h, and 72 °C for 10 min.
5. Place on ice.

3.4 Converting cDNA to dsDNA (PCR1)

1. In 100 μl PCR tubes mix the following components on ice: 36 μl H_2O, 5 μl 10× PCR buffer, 5 μl dNTPs (10 mM), 1 μl (10 μM) outer-adaptor primer, 1 μl (10 μM) homing endonuclease-specific outer primer, 5 μl RT mix (from Subheading 3.2), and 1 μl Taq DNA polymerase (5 units).
2. Bring to 50 μl and mix by pipetting or gentle vortexing and spin briefly.
3. Cycle 35 times as follows: 94 °C for 30 s, 55 °C for 30 s, 72 °C for 1 min (*see* **Note 8**).
4. Place reaction on ice.

3.5 Nested PCR (PCR2)

1. In 100 μl PCR tubes mix the following components on ice: 36 μl H_2O, 5 μl 10× PCR buffer, 5 μl dNTPs (10 mM), 1 μl (10 μM) inner-adaptor primer, 1 μl (10 μM) homing endonuclease-specific inner primer, 1 μl PCR mix (from Subheading 3.3), and 1 μl Taq DNA polymerase (5 units).
2. Bring to 50 μl with nuclease-free H_2O, mix by pipetting or gentle vortexing and spin briefly.
3. Cycle 35 times as follows: 94 °C for 30 s, 55 °C for 30 s, 72 °C for 1 min (*see* **Note 8**).
4. Place reaction on ice.
5. In the meantime, prepare a 1 % agarose gel with standard combs. You will require wells with approximately 10 μl capacity.
6. Run a small, 3–5 μl sample of the reaction on a 1 % agarose gel at 100 V to determine the success of the reaction (*see* **Note 9**).

3.6 Gel Purification

1. Precipitate the PCR products using 4.5 μl 3 M NaOAc and 90 μl 95–100 % ethanol (assuming 5 μl was removed from the sample to check the reaction success, *see* Subheading 3.5).
2. Set on ice for 30 min, then centrifuge at maximum speed for 30 min.
3. Remove the ethanol, air-dry briefly to remove ethanol traces, and resuspend the RNA pellet in 20 μl nuclease-free H_2O.
4. Mix the resuspended PCR product with 3.3 μl 6× DNA loading dye and load entire reaction onto a 1 % agarose gel. It may be necessary to leave a well blank between the samples to avoid cross-contamination.
5. Run gel at 100 V for 1 h or until the bromophenol blue indicator is approximately 3/4 to the bottom of the gel (*see* **Note 9**).
6. Carefully excise the bands from the gel and place in sterile, labelled Eppendorf tubes.
7. Use a commercially available gel extraction kit to purify the PCR products from the agarose gel (We used the QIAquick gel extraction kit from Qiagen).
8. Run 2–3 μl of the gel-purified sample on an agarose gel to confirm successful purification.

3.7 Cloning and Sequencing

1. In a sterile Eppendorf tube, mix 6 μl of the gel-purified PCR product with 2 μl of a suitable cloning vector (we used the TA Cloning® Kit (with pCR® 2.1 Vector), 1 μl of T4 DNA ligase buffer, and 400 units of T4 DNA ligase (*see* **Note 10**).
2. Bring to 10 μl with nuclease-free H_2O.
3. Incubate at room temperature (20–25 °C) for 1 h.
4. Add 100 μl competent *E. coli* cells and mix by gently flicking the tube.

5. Leave on ice for 30 min.
6. Place at 42 °C for 30 s.
7. Immediately place the cells back on ice.
8. Add 250 μl SOC media and incubate at 37 °C with shaking for 1 h.
9. Plate 50–100 μl of transformed cells onto LB plate with appropriate antibiotic.
10. Incubate the plate upside down overnight at 37 °C.
11. A minimum of ten clones should be selected for sequencing using vector-specific oligonucleotides.

4 Notes

1. The total RNA used in 5′RLM-RACE experiments should be of the highest possible quality. This ensures that the true initiation and potential processing sites are readily identified. The RNA used should show distinct and sharp rRNA bands with no smearing assessed by denaturing gel electrophoresis.
2. Prepare DTT stock solutions fresh.
3. Tobacco acid pyrophosphatase (TAP) does not require ATP.
4. This protocol uses a high concentration of adaptor RNA (1.5 μg). This is to maximize the probability of adaptor-to-transcript ligation while minimizing circularization of RNA molecules. To prevent adaptor concatenation and/or 3′ transcript-to-adaptor ligation events, the RNA adaptors should be free of 5′ phosphates.
5. The nested PCR primers can have restriction sites incorporated into their 3′ ends regardless of the sequence of the adaptor. Be sure to include a few nucleotides of overhang for the restriction enzyme to bind properly. This is crucial for downstream cloning if not using a TA cloning vector (Invitrogen).
6. While it is possible to purify the adaptor-ligated RNA from the free adaptors using sequential NH_4OAc–EtOH precipitation, this is not recommended, as contaminating adaptor RNA will interfere with downstream PCR reactions. Instead, the use of silica-based column purification is highly recommended, as this method uses size exclusion and will eliminate both small RNA fragments and the bulk of the unligated adaptor molecules.
7. The reverse transcription primer should be designed in accordance with the homing endonuclease of interest and the estimated position of the 5′ end of the transcript and with consideration to the potential for processing sites. A product of approximately 150–300 bp is ideal.

8. The annealing temperature should be adjusted to take into account the melting temperatures of the homing endonuclease-specific primers.
9. Depending on the size of the expected RLM-RACE products, the bromophenol blue may obscure the bands. While the bromophenol blue will not affect the downstream purification, this dye may be left out of the 6× DNA loading dye (prepared in 2.6) to facilitate product visualization.
10. A typical 5′RLM-RACE experiment will have a single band in the TAP+ lane and no bands in the TAP− lane. However, additional bands in either the TAP+ or TAP− lanes can indicate potential processing events or alternative initiation sites. The difference is simple to distinguish: the alternative initiation will have bands in only the TAP+ samples, while the processed RNAs will show bands in both the TAP+ and the TAP− samples. These products should be gel-purified, cloned, and sequenced.

Acknowledgement

EAG is supported by fellowships from the Canadian Institute of Health Research and the Michael Smith Foundation for Health Research.

References

1. Friedrich NC, Torrents E, Gibb EA, Sahlin M, Sjoberg BM et al (2007) Insertion of a homing endonuclease creates a genes-in-pieces ribonucleotide reductase that retains function. Proc Natl Acad Sci U S A 104:6176–6181
2. Liu Q, Belle A, Shub DA, Belfort M, Edgell DR (2003) SegG endonuclease promotes marker exclusion and mediates co-conversion from a distant cleavage site. J Mol Biol 334:13–23
3. Belle A, Landthaler M, Shub DA (2002) Intronless homing: site-specific endonuclease SegF of bacteriophage T4 mediates localized marker exclusion analogous to homing endonucleases of group I introns. Genes Dev 16:351–362
4. Sharma M, Ellis RL, Hinton DM (1992) Identification of a family of bacteriophage T4 genes encoding proteins similar to those present in group I introns of fungi and phage. Proc Natl Acad Sci U S A 89:6658–6662
5. Edgell DR (2008) Free-standing homing endonucleases of T-even phage: freeloaders or functionaries? In: Marlene Belfort BLS, Wood DW, Derbyshire V (eds) Homing endonucleases and inteins. Springer, Heidelburg, pp 147–160
6. Edgell DR, Gibb EA, Belfort M (2010) Mobile DNA elements in T4 and related phages. Virol J 7:290
7. Brok-Volchanskaya VS, Kadyrov FA, Sivogrivov DE, Kolosov PM, Sokolov AS et al (2008) Phage T4 SegB protein is a homing endonuclease required for the preferred inheritance of T4 tRNA gene region occurring in co-infection with a related phage. Nucleic Acids Res 36: 2094–2105
8. Gibb EA, Edgell DR (2007) Multiple controls regulate the expression of mobE, an HNH homing endonuclease gene embedded within a ribonucleotide reductase gene of phage Aeh1. J Bacteriol 189:4648–4661
9. Carpousis AJ, Mudd EA, Krisch HM (1989) Transcription and messenger RNA processing upstream of bacteriophage T4 gene 32. Mol Gen Genet 219:39–48
10. Triezenberg SJ (2001) Primer extension. In: Ausubel FM et al (eds) Current protocols in molecular biology. Chapter 4: Unit4 8. PMID: 18265242
11. Emery P (2007) RNase protection assay. Meth Mol Biol 362:343–348
12. Fromont-Racine M, Bertrand E, Pictet R, Grange T (1993) A highly sensitive method for mapping the 5′ termini of mRNAs. Nucleic Acids Res 21:1683–1684

Chapter 6

PCR Analysis of Chloroplast Double-Strand Break (DSB) Repair Products Induced by I-CreII in Chlamydomonas and Arabidopsis

Taegun Kwon, Obed W. Odom, Weihua Qiu, and David L. Herrin

Abstract

Homing endonuclease I-CreII has been used to study the consequences and repair of a double-strand break (DSB) in the chloroplast genome of *Chlamydomonas* and *Arabidopsis*. Since I-CreII is from a mobile *psbA* intron of *Chlamydomonas*, it cleaves the *psbA* gene of an intronless-*psbA* strain of *Chlamydomonas*. And it cleaves specifically in the *psbA* gene of *Arabidopsis*, which is naturally intronless. We have shown further that most of the repair products of an I-CreII-induced break in chloroplast DNA can be defined by PCR analysis with total nucleic acids and the appropriate primers. Here, we provide protocols for small-scale preparation of nucleic acids from *Chlamydomonas* and *Arabidopsis*, as well as guidelines for the subsequent PCR analysis.

Key words DNA amplification, Double-strand break-repair (DSBR), Homing endonuclease, Microhomology-mediated end-joining (MMEJ), Plastid DNA, Single-strand annealing (SSA)

1 Introduction

The chloroplast genome (CpDNA) is typically circular, well conserved, and encodes either ~100 or ~250 genes, depending on whether it is from the green or red lineage of this semiautonomous organelle. Although CpDNA is highly polyploid, there is very little genetic redundancy among the genes themselves (i.e., there are no protein-gene families). Moreover, there is a strong tendency for the genome to be homoplasmic (all copies have the same genes and alleles). The requirement for radiant energy to drive photosynthesis adds to the stress on plant genomes. And this is especially true for CpDNA, which spends much of its life attached to oxygen-producing membranes [1]. Although a number of processes that help protect chloroplasts have been identified, relatively little is known about its DNA repair mechanisms. In recent years, however, it has become clear that the chloroplast has multiple, efficient pathways for repairing double-strand breaks (DSBs).

David R. Edgell (ed.), *Homing Endonucleases: Methods and Protocols*, Methods in Molecular Biology, vol. 1123, DOI 10.1007/978-1-62703-968-0_6,

This line of research started with an investigation of group I introns—and their encoded proteins—in the CpDNA of *Chlamydomonas reinhardtii* (*Chlamydomonas*) [2]. Group I intron homing is essentially a form of DSB repair that inserts a mobile intron into its intron-minus allele, such that it disrupts the target site of the intron-encoded endonuclease [3]. To take this line of research beyond intron mobility, we used the homing endonuclease I-CreII, which is encoded by the *Cr.psbA4* intron of *Chlamydomonas* [4, 5]. The native target for I-CreII is an ~30-bp region of intron-minus *psbA* that spans the exon 4–exon 5 junction [6, 7]. A *Chlamydomonas* strain with an intronless *psbA* allele at both locations (1 for each inverted repeat) was used to examine the consequences of 2 transient DSBs in the genome induced by plasmid-borne I-CreII [8]. Since the DNA analysis was performed on clones that had segregated after the transformation with I-CreII, there was generally only one repair product per colony clone (Fig. 1); a process termed "copy correction" insures that both copies of the large inverted repeat are identical (unless one copy has a deletion that extends into the single-copy region) [8]. Thus, the clonal analysis greatly facilitates the molecular analysis by PCR, which in turn allows one to use relatively small amounts of cells and/or tissue.

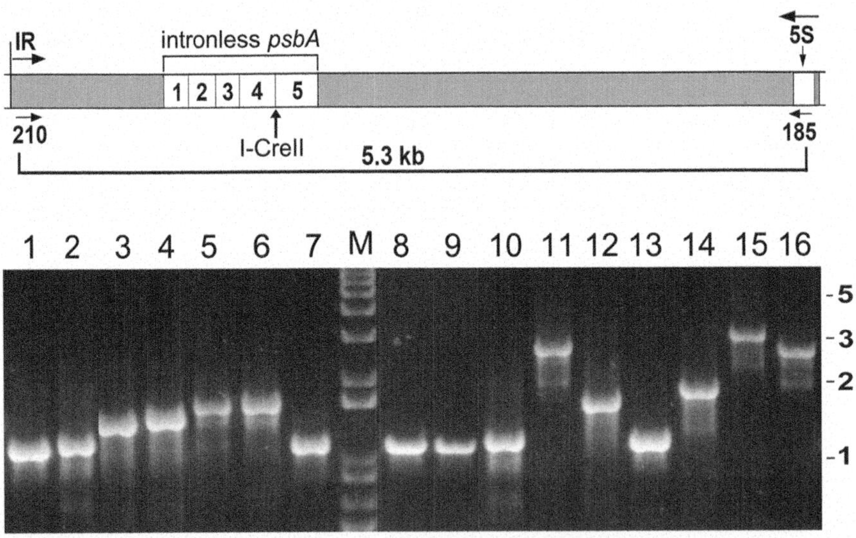

Fig. 1 PCR analysis of *Chlamydomonas* clones whose chloroplast had been transiently transformed with I-CreII. Above the agarose gel is a simple diagram of the *psbA* region in the intronless-*psbA* strain that received I-CreII. *PsbA* is in the large inverted repeat, whose beginning is indicated with IR (and an *arrow*); 5S refers to the 5S rRNA gene. The locations of the PCR primers 210 (5′-aggacgttagtcgatatttatacactc-3′) and 185 (5′-tggaatctcagttctagtgctaggg-3′) are indicated, as is the size of the product (5.3 kb) obtained with the recipient strain. TNA from 16 independent clones (1–16) was used for PCR; lane M contained DNA markers, and the sizes of several are indicated to the right. (Modified from Odom et al. [8])

Since *psbA* encodes a highly conserved protein of photosystem II, I-CreII can cleave the *psbA* gene of a number of different plant systems, including the model angiosperm, *Arabidopsis thaliana* (*Arabidopsis*) [6]. To study how angiosperm plastids respond to a DSB in the chromosome, I-CreII was engineered for inducible expression from the nucleus, and with a chloroplast targeting peptide so that it would cleave at *psbA*. The I-CreII fusion gene was integrated into the nuclear genome and induced with estradiol, which inhibited shoot growth in proportion to the fraction of CpDNA that had undergone mutagenic DSB repair [9]. The repair events were defined by using PCR [10] to amplify, from total seedling DNA, the CpDNA around *psbA* (Fig. 2). And the heteroplasmic mix of repair products was sorted out by cloning and sequencing the PCR products [9].

2 Materials

Prepare the solutions for nucleic acid purification with molecular biology-grade chemicals and ultrapure water (resistivity of 18 megohms). Sterilize the salt and buffer solutions by autoclaving; prepare the others with ultrapure, autoclaved water. Use disposable microtubes and pipet tips that are either RNase- and DNase-free, or sterile. For PCR, pipet tips with a filter are used to prevent cross-contamination.

2.1 Materials for the Nucleic Acid Isolations

1. *Chlamydomonas* Extraction Buffer: 100 mM NaCl, 50 mM EDTA, 20 mM Tris–HCl pH 8 (needed only for the *Chlamydomonas* protocol).
2. *Arabidopsis* Extraction Buffer: 200 mM Tris–HCl pH 7.5, 250 mM NaCl, 25 mM EDTA, 0.5 % SDS; stable at RT for at least 6 months.
3. Proteinase K: make fresh for each set of extractions by dissolving the powder in cold water to a final concentration of 5 mg/mL; keep on ice (needed only for the *Chlamydomonas* protocol).
4. SDS: 20 % (w/v) solution, prepared with autoclaved water and stored at RT.
5. *N*-lauryl sarcosine: 20 % (w/v) solution, prepared with autoclaved water and stored at RT (needed only for the *Chlamydomonas* protocol).
6. KOAc: 4 M, the pH is not adjusted, store at 4 °C (needed only for the *Chlamydomonas* protocol).
7. Phenol: to liquefied phenol, add hydroxyquinoline to 0.1 %, and adjust the pH to >7.5 by repeated extractions with 0.1 M Tris–HCl pH 8.0; the aqueous phase is removed by aspiration

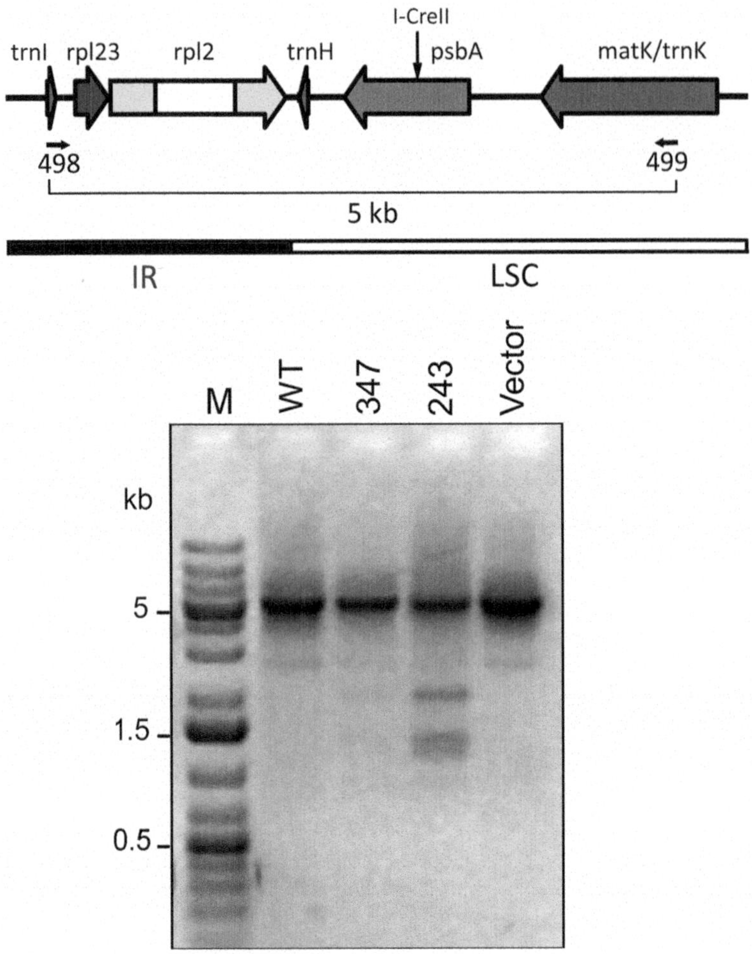

Fig. 2 PCR analysis at the *psbA* locus in transgenic *Arabidopsis* expressing chloroplast-targeted I-CreII. Above the gel is a map of the *psbA* region of CpDNA, which shows the locations of PCR primers and the I-CreII cleavage site. MatK is the protein encoded within the *trnK* intron. LSC refers to the large single-copy region, and IR to one of the two copies of the large inverted repeat. The agarose gel contained the PCR products from the indicated plants plus size markers in lane M; the ethidium-DNA fluorescence image was inverted. The plant lines were: WT, wild-type; 347, a weak I-CreII expresser; 243, a strong I-CreII expresser; Vector, transformed with the vector only. All the plants were grown on β-estradiol to induce the I-CreII *trans*gene, and TNA was used for PCR with primers 498 (5′-ccatgtacgaggatccccac-3′) and 499 (5′-cggcaatcctagggttgctc-3′), which amplify nt 152241-2786 of chloroplast DNA. (Modified from Kwon et al. [9])

with a glass pipet and an appropriate vacuum trap. Store the saturated phenol at 4 °C in a dark bottle [11]. (*Caution*: phenol is caustic and should be handled with great care, e.g., by wearing two pair of form-fitting gloves).

8. Chloroform (reagent grade): used as purchased, stored at RT in a dark bottle.

9. NaCl: 4 M, store at RT.
10. Isopropanol (reagent grade): used as purchased, stored at RT in a dark bottle.
11. 100 % EtOH: used as purchased, stored at −20 °C for *Chlamydomonas* or RT for *Arabidopsis.*
12. 70 % EtOH: prepared by adding water (0.3 vol) to 100 % EtOH (0.7 vol), store at −20 °C.
13. TE: 10 mM Tris–HCl pH 8.0, 1 mM EDTA; prepared by diluting 1 M Tris–HCl (pH 8.0 at RT) and 0.5 M Na_2EDTA (pH 8.0 at RT) with water; store at 4 °C.
14. Micropestles for 1.5-mL microfuge tubes. (Needed only for the *Chlamydomonas* protocol).

2.2 Materials for PCR and Product Analysis

1. dNTP mix: containing 2 mM each, of dATP, dCTP, dGTP, and dTTP.
2. 10× Taq DNA Polymerase buffer: 100 mM Tris–HCl, pH 8.3 at 25 °C, 500 mM KCl, 15 mM $MgCl_2$.
3. Taq DNA polymerase.
4. QIAquick PCR Purification Kit.
5. Preparative grade of agarose.
6. QIAquick Gel Extraction Kit.
7. pGC-Blue vector, ready for G-C cloning (Lucigen).

3 Methods

3.1 Isolation of Total Nucleic Acids (TNA) from Chlamydomonas

1. Suspend cells scraped from a 2 cm × 2 cm area of an agar plate, or pelleted from 10 to 15 mL of a liquid culture, in 350 μL of *Chlamydomonas* Extraction Buffer.
2. In a microfuge tube, add 40 μL of proteinase K and 50 μL of SDS to the cell suspension, and mix for 1–2 min on a rocker platform.
3. Add 50 μL of N-lauryl sarcosine, mix by inversion, and place on the rocker for 1–1.5 h (*see* **Note 1**).
4. Cool on ice, add 62 μL of KOAc, vortex, and place on ice for 15–30 min.
5. Centrifuge at 12,000 × *g* for 15 min (4 °C) and transfer the supernatant to a new microfuge tube.
6. Add 200 μL of phenol, vortex well, and mix on the rocker for at least 10 min.
7. Add 200 μL of chloroform, vortex, let stand for 1 min, and vortex again.

8. Centrifuge at 10,000 × *g* (RT) for 4–5 min and remove the aqueous phase to a new tube.
9. Repeat **steps 6–8**, except shorten the mixing time for the phenol extraction (**step 6**) to 1 min.
10. Fill the microfuge tube (containing the twice-extracted aqueous phase) with cold 100 % EtOH, and place at −20 °C for 0.5–1 h.
11. Centrifuge at 14,000 × *g* (4 °C) for 15 min, remove the supernatant, and add 0.5–1 mL of cold 70 % EtOH (carefully) to the pellet. Remove as much liquid as possible, and dry the pellet under mild vacuum for 5–10 min (or in air for 15–30 min).
12. Add cold TE (50 μL) to the pellet and resuspend the TNA with gentle mixing on ice. Store at −70 °C, or if DNA only will be analyzed, it can be stored at −20 °C (*see* **Note 2**).

3.2 Isolation of TNA from Arabidopsis for PCR

This protocol was obtained from Enamul Huq (University of Texas at Austin) and is a modification of the protocol of Edwards et al. [12]. After the initial tissue grinding (**step 1**), the rest of the procedure is performed at RT.

1. Pulverize 100 mg (fresh weight) of tissue in liquid nitrogen using a 1.5-mL microfuge tube and a micropestle (*see* **Note 3**).
2. Add 0.25 mL of *Arabidopsis* Extraction Buffer and grind further with the micropestle (at RT) (*see* **Note 4**).
3. Add another 0.25 mL of Extraction Buffer and continue grinding until the solution becomes uniformly green and translucent (*see* **Note 5**).
4. Centrifuge for 5 min at 14,000 × *g*.
5. Remove 0.6 mL of the supernatant, avoiding the pellet of plant debris, and deposit it into a new microfuge tube (*see* **Note 6**).
6. Add 0.5 mL of isopropanol (RT) and mix by inversion.
7. Centrifuge at 14,000 × *g* for 5 min (RT) and gently wash the pellet with 70 % ethanol.
8. Remove as much liquid as possible and dry the pellet under mild vacuum for 5–10 min (or in air for 15–30 min).
9. Dissolve the pellet in 0.1 mL of TE and store at −20 °C. The typical yield is 5–10 μg of TNA, and 1–2 μL is sufficient for PCR (*see* **Note 1**).

3.3 PCR and Analysis of the Products

Most of the DSB repair products can be detected using PCR as follows (*see* **Note 7**).

A typical 1× PCR reaction (30 μL) contained the following:

0.5–2 μL of TNA (or H_2O, for a no-DNA control rxn).

1 μL of each forward and reverse primer (10 μM stock).

3 μL of 10× Taq DNA Polymerase buffer (100 mM Tris–HCl, pH 8.3 at 25 °C, 500 mM KCl, 15 mM $MgCl_2$).

3 μL of dNTP mix (stock = 2 mM each, of dATP, dCTP, dGTP, and dTTP).

0.3 μL of Taq DNA polymerase (5 U/μL).

H_2O to bring the total vol to 30 μL.

A typical temperature regimen is: 94 °C for 5 min, followed by 30–35 cycles of 94 °C for 30 s, 60 °C (or 5 °C below the predicted T_m of the primers) for 30 s, and 70 °C for 5 min. A final extension step of 70 °C for 10 min terminated the regimen.

A portion of each reaction (3–10 μL) is separated on a standard analytical agarose gel (1–1.5 % agarose) containing ethidium bromide (1 μg/mL). Larger amounts are loaded if it is a preparative agarose gel (*see* **Note 7**) used to obtain DNA size fractions.

4 Notes

1. The solution becomes highly viscous when the cells are first lysed, but the incubation with proteinase K should decrease the viscosity significantly (which is necessary for a quantitative extraction).
2. If the RNA in the preparation interferes with visualizing the subsequent PCR products on an agarose gel, you can try using threefold to fivefold less of the TNA in the PCR reaction. Alternatively, the RNA can be removed by digestion with RNase A, which is prepared by dissolving pancreatic RNase A powder in 0.01 M NaOAc (pH 5.2 or 6.0), heating to 100 °C for 10 min, and cooling slowly to RT. Then, adjust the pH to 7.5 by adding Tris–HCl pH 7.5–0.05 M [11].

   To remove most of the RNA from the TNA, use these steps:

   (a) Add 5 μL of RNase A and incubate at 37 °C for 30 min.

   (b) Extract with an equal volume of phenol–chloroform (1:1), centrifuge at 10,000 ×*g* (RT) for 5 min, and remove the aqueous phase to a new tube.

   (c) Add 5 μL of 4 M NaCl and 0.2 mL of 100 % ethanol. Mix by vortexing gently.

   (d) Centrifuge at 14,000 ×*g* (4 °C) for 15 min and wash the pellet with 70 % ethanol.

   (e) Air-dry the pellet and then dissolve it in 50 μL of TE; 1 μL is sufficient for PCR.
3. This protocol works with as little as a single cotyledon from a ~2-week-old plant. On the other hand, >100 mg of tissue is problematic to grind in a microfuge tube and results in poor

DNA yield and quality. If more DNA is desired, the plant material should be split among multiple microfuge tubes, or a large-scale extraction protocol should be used.

4. This step can be skipped if only one cotyledon is used for the extraction.
5. When processing multiple tubes, it is okay to leave the ground samples in this buffer for 30 min.
6. An optional extraction with phenol–chloroform will give a higher purity of DNA, which may be helpful for long-term storage, but is usually not necessary for PCR. To do this, add 0.6 mL of phenol–chloroform (1:1), and mix by rocking the tube for several min. Centrifuge at 14,000 × *g* for 5 min, and remove the aqueous phase to a new tube.
7. The conditions for PCR amplification of the products of DSB repair in both *Chlamydomonas* and *Arabidopsis* are fairly standard, except for the long extension time (5 min) that is needed to amplify products of at least 5 kb. The most important factor is the choice of primers. In *Chlamydomonas*, we had to use multiple pairs of primers located at different distances from the I-CreII cleavage site to get a product in all 50 transformants that we knew had alterations at the *psbA* locus. Although the majority of the clones had a deletion, or a deletion plus a small insertion, and could be amplified with the primers used in Fig. 1, some had a large insertion that required us to do a form of primer walking to determine its structure [8]. The insertion was the same in all of those clones, and included plasmid vector sequences, which had integrated because of unanticipated homology to the 5′ and 3′ flanking regions of *psbA*. We did not obtain such insertions in *Arabidopsis*, because I-CreII was expressed from the nucleus (i.e., the chloroplast was not transformed), and primers that annealed ~2.5 kb, each, from the cleavage site were the most informative (*see* Fig. 2). In both organisms, the deletions that accompanied repair of the DSB by the mutagenic pathways (i.e., SSA-, MMEJ-, and NHEJ-like), which were the only ones we could track in *Arabidopsis*, were relatively large, ranging from 2.4 to 5 kb; this should be kept in mind when designing primers.

   There are two further considerations specific to *Chlamydomonas*; one is the relatively large number of repeated sequences in its chloroplast DNA (compared to land plants), which should be avoided when designing PCR primers. A simple way to do that is to use coding regions, which lack repeats, as priming spots, but when that is not possible, computer programs that identify repeats [13, 14] should be used to avoid them. The second consideration is positive in that only one DSB repair product is obtained for each clonal cell line, so the resulting PCR product is sequenced directly without subcloning.

There are a number of ways to clean up PCR products for sequencing, and most work well. The purified PCR products were sequenced at the University of Texas DNA Center employing the same primers used to make the PCR product, as well as internal primers to get the overlapping sequence.

In contrast, *Arabidopsis* seedlings contain many DSB repair products induced by I-CreII, so the PCR products must be cloned before sequencing. Since we employ *Taq* DNA polymerase for the amplification (see below), the mix of PCR products is cloned easily in the C-tailed vector pGC-Blue (Lucigen). To reduce the sequencing of clones that have parental (i.e., unchanged) DNA, the PCR products that are smaller (or larger) than wild-type DNA can be recovered from an agarose gel like that in Fig. 2, except it is poured with a preparative grade of agarose. The recovered DNAs are cloned into pGC-Blue similar to the unfractionated PCR products.

Acknowledgments

This work was supported by grants (to DLH) from the Dept. of Energy (DE-FG03-02ER15352), the R.A. Welch Foundation (F-1164), and the Texas Advanced Research Program (ARP 003658-0144-2007).

References

1. Sakai A, Takano H, Kuroiwa T (2004) Organelle nuclei in higher plants: structure, composition, function, and evolution. Int Rev Cytol 238:59–118
2. Herrin DL, Kuo T-C, Goldschmidt-Clermont M (1998) RNA splicing in the chloroplast. In: Rochaix J-D, Goldschmidt-Clermont M, Merchant S (eds) The molecular biology of chloroplasts and mitochondria in *Chlamydomonas*. Kluwer Academic Press, Dordrecht, The Netherlands, pp 183–195
3. Durrenberger F, Thompson AJ, Herrin DL, Rochaix J-D (1996) Double strand break-induced recombination in chloroplasts of *Chlamydomonas reinhardtii*. Nucleic Acids Res 24:3323–3331
4. Holloway SP, Deshpande NN, Herrin DL (1999) The catalytic group I introns of the *psbA* gene of *Chlamydomonas reinhardtii*: secondary structures, ORFs and evolutionary implications. Curr Genet 36:69–78
5. Odom OW, Holloway SP, Deshpande NN, Lee J, Herrin DL (2001) Mobile introns from the *psbA* gene of *Chlamydomonas reinhardtii*: highly efficient homing of an exogenous intron containing its own promoter. Mol Cell Biol 21:3472–3481
6. Kim H-H, Corina L, Suh J-K, Herrin DL (2005) Expression, purification, and biochemical characterization of the intron-encoded endonuclease, I-CreII. Protein Expr Purif 44:162–172
7. Corina LE, Qiu W, Desai A, Herrin DL (2009) Biochemical and mutagenic analysis of I-CreII reveals distinct but important roles for both the H-N-H and GIY-YIG motifs. Nucleic Acids Res 37:5810–5821
8. Odom OW, Baek K-H, Dani RN, Herrin DL (2008) *Chlamydomonas* chloroplasts can use short dispersed repeats (SDRs) and multiple pathways to repair a double-strand break in the genome. Plant J 53:842–853
9. Kwon T, Huq E, Herrin DL (2010) Microhomology-mediated and nonhomologous repair of a double-strand break in the chloroplast genome of *Arabidopsis*. Proc Natl Acad Sci U S A 107:13954–13959
10. Mullis K, Faloona F, Scharf S, Saiki R, Horn G, Erlich H (1986) Specific enzymatic amplification of DNA in vitro: the polymerase chain

reaction. Cold Spring Harb Symp Quant Biol 51:263–273
11. Sambrook J, Russell DW (2001) Molecular cloning: a laboratory manual, 3rd edn. Cold Spring Harbor Laboratory Press, Cold Spring Harbor, NY
12. Edwards K, Johnstone C, Thompson C (1991) A simple and rapid method for the preparation of plant genomic DNA for PCR analysis. Nucleic Acids Res 19:1349
13. Kurtz S, Choudhuri JV, Ohlebusch E, Schleiermacher C, Stoye J, Giegerich R (2001) REPuter: the manifold applications of repeat analysis on a genomic scale. Nucleic Acids Res 29:4633–4642
14. Betley JN, Frith MC, Graber JH, Choo S, Deshler JO (2002) A ubiquitous and conserved signal for RNA localization in chordates. Curr Biol 12:1756–1761

Chapter 7

A Two-Plasmid Bacterial Selection System for Characterization and Engineering of Homing Endonucleases

Ning Sun and Huimin Zhao

Abstract

Homing endonucleases recognize long DNA sequences and generate site-specific DNA double-stranded breaks. They can serve as a powerful genomic modification tool in various industrial and biomedical applications. Here, we describe a two-plasmid bacterial selection system for characterization and engineering of homing endonucleases. This selection system couples the DNA cleavage activity of a homing endonuclease with the survival of host cells. Therefore, it can be used for assaying in vivo activity of homing endonucleases. Moreover, due to its high sensitivity, it can be applied for directed evolution of homing endonucleases with altered sequence specificity.

Key words Homing endonucleases, Protein engineering, Directed evolution, Gene targeting, Gene therapy

1 Introduction

Homing endonucleases, also known as meganucleases, represent a family of naturally occurring rare-cutting endonucleases. Homing endonucleases recognize long DNA sequences (14–40 bp) and can be used to generate site-specific double-strand breaks (DSBs) in the chromosome [1]. Subsequent repair of the DSBs by nonhomologous end joining or homologous recombination results in desired genetic modifications such as gene deletion, gene insertion or gene replacement [2]. Therefore, homing endonucleases represent a promising tool for targeted genome engineering in systems biology, synthetic biology, and human gene therapy. However, naturally occurring homing endonucleases have a limited repertoire of recognition sequences, which severely hampers their application. Directed evolution [3] serves as a powerful tool for engineering homing endonucleases with altered specificity to recognize novel sequences [4, 5].

David R. Edgell (ed.), *Homing Endonucleases: Methods and Protocols*, Methods in Molecular Biology, vol. 1123, DOI 10.1007/978-1-62703-968-0_7, © Springer Science+Business Media New York 2014

To study and engineer homing endonucleases in vivo, we developed a highly sensitive selection system in *Escherichia coli* (Fig. 1) [6]. The system comprises two plasmids. The reporter plasmid encodes the *ccdB* toxic gene under an arabinose-inducible promoter, followed by a homing endonuclease cleavage site. The expression plasmid contains a homing endonuclease gene under an isopropyl-β-d-thiogalactopyranoside (IPTG)-inducible promoter. To carry out the assay, both the reporter plasmid and the expression plasmid are co-transformed into *E. coli* cells. Homing endonucleases are expressed following IPTG induction. An inactive homing endonuclease cannot recognize the cleavage site on the reporter plasmid, leaving the reporter plasmid intact, which results in cell death after CcdB toxin expression is induced by arabinose. On the other hand, the cleavage of the target DNA sequence by the active homing endonuclease before arabinose induction eliminates the cytotoxic reporter plasmid and leads to cell survival. This assay system couples the enzymatic DNA cleavage of homing

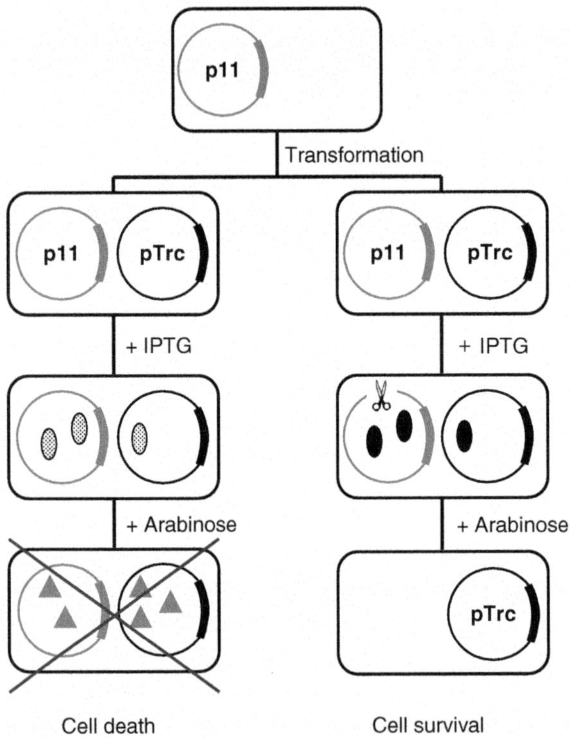

Fig. 1 Schematic of the bacterial two plasmid selection system. Details are provided in the text. "p11" represents the reporter plasmid containing the *ccdB* toxic gene and the recognition site of a homing endonuclease. "pTrc" represents the expression plasmid encoding a homing endonuclease gene. *Dotted ovals* represent the non-active homing endonucleases. *Black ovals* represent the active homing endonucleases. *Triangles* represent the CcdB toxins

endonucleases with the survival of the host *E. coli* cells. Therefore, it can be applied to assay the homing endonuclease activity in vivo with ease. Here, we use I-SceI [7] as an example to illustrate the experimental procedures. Due to its high sensitivity, this system can also be applied to directed evolution of homing endonucleases with altered target sequences [8]. Moreover, with appropriate modifications, this system can be used to characterize and engineer other groups of rare-cutting endonucleases such as zinc finger nucleases (ZFNs) [9] and transcription activator-like effector nucleases (TALENs) [10].

2 Materials

Prepare all solutions using ultrapure water, prepared by purifying deionized water to attain a resistivity of 18.2 MΩ cm at 25 °C. Prepare and store all reagents at room temperature unless indicated otherwise.

2.1 Preparation of Electrocompetent E. coli BW25141

1. *E. coli.* BW25141 ($lacI^q$ $rrnB_{T14}$ $\Delta lacZ_{WJ16}$ $\Delta phoBR580$ *hsdR514* $\Delta araBAD_{AH33}$ $\Delta rhaBAD_{LD78}$ *galU95* $endA_{BT333}$ *uidA*(Δ*Mlu*I)::pir^+ *recA1*) [11, 12].
2. LB medium: Add 20 g of LB Broth (Fisher, Fair Lawn, NJ) into 1 L of deionized water. Autoclave at 121 °C for 15 min.
3. LB agar plates: LB medium and 20 g/L agar.
4. 10 % glycerol: Mix 100 mL of glycerol and 900 mL of deionized water. Autoclave at 121 °C for 15 min.
5. Benchtop centrifuges to separate cells and supernatant.

2.2 Construction of Reporter Plasmid p11-LacY-wt1 and Preparation of Electrocompetent E. coli BW25141 Harboring p11-LacY-wt1

1. T4 Polynucleotide Kinase (New England Biolabs, Beverly, MA).
2. T4 DNA Ligase with 10× T4 DNA Ligase Buffer (New England Biolabs, Beverly, MA).
3. Plasmid p11-LacY was constructed previously [6].
4. *Xba*I and *Sph*I restriction enzymes with 10× NEBuffer 4 and 100× Bovine Serum Albumin (BSA, New England Biolabs, Beverly, MA).
5. Antarctic Phosphatase with 10× Antarctic Phosphatase Buffer (New England Biolabs, Beverly, MA).
6. 1-Butanol (Fisher, Fair Lawn, NJ).
7. 1 M glucose solution: Dissolve 90 g of d-glucose in 400 mL of deionized water and adjust to a final volume of 500 mL. Filter-sterilize.
8. SOC medium: Add 20 g of tryptone, 5 g of yeast extract, 0.5 g of NaCl, 0.186 g of KCl, 0.95 g of $MgCl_2$, 1.2 g of $MgSO_4$

into 980 mL of deionized water. Autoclave at 121 °C for 15 min. After the solution cools down to 60 °C, add 20 mL of sterile 1 M glucose solution.

9. Ampicillin stock solution: Dissolve 1 g of ampicillin powder in 10 mL of deionized water and filter-sterilize it.
10. LB-Amp$^+$ medium: LB medium plus 100 μg/mL ampicillin.
11. LB-Amp$^+$ agar plates: LB-Amp$^+$ medium and 20 g/L agar.
12. QIAprep Miniprep Kit (QIAGEN, Valencia, CA).

2.3 Construction of Expression Plasmid pTrc-IScel

1. Plasmid pSCM525 [7] and plasmid pTrc-p15a [6].
2. Phusion High-Fidelity DNA Polymerase with 5× Phusion HF Buffer (New England Biolabs, Beverly, MA).
3. 10× dNTPs solution: 2.5 mM each of dATP, dCTP, dGTP, and dTTP.
4. Concentrated stock solution of TAE (50×): Weigh 242 g of Tris base (MW = 121.14) and dissolve it in approximately 750 mL of deionized water. Carefully add 57.1 mL of glacial acetic acid and 100 mL of 0.5 M EDTA, and adjust the solution to a final volume of 1 L. This stock solution can be stored at room temperature. The pH of this buffer is not adjusted and should be about 8.5.
5. Working solution of TAE buffer (1×): Dilute the stock solution by 50-fold with deionized water. Final solute concentrations are 40 mM Tris acetate and 1 mM EDTA.
6. 1 % Agarose gel in 1× TAE buffer: Add 1 g of agarose into 100 mL of 1× TAE buffer and microwave until agarose is completely melted. Cool the solution to approximately 70–80 °C. Add 5 μL of ethidium bromide into the solution and mix well. Pour 25–30 mL of solution onto an agarose gel rack with a 2-well comb.
7. QIAquick Gel Extraction Kit (QIAGEN, Valencia, CA).
8. QIAquick PCR Purification Kit (QIAGEN, Valencia, CA).
9. *Eco*RI and *Kpn*I restriction enzymes (BSA, New England Biolabs, Beverly, MA).
10. Kanamycin stock solution: Dissolve 0.5 g of kanamycin powder in 10 mL of deionized water and filter-sterilize.
11. LB-Kan$^+$ medium: LB medium plus 50 μg/mL kanamycin.
12. LB-Kan$^+$ agar plates: LB-Kan$^+$ medium and 20 g/L agar.

2.4 In Vivo Activity Assay

1. 0.5 M IPTG: Dissolve 5.96 g of IPTG into 50 mL of deionized water. Filter-sterilize.
2. LB-Kan$^+$-Ara$^+$ agar plates: LB-Kan$^+$ agar plate plus 10 mM of L-arabinose.

3 Methods

3.1 Preparation of Electrocompetent E. coli BW25141

1. Streak *E. coli* strain BW25141 on an LB agar plate from frozen stock. Incubate overnight at 37 °C.
2. Pick up a single colony to inoculate a 4 mL overnight culture in LB medium at 37 °C.
3. Inoculate 1 mL of overnight culture into 400 mL LB medium. Grow at 37 °C with shaking for 2–4 h (*see* **Note 1**). When OD_{600} reaches 0.6–0.8, immediately put the cells on ice (or 4 °C).
4. Chill the culture for 15–30 min. Keep the cells on ice (or 4 °C) for the remainder of the procedure. Prechill the centrifuge and centrifuge bottles at 4 °C.
5. Harvest the cells by centrifugation at 6,000 × *g* for 10 min at 4 °C. Decant the supernatant and resuspend the cell pellet in 400 mL of prechilled sterile ddH_2O.
6. Harvest the cells by centrifugation at 6,000 × *g* for 10 min at 4 °C. Decant the supernatant and resuspend the cell pellet in 200 mL of prechilled sterile ddH_2O.
7. Harvest the cells by centrifugation at 6,000 × *g* for 10 min at 4 °C. Decant the supernatant and resuspend the cell pellet in 40 mL of prechilled 10 % glycerol. Transfer to a 50 mL centrifuge tube.
8. Harvest the cells by centrifugation at 3,000 × *g* for 10 min at 4 °C. Decant the supernatant and resuspend the cell pellet in 20 mL of prechilled 10 % glycerol.
9. Harvest the cells by centrifugation at 3,000 × *g* for 10 min at 4 °C. Carefully aspirate the supernatant and resuspend the cell pellet in 1 mL of prechilled 10 % glycerol.
10. Aliquot 50 μL of the cells into sterile 0.5 mL microfuge tubes and snap-freeze with dry ice or liquid nitrogen. Store frozen cells at −80 °C.

3.2 Construction of Reporter Plasmid p11-LacY-wt1 and Preparation of Electrocompetent E. coli BW25141 Harboring p11-LacY-wt1

1. Mix 50 μL (100 pmol/μL) each of Oligo1 and Oligo2, which contain the recognition sequence of I-SceI and restriction enzymes sites, respectively. Sequences of the oligos are shown below. Recognition sequence of I-SceI is shown in capital letters while *Xba*I and *Sph*I sites are shown in italics (*see* **Note 2**).

Oligo1: 5′-*ctagc* attacgc TAGGGATAACAGGGTAAT atcacgc *tctaga* catacg *gcatg*-3′.

Oligo2: 5′-*c* cgtatg *tctaga* gcgtgat ATTACCCTGTTA TCCCTA gcgtaat *g*-3′.

2. Anneal the oligo mix in a thermocycler. Reaction condition: denature at 95 °C for 5 min; decrease the temperature to 4 °C at the rate of 0.1 °C/s; keep at 4 °C.

3. Phosphorylate the annealed oligos at 37 °C for 30 min followed by heat inactivation at 65 °C for 10 min. Reaction mixture contains 10 μL of annealed oligos, 5 μL of 10× T4 DNA Ligase Reaction Buffer, 1 μL of T4 Polynucleotide Kinase, and 34 μL of ddH_2O (*see* **Note 3**).
4. Digest p11-LacY plasmid by *Xba*I and *Sph*I at 37 °C for 2.5 h. Digestion condition: 6 μL of 10× NEBuffer 4, 0.6 μL of 100× BSA, 15 U of *Xba*I, 15 U of *Sph*I, 1 μg of p11-LacY. Add ddH_2O to a final volume of 60 μL. After digestion, add 7 μL of 10× Antarctic Phosphatase Reaction Buffer and 1 μL of Antarctic Phosphatase to the digestion mixture and incubate at 37 °C for 30 min followed by heat inactivation at 65 °C for 10 min (*see* **Note 4**).
5. Insert phosphorylated oligos into p11-LacY plasmid through *Xba*I and *Sph*I sites. Ligation reaction: 50 ng of digested p11-LacY plasmid, 1 μL of phosphorylated oligos, 2 μL of 10× T4 DNA Ligase Reaction Buffer and 1 μL of T4 DNA Ligase. Adjust the volume to 20 μL with ddH_2O. Incubate overnight at 16 °C.
6. Transfer the ligation mixture into 980 μL of 1-butanol and vortex vigorously. Centrifuge for 20 min at top speed in a tabletop microcentrifuge (*see* **Note 5**). Carefully aspirate the supernatant and air-dry the pellet for 5–10 min. Resuspend the pellet in 4 μL of ddH_2O.
7. Mix 2 μL of the redissolved DNA with 50 μL of electrocompetent *E. coli* BW25141 cells and transfer the mixture into a pre-chilled 0.2 cm electroporation cuvette.
8. Electroporate the cells at 2.5 kV (*see* **Note 6**). Quickly add 1 mL of SOC medium and resuspend the cells gently. Transfer to a round-bottom culture tube.
9. Shake at 250 rpm at 37 °C for 1 h.
10. Dilute the cells as appropriate and spread 100–200 μL cells onto a LB-Amp^+ agar plate. Incubate at 37 °C overnight.
11. Inoculate single colonies to 4 mL of LB-Amp^+ medium and grow with shaking at 37 °C overnight.
12. Purify plasmids from each 4 mL of culture using the QIAprep Spin Miniprep Kit.
13. Confirm the identity of the plasmids by DNA sequencing (Fig. 2a).
14. Electroporate p11-LacY-wt1 plasmid into electrocompetent BW25141. Shake at 250 rpm at 37 °C for 1 h. Dilute the cells as appropriate and spread 100–200 μL cells onto a LB-Amp^+ agar plate. Incubate at 37 °C for 10–12 h until colonies become visible.

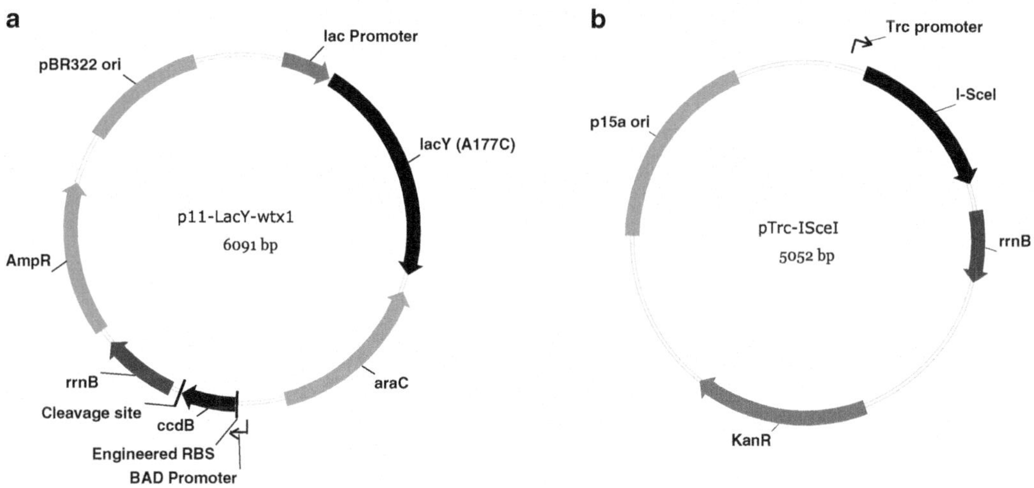

Fig. 2 The two plasmids used in this bacterial two plasmid selection system. (**a**) The reporter plasmid p11-LacY-wt1 encodes the toxic *ccdB* gene under the arabinose-inducible *BAD* promoter and a single copy of the I-SceI cleavage site. It also encodes an arabinose transporter gene *lacY (A177C)* under the IPTG-inducible *lac* promoter to facilitate the induction of *ccdB* by arabinose. The ribosomal binding site (RBS) of *ccdB* gene was engineered to increase system sensitivity [6]. (**b**) The expression plasmid pTrc-ISceI encodes I-SceI under the IPTG-inducible *Trc* promoter

15. Inoculate a single colony directly into 400 mL LB-Amp⁺ medium. Grow at 37 °C with shaking for 7–10 h (*see* **Note 7**). When OD_{600} reaches 0.4–0.6, immediately put the cells on ice (or 4 °C).
16. Follow **steps 4–10** of Subheading 3.1.

3.3 Construction of Expression Plasmid pTrc-I-SceI

1. PCR-amplify the homing endonuclease I-SceI gene from plasmid pSCM525 [7] with primers EcoRI-SceI', atcagt *gaattc* aggaaactcgagatgaaaaatattaaaaaaaa (*Eco*RI site is shown in italics), and KpnI-Isce-2-C, atgccg *ggtacc* ttattttaaaaaagtttcgg (*Kpn*I site is shown in italics). PCR mixture: 20 μL of 5× Phusion HF Buffer, 10 μL of 10× dNTPs solution, 50 pmol of EcoRI-SceI', 50 pmol of KpnI-Isce-2-C, 10 ng of pSCM525, 1 μL of Phusion High-Fidelity DNA Polymerase. Adjust the volume to 100 μL with ddH_2O.
2. PCR condition: Fully denature at 98 °C for 2 min, followed by 25 cycles of 98 °C for 30 s, 57 °C for 30 s, and 72 °C for 20 s, with a final extension at 72 °C for 10 min.
3. Load all the PCR products onto a 1 % agarose gel and perform electrophoresis at 120 V for 15 min.
4. Gel-purify PCR products using the QIAquick Gel Extraction Kit.
5. Digest PCR product and pTrc-p15a by *Eco*RI and *Kpn*I at 37 °C for 3 h. Digestion condition: 6 μL of 10× NEBuffer 4,

0.6 μL of 100× BSA, 15 U of *Eco*RI, 15 U of *Kpn*I, 1 μg of PCR product or pTrc-p15a. Add ddH_2O to a final volume of 60 μL (*see* **Note 8**).

6. Purify the digestion product by QIAquick PCR Purification Kit.
7. Set up a ligation reaction: 50 ng of digested pTrc-p15a, 50 ng of digested PCR product containing the I-SceI gene, 2 μL of 10× T4 DNA Ligase Reaction Buffer, and 1 μL of T4 DNA Ligase. Adjust the volume to 20 μL with ddH_2O. Incubate overnight at 16 °C.
8. Transfer the ligation mixture into 980 μL of 1-butanol and vortex vigorously. Centrifuge for 20 min at top speed in a tabletop microcentrifuge. Carefully aspirate the supernatant and air-dry the pellet for 5–10 min. Resuspend the pellet in 4 μL of ddH_2O.
9. Mix 2 μL of the redissolved DNA with 50 μL of electrocompetent *E. coli* BW25141 cells (*see* **Note 9**) and transfer the mixture into a prechilled 0.2 cm electroporation cuvette.
10. Electroporate the cells at 2.5 kV (*see* **Note 6**). Quickly add 1 mL of SOC medium and resuspend the cells gently. Transfer to a round-bottom culture tube.
11. Shake at 250 rpm at 37 °C for 1 h.
12. Dilute the cells as appropriate and spread 100–200 μL cells onto a LB-Kan^+ agar plate. Incubate at 37 °C overnight.
13. Inoculate single colonies to 4 mL of LB-Kan^+ medium and grow with shaking at 37 °C overnight.
14. Purify plasmids from each 4 mL of culture using the QIAprep Spin Miniprep Kit.
15. Confirm the identity of the plasmids by DNA sequencing (Fig. 2b).

3.4 *In Vivo* Activity Assay

1. Mix 1–100 ng of pTrc-ISceI with 50 μL of electrocompetent *E. coli* BW25141 harboring p11-LacY-wtl and transfer the mixture into a prechilled 0.2 cm electroporation cuvette.
2. Electroporate the cells at 2.5 kV. Quickly add 1 mL of SOC medium and resuspend the cells gently. Transfer to a round-bottom culture tube.
3. Recover the culture by shaking at 250 rpm at 37 °C for 5 min.
4. Add 4 mL of SOC medium and 5 μL of 0.5 M IPTG to the culture. Grow with shaking at 250 rpm at 37 °C for 70 min (*see* **Note 10**).
5. Shake at 250 rpm at 30 °C for 1 h (*see* **Note 11**).
6. Dilute the cells as appropriate and spread 100–200 μL of cells onto a LB-Kan^+ agar plate. Spread a second aliquot onto a LB-Kan^+-Ara^+ agar plate (*see* **Note 12**).

7. Incubate at 37 °C overnight until colonies become clearly visible.
8. Count the colonies on the LB-Kan$^+$ agar plate and the LB-Kan$^+$-Ara$^+$ agar plate.
9. Calculate the survival rate by dividing the number of colonies on the LB-Kan$^+$-Ara$^+$ agar plate by the number of colonies on the LB-Kan$^+$ agar plate, accounting for dilution factors (*see* **Note 13**).

4 Notes

1. Normally, the doubling time for a BW25141 strain is approximately 30 min.
2. The internal *Xba*I site is used to insert additional recognition sequences if necessary.
3. 10× T4 DNA Ligase Reaction Buffer is used instead of 10× T4 Polynucleotide Kinase Reaction Buffer. Therefore, the phosphorylated oligos can be directly used for the ligation reaction in **step 5** of Subheading 3.2 without changing the reaction buffer.
4. This step helps to decrease the vector background due to self-ligation.
5. This step removes salt from the ligation buffer to increase electroporation efficiency and eliminate arcing of cuvettes.
6. For an efficient electroporation, a time constant of 4.8–5.2 ms should be obtained.
7. It is important that cells do not grow to log phase at any stage during competent cells preparation.
8. Antarctic Phosphatase can be used as in **step 4** of Subheading 3.2 to decrease the vector background due to self-ligation.
9. Any other *E. coli* strain suitable for DNA cloning, such as DH5α and JM109, can be used.
10. A final concentration of 0.5 mM IPTG is used to induce the I-SceI expression under the *Trc* promoter and the LacY (A177C) expression under the *lac* promoter.
11. This step allows I-SceI to cut its recognition sequence on p11-LacY-wt1 efficiently.
12. A final concentration of 10 mM l-arabinose on agar plates is used to induce the *ccdB* expression under the *BAD* promoter.
13. The survival rate of wild-type I-SceI is typically >90 %. The survival rate of a non-active mutant of I-SceI (D44A) is typically <0.2 %.

References

1. Hafez M, Hausner G (2012) Homing endonucleases: DNA scissors on a mission. Genome 55:553–569
2. Sun N, Abil Z, Zhao H (2012) Recent advances in targeted genome engineering in mammalian systems. Biotechnol J 7:1074–1087
3. Cobb RE, Sun N, Zhao H (2013) Directed evolution as a powerful synthetic biology tool. Methods 60:81–90
4. Arnould S, Perez C, Cabaniols JP, Smith J, Gouble A, Grizot S, Epinat JC, Duclert A, Duchateau P, Paques F (2007) Engineered I-CreI derivatives cleaving sequences from the human XPC gene can induce highly efficient gene correction in mammalian cells. J Mol Biol 371:49–65
5. Smith J, Grizot S, Arnould S, Duclert A, Epinat JC, Chames P, Prieto J, Redondo P, Blanco FJ, Bravo J, Montoya G, Paques F, Duchateau P (2006) A combinatorial approach to create artificial homing endonucleases cleaving chosen sequences. Nucleic Acids Res 34:e149
6. Chen Z, Zhao H (2005) A highly sensitive selection method for directed evolution of homing endonucleases. Nucleic Acids Res 33:e154
7. Perrin A, Buckle M, Dujon B (1993) Asymmetrical recognition and activity of the I-SceI endonuclease on its site and on intron-exon junctions. EMBO J 12:2939–2947
8. Chen Z, Wen F, Sun N, Zhao H (2009) Directed evolution of homing endonuclease I-SceI with altered sequence specificity. Protein Eng Des Sel 22:249–256
9. Guo J, Gaj T, Barbas CF 3rd (2010) Directed evolution of an enhanced and highly efficient FokI cleavage domain for zinc finger nucleases. J Mol Biol 400:96–107
10. Sun N, Liang J, Abil Z, Zhao H (2012) Optimized TAL effector nucleases (TALENs) for use in treatment of sickle cell disease. Mol Biosyst 8:1255–1263
11. Datsenko KA, Wanner BL (2000) One-step inactivation of chromosomal genes in Escherichia coli K-12 using PCR products. Proc Natl Acad Sci U S A 97:6640–6645
12. Lessard IA, Pratt SD, McCafferty DG, Bussiere DE, Hutchins C, Wanner BL, Katz L, Walsh CT (1998) Homologs of the vancomycin resistance D-Ala-D-Ala dipeptidase VanX in *Streptomyces toyocaensis*, *Escherichia coli* and *Synechocystis*: attributes of catalytic efficiency, stereoselectivity and regulation with implications for function. Chem Biol 5: 489–504

Chapter 8

Rapid Screening of Endonuclease Target Site Preference Using a Modified Bacterial Two-Plasmid Selection

Jason M. Wolfs, Benjamin P. Kleinstiver, and David R. Edgell

Abstract

Homing endonucleases and other site-specific endonucleases have potential applications in genome editing, yet efficient targeting requires a thorough understanding of DNA-sequence specificity. Here, we describe a modified two-plasmid genetic selection in *Escherichia coli* that allows rapid profiling of nucleotide substitutions within a target site of given endonucleases. The selection utilizes a toxic plasmid (pTox) that encodes a DNA gyrase toxin in addition to the endonuclease target site. Cleavage of the toxic plasmid by an endonuclease expressed from a second plasmid (pEndo) facilitates growth under selective conditions. The modified protocol utilizes competent cells harboring the endonuclease expression plasmid into which target site plasmids are transformed. Replica plating on nonselective and selective media plates identifies cleavable and non-cleavable targets. Thus, a library of randomized target sites, or many individual target sites, can be analyzed using a single transformation. Both cleavable and non-cleavable targets can be analyzed by DNA sequencing to gain information about nucleotide preference in the endonuclease's target site.

Key words Homing endonuclease, DNA target sequence preference, Two-plasmid genetic selection, DNA gyrase toxin

1 Introduction

Homing endonucleases and other engineered site-specific DNA endonucleases target long DNA sequences of 14–40 base pairs to generate a double-strand break (DSB) [1, 2]. These site-specific endonucleases possess the ability to introduce DSBs at precise locations for genome-editing applications [3, 4]. Targeting of site-specific endonucleases requires a synthesis of information on nucleotide preference derived from in vitro and in vivo methods [5–8]. Towards this goal, we have modified a two-plasmid bacterial selection system to facilitate rapid screening of the sequence tolerance of homing endonucleases (and other site-specific endonucleases) [9, 10]. Briefly, individual mutant target sites (or a library of sites) are cloned into a toxic reporter plasmid containing a *ccdB* gene that encodes a DNA gyrase toxin, whereby cell survival is

David R. Edgell (ed.), *Homing Endonucleases: Methods and Protocols*, Methods in Molecular Biology, vol. 1123, DOI 10.1007/978-1-62703-968-0_8,

dependent on cleavage of the target site by the homing endonuclease expressed from a separate plasmid (Fig. 1). Our protocol diverges from the standard two-plasmid selection in that the initial transformation is into competent cells harboring the pEndo expression plasmid (Fig. 1, *right*) rather than into cells harboring the pTox plasmid (Fig. 1, *left*). This change means that only a single strain harboring the pEndo plasmid can be made competent, which can be used to screen a library of substrates or many individual substrates. In contrast, the standard protocol would require making competent cells for every target site to be tested. In our modified protocol, transformed colonies containing individual target site and endonuclease-expressing plasmids are replica-gridded on

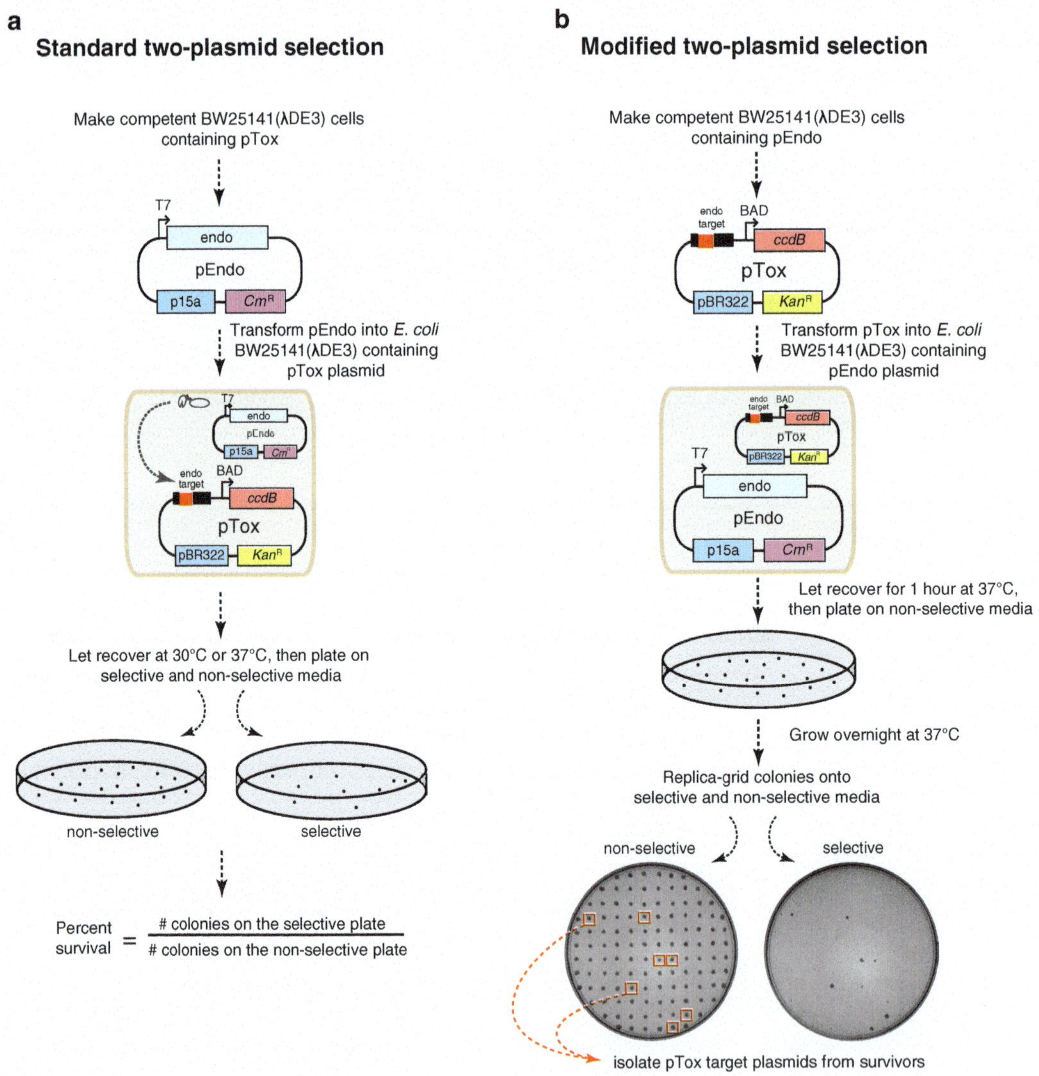

Fig. 1 (**a**) Schematic for the in vivo two-plasmid screen for randomized target sites (randomized region shown in *red*). (**b**) Work flow for two-plasmid screen and target site validation

nonselective and selective plates to repress or express the toxic gene, respectively. Cells harboring target sites cleavable by the endonuclease survive on both plates, while cells harboring non-cleavable target sites only survive on the nonselective plates (Fig. 1, *right*). This method enables the isolation of cleavable target sites from the nonselective plates for DNA sequencing. In some cases, the non-cleavable target sites may be of equal interest and can be isolated from the nonselective plates. Furthermore, when dealing with a large pool of target sites, or a randomized target site library, this method can be used as an initial screen as it efficiently identifies cleavable from non-cleavable target sites.

2 Materials

2.1 Plasmids and E. coli Strains

1. *E. coli* strain: BW25141(λDE3) modified from strain BW25141 (*see* **Note 1**).
2. Toxic reporter plasmid: pTox modified from pll-lacY-wtx1 (Fig. 2, and *see* **Note 2**).
3. Expression vector: pEndo created from pACYCDeut-1 with endonuclease ORF.

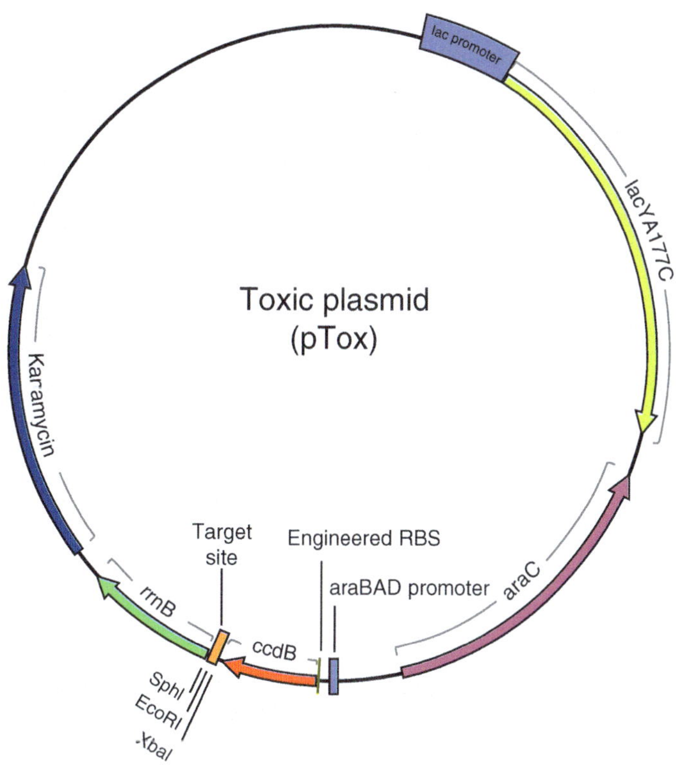

Fig. 2 Plasmid map of pTox

2.2 $CaCl_2$ Competent BW25141(λDE3)

1. SOC media (1 L): 20 g tryptone, 5 g yeast extract, 0.5 g NaCl, and 2.5 mM KCl. Bring volume to 1 L with dH_2O, then autoclave. Let media cool and then add 10 mM $MgCl_2$ and 20 mM glucose.
2. Ice-cold sterile 100 mM $CaCl_2$ solution and 100 mM $CaCl_2$ solution containing 15 % glycerol.
3. Sterile 500 mL centrifugation bottles.
4. Media/plates: LB plus chloramphenicol (25 μg/mL) and glucose (0.2 %); for plates add agar (15 g/L).
5. Liquid nitrogen.

2.3 Library Construction

1. Enzymes: Klenow fragment (-exo) polymerase, T4 DNA ligase, and restriction enzymes—EcoRI-HF and SphI-HF (XbaI could be used as an alternative).
2. PCR cleanup kits.
3. SEM competent NEB5α *E. coli* cells.
4. SOC media.
5. Media/plates: LB plus kanamycin (50 μg/mL) and glucose (0.2 %); for plates add agar (15 g/L).

2.4 Target Site Two-Plasmid Screen

1. SOC media.
2. Media: LB plus kanamycin (50 μg/mL) and glucose (0.2 %).
3. Nonselective plates: LB plus chloramphenicol (25 μg/mL), kanamycin (50 μg/mL), and glucose (0.2 %) with agar (15 g/L).
4. Selective plates: LB plus chloramphenicol (25 μg/mL) and L-(+)-arabinose (10 mM) with agar (15 g/L).
5. Bar-coded primers for sequencing target sites using next-generation sequencing.

3 Methods

3.1 Making $CaCl_2$ Competent BW25141(λDE3) Cells Containing pEndo

1. Day 1, transform pEndo into BW25141(λDE3) cells.
2. Add ~70 ng of pEndo plasmid to 50 μL of $CaCl_2$ competent BW25141(λDE3) cells. Incubate on ice for 30 min.
3. Heat shock at 42 °C for 45 s; place on ice for 2 min. Add 400 μL of SOC media to BW25141(λDE3) cells.
4. Incubate in a 37 °C incubator shaking at 200 rpm for 1 h.
5. Plate 100 μL of cells on plates containing LB plus chloramphenicol (25 μg/mL) and glucose (0.2 %). Incubate overnight at 37 °C (~16 h).

6. Day 2, start overnight culture for BW25141(λDE3) cells containing pEndo vector.
    (a) Add 5 mL of LB plus chloramphenicol (25 μg/mL) and glucose (0.2 %) media to a sterile test tube.
    (b) Inoculate media with a single colony of BW25141(λDE3) cells containing pEndo.
    (c) Incubate overnight in a 37 °C incubator shaking at 200 rpm (~16 h).
7. Day 3, make $CaCl_2$ competent BW25141(λDE3) cells containing pEndo.
    (a) Add 5 mL overnight culture to flask containing 500 mL of LB plus chloramphenicol (25 μg/mL) and glucose (0.2 %).
    (b) Grow in 37 °C shaking at 200 rpm to an A600 of ~0.2.
    (c) Incubate culture on ice for 10 min.
    (d) Pour culture into sterile centrifuge bottles and spin at 4,300 × *g* for 10 min.
    (e) Pour off supernatant; gently resuspend cells in 100 mL of sterile, ice-cold 100 mM $CaCl_2$ solution.
    (f) Spin cultures at 4,300 × *g* for 10 min.
    (g) Pour off supernatant; gently resuspend cells in 1.5 mL of sterile, ice-cold 100 mM $CaCl_2$ solution containing 15 % glycerol.
    (h) Aliquot cells into 1.5 mL microfuge tubes, freeze using liquid nitrogen and store cells at −80 °C.

3.2 Construction of pTox Target Site Library (pToxLib)

1. Add 4 μL of 10× NEBuffer 4 (from NEB), 2 μL of 100 μM target site oligo, 4 μL of 100 μM extension oligo, 10 μL of 2 mM dNTPs, 0.5 μL of Klenow polymerase (5,000 U/mL), and 19.5 μL of ddH_2O.
2. Incubate at 37 °C for 30 min.
3. PCR cleanup and elute in 35 μL of ddH_2O.
4. Add 4 μL of 10× NEBuffer 4 (from NEB), 1 μL of EcoRI-HF, and 1 μL of SphI-HF to PCR cleaned-up extension reaction. In addition, digest (EcoRI/SphI) and dephosphorylate the pTox plasmid.
5. Incubate at 37 °C for 1 h.
6. PCR cleanup and elute in 35 μL of ddH_2O.
7. Ligation: set up reaction with a 1:3 M ratio of EcoRI/SphI-digested pTox plasmid to insert (digested randomized target site). Set up a large enough reaction for 20 or more transformation (need 5 μL per transformation).
8. Incubate at 25 °C for 20 min.

9. Add 5 μL to 60 μL of SEM competent NEB5α (set up 20 or more transformations depending on desired library complexity).
10. Incubate on ice for 30 min.
11. Heat shock at 42 °C for 45 s; then place on ice for 2 min.
12. Add 400 μL of SOC media to NEB5α cells.
13. Incubate in a 37 °C incubator shaking at 200 rpm for 1 h.
14. Combine all the transformations into a 250 mL flask, plate 20 μL on LB kanamycin (50 μg/mL) with glucose (0.2 %).
15. Add 100 mL of LB plus kanamycin (50 μg/mL) and glucose (0.2 %) to the remaining transformation and incubate overnight at 37 °C (~16 h).
16. The next day, make culture stocks of your library and freeze at −80 °C, then miniprep remaining O/N culture.
17. The library complexity can be estimated by counting the number of colonies on the plate, then multiplying it by the dilution factor and dividing by 4 for the number of doubling times.

3.3 Target Site Two-Plasmid Screen

1. Transform ~200 ng of pToxLib plasmid into BW25141(λDE3) cells harboring the pEndo plasmid; in addition transform pTox plasmid containing a positive and negative control target site.
2. Incubate on ice for 30 min.
3. Heat shock at 42 °C for 45 s; then place on ice for 2 min.
4. Add 400 μL of SOC media to BW25141(λDE3) cells.
5. Incubate in a 37 °C incubator shaking at 200 rpm for 1 h.
6. Plate 100 μL on plates containing LB plus chloramphenicol (25 μg/mL), kanamycin (50 μg/mL), and glucose (0.2 %).
7. Incubate overnight at 37 °C (~16 h).
8. The next day, replica-grid individual colonies onto selective then nonselective plates (*see* **Note 3**); grid positive and negative target site control colonies onto every plate (*see* **Note 4**).
9. Incubate at 37 °C for 16 h.
10. Identify cleavable target site "survivors" (*see* **Note 5**) on the selective media and mark them on the nonselective plate. Pick the marked colonies from the nonselective plate and grow overnight (~16 h) in 96-well plates containing 1 mL of LB media plus kanamycin (50 μg/mL) and glucose (0.2 %), shaking at 300 rpm. Grow the desired number of cleavable target "survivors" and non-cleavable target "dead" colonies for sequencing (Fig. 2).
11. Combine 100 μL of culture from each well (each individual target site) and miniprep to isolate the target site pToxLib plasmids.

12. PCR conditions: Using bar-coded primers, PCR amplify the pToxLib library cleavable target sites ("survivors") and non-cleavable target sites ("dead"). Perform 25 cycles using ~100 ng of pTox plasmid. Denature 94 °C for 20 min, 25× (Denature: 94 °C for 30 s, anneal: 54 °C for 30 s, extension: 72 °C for 45 s) and final extension 72 °C for 2 min. Combine ~50 ng of PCR product from each reaction and sequence using next-generation sequencing platform of choice.

4 Notes

1. λDE3 lysogen kit was used to construct a BW25141 strain that carries an inducible T7 RNA polymerase BW25141(λDE3).
2. To create pTox (Fig. 2), pll-lacY-wtx1 was modified by removing the EcoRI site and cloning an oligo cassette containing an EcoRI site into XbaI/SphI. A kanamycin-resistant gene was cloned into the ScaI site of pTox, eliminating the ampicillin-resistant gene in the process. Target sites can be cloned utilizing EcoRI, XbaI, or SphI sites in pTox.
3. Replica-grid controls onto every plate; grid colonies to selective then nonselective plate to ensure that glucose from the nonselective plate is not transferred to the selective plates granting colony survival. *Important*: Mark the top of the plate so that cleavable target site "survivors" can be referenced back to the nonselective plate for sequencing.
4. The number of transformants to be screened is based on the complexity of the pTox target site library, and the degree of confidence that all possible variants of the library have been screened at least once (Fig. 3). For instance, to screen all individual target sites in a library consisting of four randomized positions (256 possible variants) with 90 % confidence, 588 transformants must be replica plated.
5. Robustly cleaved target sites known as "strong" survivors are colonies that have growth comparable in diameter to the wild-type target site, while poorly cleaved target sites are called "weak" survivors grow to a smaller diameter as compared to the wild-type target site and often with a "cauliflower" morphology. Non-cleavable target sites have no growth on the selective media (Fig. 1). In a test study to quantify "strong" and "weak" survivors [10], we isolated multiple pTox plasmids from nonselective cells and found that "weak" uniformly corresponds to <0.5 % survival in the standard two-plasmid assay, while "strong" corresponds between 5 and 100 % survival, depending on the individual target site sequence. Before retransformation of isolated pTox plasmids, we incubate the DNA preparation with NcoI/HpaI to digest any co-purified pEndo plasmid.

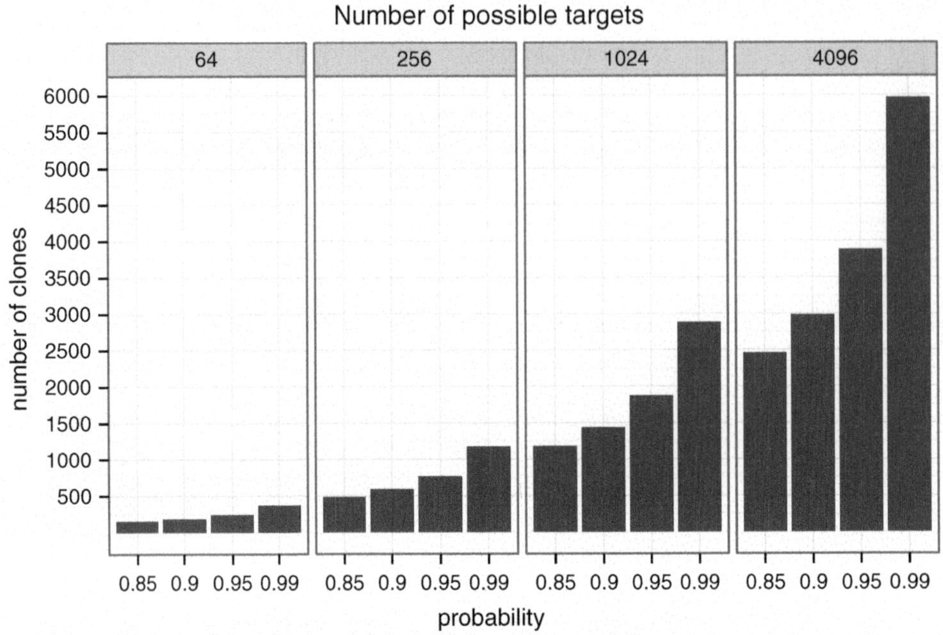

Fig. 3 Graphical representation of the number of colonies required to be screened to achieve a desired confidence that all possible target sites from a random library of *N* nucleotides in length have been sampled at least once

References

1. Belfort M, Derbyshire V, Cousineau B, Lambowitz A (2002) Mobile introns: pathways and proteins. In: Craig N et al (eds) Mobile DNA II. ASM, New York, pp 761–783
2. Stoddard BL (2005) Homing endonuclease structure and function. Q Rev Biophys 38(1): 49–95
3. Takeuchi R, Lambert AR, Mak AN, Jacoby K, Dickson RJ, Gloor GB et al (2011) Tapping natural reservoirs of homing endonucleases for targeted gene modification. Proc Natl Acad Sci U S A 108(32):13077–13082
4. Bogdanove AJ, Voytas DF (2011) TAL effectors: customizable proteins for DNA targeting. Science 333(6051):1843–1846
5. Zhao L, Pellenz S, Stoddard BL (2009) Activity and specificity of the bacterial PD-(D/E)XK homing endonuclease I-Ssp6803I. J Mol Biol 385(5):1498–1510
6. Argast GM, Stephens KM, Emond MJ, Monnat RJ Jr (1998) I-PpoI and I-CreI homing site sequence degeneracy determined by random mutagenesis and sequential in vitro enrichment. J Mol Biol 280(3):345–353
7. Jarjour J, West-Foyle H, Certo MT, Hubert CG, Doyle L, Getz MM et al (2009) High-resolution profiling of homing endonuclease binding and catalytic specificity using yeast surface display. Nucleic Acids Res 37(20): 6871–6880
8. Arnould S, Chames P, Perez C, Lacroix E, Duclert A, Epinat JC et al (2006) Engineering of large numbers of highly specific homing endonucleases that induce recombination on novel DNA targets. J Mol Biol 355(3): 443–458
9. Chen Z, Zhao H (2005) A highly sensitive selection method for directed evolution of homing endonucleases. Nucleic Acids Res 33(18):e154
10. Kleinstiver BP, Wolfs JM, Edgell DR (2013) The monomeric GIY-YIG homing endonuclease I-BmoI uses a molecular anchor and a flexible tether to sequentially nick DNA. Nucleic Acids Res 41(10):5413–5427

Chapter 9

A Yeast-Based Recombination Assay for Homing Endonuclease Activity

Jean-Charles Epinat

Abstract

Homing endonucleases (HEs) are natural enzymes that cleave long DNA target with a high specificity and trigger homologous recombination at the exact site of the break. Such mechanisms can thus be used for all the applications covered today by the generic name of "genome engineering": targeted sequence insertion, removal, or editing. However, before being able to address those applications, the engineering of HEs must be mastered so that any potential target would be efficiently and specifically recognized and cleaved. Working on the I-CreI model, we have developed a very powerful platform to generate HEs with new tailored specificity. We have put in place the first in vivo, functional, high throughput assay to generate I-CreI variants and measure their activity. We use semi-rational design combined with proprietary in silico predictions to design and synthesize I-CreI mutants that are tested for their capacity to induce homologous recombination in a yeast cell. The process has been standardized and robotized so that we can generate thousands of I-CreI derivatives, characterize their cleavage profile, and deliver them for further applications in the research, therapeutic, or agrobusiness fields.

Key words Homing endonucleases, High throughput screening, Homologous recombination, Genome engineering, Yeast

1 Introduction

Genome engineering is the most powerful approach to modify a cell today. Being able to chirurgically insert, delete, or modify one sequence at a chosen locus either by triggering homologous recombination (HR) or nonhomologous end joining (NHEJ) is a goal for anyone interested by studying gene function, protein expression, or "trait" development. The challenge is then to define the tool that is going to be the most successful to enhance site-directed genome modification. Over the last decade, several classes of proteins have been studied and developed to reach this goal. Among them, we can include natural endonucleases as well as artificial enzymes: chemical nucleases [1], zinc finger nucleases (ZFNs) [2], homing endonucleases (also known as Meganucleases) [3], and more recently transcription activator-like nucleases (TALEN™) [4, 5].

David R. Edgell (ed.), *Homing Endonucleases: Methods and Protocols*, Methods in Molecular Biology, vol. 1123, DOI 10.1007/978-1-62703-968-0_9,

Homing endonucleases were discovered in the 1980s as proteins capable of driving the mobility of introns in yeast [6–14]. Today, a considerable body of biochemical, genetic, and structural data showed that they display a high degree of modularity and has also established that these endonucleases could be used as molecular tools leading to targeted mutagenesis (TM) (by NHEJ) or to homologous gene targeting (HGT) (by HR mechanisms such as gene conversion (GC) or single-strand annealing (SSA)) (for general reviews: [3, 5, 15–19]).

However, before starting to introduce any kind of modification into a chosen locus induced by Meganucleases, the large capacity of engineering such enzymes needed to be proven. In 2000, neither molecular tools nor experimental procedures and even less selection/screening methods were in place. During the last 10 years, a variety of approaches have been developed and used for studying and identify new Meganucleases with new specificities [20–33]. Among those techniques, none represents the perfect tool. They belong to the in vitro world; they involve binding only, but not cleavage; they take cleavage into account, but in prokaryotic cell or on the surface of a "displayer." The throughput is not adapted, and only hundreds of mutants can be tested on few targets only; or it allows very large libraries' selection, but the identification of the positive clones is much less efficient. With most of those techniques, it is often difficult to link the binding and cleavage activity of the enzyme since the assay measures only one of them. Moreover, those methods are often not able to quantitatively measure activity levels. Most importantly, those assays are only able to address the activity of monomeric proteins or homodimeric variants, but never heterodimers' activity. The in vivo functional screening assay that we developed and that is described below has the advantage to cover all those aspects. Originally, we designed an assay such as we would be able to easily measure Meganuclease activity in vivo. The goal was to implement a functional assay (mimicking the whole "DSB-induced homologous recombination" pathway) that could be simple, versatile, sensitive, and suitable for high throughput automation. The baker's yeast *Saccharomyces cerevisiae* was immediately chosen due to its powerful genetics tools (easy to transform, strains with different auxotrophic selectable markers, inducible and constitutive promoters (strong and weak), shuttle vectors with high- or low-copy numbers, growth range at various temperatures, several carbon source with known effects, etc.) and because it is the perfect organism for measuring homologous recombination [34, 35].

The goal was set to develop all needed laboratory practices so that a Meganuclease engineering platform could arise. In one hand, the I-CreI Meganuclease was chosen as a semi-rational engineering scaffold. In another hand, a qualitative and quantitative in vivo and functional yeast-based assay for DSB-induced HR was developed that allowed us to genetically link binding and cleavage activity of one enzyme (monomeric, homodimeric, or heterodimer) to its target. The common denominator was that all together a high throughput industrial process had to be implemented [26, 31, 36–42].

2 Materials

2.1 Yeast and Bacteria Culture Media

1. Yeast-complete media (YP) are composed of Bacto Peptone (10 g/l) and Bacto Yeast Extract (10 g/l). Carbon source is added to final concentration of 2 % for D-glucose (added before sterilization) (*see* **Note 1**), 3 % for glycerol (added before sterilization) (*see* **Note 1**), and 2 % for galactose (sterilized by filtration and added after autoclaving (*see* **Notes 2** and **3**) [43]).
2. Yeast minimum synthetic media (MM) are composed of Yeast Nitrogen Base without amino acids and ammonium sulfate (1.7 g/l); ammonium sulfate (5 g/l); desired dropout mix (1 g/l) and carbon source as above (*see* **Note 4**).
3. Complete dropout mix for yeast MM is composed of adenine hemisulfate salt (2 g); uracil (2 g); L-arginine HCl (2 g); L-HISTIDINE HCl monohydrate (2 g); L-isoleucine (2 g); L-lysine HCl (2 g); L-methionine (2 g); L-serine (2 g); L-threonine (2 g); L-tyrosine (2 g); L-phenylalanine (3 g); L-tryptophan (3 g); L-valine (9 g); and L-leucine (4 g). Each particular dropout mix is based on the same quantities except that it misses the amino acid(s) you are selecting prototrophy for. Dropout mixes are added before autoclaving (*see* **Note 5**).
4. For yeast media, G418 is added (after autoclaving media) when needed at final concentration of 200 mg/l. Stock solution at 50 mg/ml (active units) in deionized water and store at −20 °C.
5. For yeast media, pH of all media is brought to 5.6.
6. For yeast and bacteria cultures, stock solution for −80 °C freezing is composed of regular culture media (complete or synthetic) and glycerol at 12.5 % final concentration. It can also be made of 1 volume of yeast culture plus 1 volume of stock solution (Bacto Peptone 20 g/l, Bacto Yeast Extract 20 g/l, glucose 40 g/l, and glycerol 50 %) (*see* **Note 1**).
7. For yeast plates, tetracycline can be added to avoid bacterial contamination. Final concentration is 50 mg/l (added after autoclaving). Tetracycline stock solution is 10 mg/ml in 50 % ETOH (keep at −20 °C).
8. X Gal staining plates are composed of 0.5 M sodium phosphate buffer at pH 7, 0.1 % SDS, 1,5 % w/v agarose, and 0.04 %

X-Gal. X-Gal stock solution is composed of 2 or 10 % w/v in *NA N*-dimethylformamide (DMF) (*see* **Note 6**). Keep at −20 °C in the dark. X-Gal is added after autoclaving (temperature less than 55 °C).

9. *E. coli* LB (Luria-Bertani) growing medium is composed of Bacto tryptone (10 g/l), Bacto yeast extract (5 g/l), NaCl (5 g/l), and pH 7.5.
10. For *E. coli* media, antibiotics can be added after autoclaving: ampicillin (100 μg/ml) or kanamycin (25 mg/ml). Ampicillin and kanamycin stock solutions are, respectively, 100 and 25 mg/ml in deionized water and kept at −20 °C.
11. For yeast and bacteria plates, agar is added to liquid media at final concentration of 20 g/l before autoclaving.

2.2 Yeast Transformation Media

1. Single-stranded salmon sperm DNA (ssssDNA) stock solution at 2 mg/ml in TE pH 8 (*see* **Note 7**).
2. Lithium acetate 1 M, filter sterilized, and kept at room temperature for a month.
3. PEG (polyethylene glycol) 3350 50 % w/v in water, filter sterilized, and kept at 4 °C.
4. TE (Tris base 10 mM, EDTA 1 mM, pH 8).
5. Transformation mix (for ten high-efficiency library transformations): 5.28 ml PEG 50 %, 792 μl LiAc 1 M, and 1.1 ml ssssDNA.

2.3 DNA Extractions Buffers

1. Plasmid DNAs from *E. coli* are extracted manually with Qiagen or NucleoBond kits using manufacturer's protocol.
2. In case of high throughput, *E. coli* are grown in LB medium (with ampicillin or kanamycin depending on the plasmid resistance maker, see above) in 96-deepwell plates (1.5 ml per well) for 48 h, and extraction is performed on a Beckman BioMek FX platform using Wizard MagneSil TFx System from Promega.
3. Yeast cultures for plasmid isolation are grown in 96-deep-well plates (2 ml per well) in the proper selective media (depending on the auxotrophic marker of the plasmid, see above) for 48 h.
4. For yeast DNA preparation: KH_2PO_4 33.5 mM pH 7.5 (1 M stock solution), 20 % SDS, and Lyticase (Arthrobacter luteus; SIGMA Ref: L-2524) at 2.5 U/ μl in KH_2PO_4 33.5 mM pH 7.5 (keep at −20 °C) are needed.

2.4 Robotic Equipment

1. Colony gridders QPix and QPix II XT from Genetix (maintained by Proteigene) are used for the gridding of yeast strains as well as for colony picking, plate replicating, and rearraying.
2. High throughput DNA extraction from bacteria and yeast are performed on Beckman BioMek FX platform which is also used for cherry picking, dilution, etc.

3 Methods

We designed a mating assay where one strain will express Meganuclease genes and another strain would carry an inactive reporter plasmid. After mating, the transformed diploids would be selected, and if the Meganuclease recognized and cleaved its target site, the reporter would be converted into an active form (Fig. 1). Both strains can be used for screening and kept for future experiments. We thus started to build a very large library of Meganuclease expression and target screening strains that could be mated at any time.

3.1 Plasmids Constructs

1. Target vectors: We designed a reporter vector with two inactive direct repeats of the LacZ gene (*see* **Note 8**) that would be susceptible to SSA after cleavage (Fig. 2 and Table 1). The DNA targets (24 base pairs up to 5 kb length) used in the yeast screening assay are inserted into the yeast vector pFL39-ADH-LACURAZ by in vivo homologous recombination [36] (*see* **Notes 9** and **10**).

   Yeast reporter vectors are used to transform *Saccharomyces cerevisiae* strain FYBL2-7B (*MAT a*, *ura3Δ851*, *trp1Δ63*, *leu2Δ1*, *lys2Δ202*) [44]. Transformants are selected on synthetic medium lacking uracil and validated (prototroph for

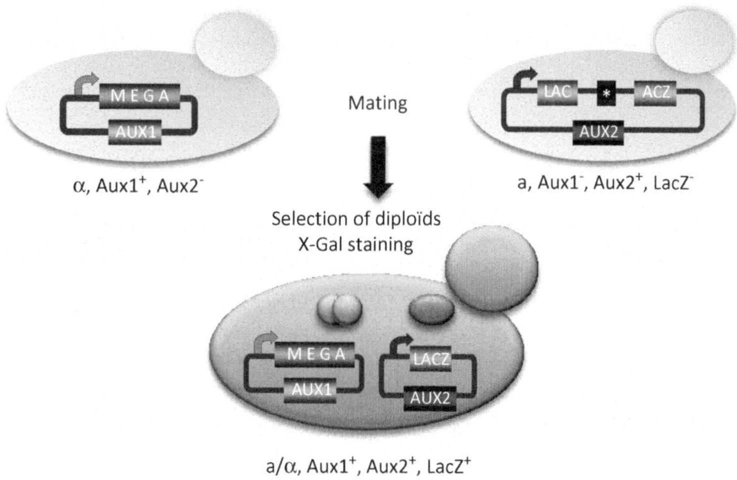

Fig. 1 Screening assay in yeast. One yeast strain is used for Meganuclease expression, and another strain with the opposite mating type receives the target reporter vector. After mating and transformed diploid selection, the expression of the enzyme is induced on galactose medium. If the enzyme recognizes and cleaves its target sequence, the LacZ reporter recombines by a single-strand annealing event. The reconstituted gene allows LacZ expression which is measured by X-Gal staining

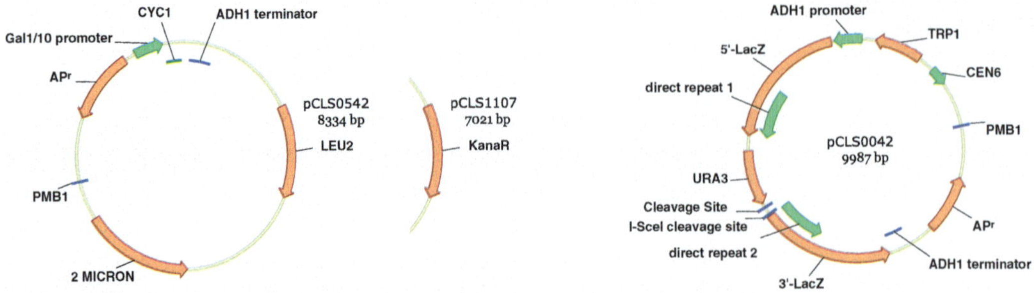

Fig. 2 Schematic representation of expression and target vectors. *Left*: The Meganuclease expression vector is a shuttle plasmid with a 2 μm origin of replication, which allows a high copy number after yeast transformation. The expression cistron is induced when yeast grow on a galactose-based carbon source. We have developed a couple of plasmids that allow the expression of heterodimers. One plasmid can be selected on media without leucine, the second for G418 resistance. *Right*: The target vector is a centromeric-based, low-copy, plasmid. It is composed of a partial LacZ direct repeat driven by a constitutive ADH1 promoter. In between the tandem repeats, we find an I-SceI cleavage site, the desired target site, and a URA3 cassette that allow the selection of "non-recombined" targets after yeast transformation on media lacking uracil. Later on, the presence of the vector can be selected on media lacking tryptophan

Table 1
Yeast genetic tools for in vivo functional assay

		Nuclease	Target
Yeast strain		FYC2-6A : *MATα*, *trp1Δ63*, *leu2Δ1*, *his3Δ200*	FYBL2-7B : *MAT a*, *ura3Δ851*, *trp1Δ63*, *leu2Δ1*, *lys2Δ202*
Plasmid	ORI (copy nb /cell)	2 μm (10–50)	CEN6 (1–5)
	Promoter; terminator	Inducible P_{GAL10}; ADH1 term	Constitutive P_{ADH1}; ADH1 term
	Marker	LEU2 and/or Kan^R	URA3 (selection of "original" transformants : before validation) and TRP1 (selection transformants during screening, after validation)

The system is based on mating a yeast strain that expresses a Meganuclease gene (or two) with a yeast strain containing a LacZ reporter. The plasmids have distinct selectable markers. The expression vector is a high-copy-number plasmid and the expression of the enzyme is galactose dependant. The reporter vector is a low-copy plasmid, and the LacZ gene is constitutively expressed after cleavage induced reconstitution

uracil, white and cleavable by I-SceI) before entering the screening process (*see* **Note 11**). Constructs are always confirmed by sequencing.

2. Expression vectors: DNA sequences encoding the various Meganucleases were generated by PCR (*see* [3, 15, 16, 36–38, 41] for details). Genes are inserted by in vivo cloning (*see* **Note 9**) into the 2 μm-based replicative vector pCLS0542, which contains

the LEU2 gene for selective growth or into pCLS1107 that contains a kanamycin-resistant cassette (Fig. 2 and Table 1). *Saccharomyces cerevisiae* strain FYC2-6A (*MATα*, *trp1Δ63*, *leu2Δ1*, *his3Δ200*) [44] was transformed with these vectors using a high-efficiency lithium acetate transformation protocol (see below and ref. 45) (*see* **Note 12**).

3.2 Transformations

1. Every screening campaign involves Meganuclease expression vectors and target reporter plasmids transformation in yeast strains. The initial step is always manual transformation and plating. Since we often have libraries to be screened, the efficiency of the transformation procedure is crucial. The following protocol consistently gives 10^6 transformants per microgram of DNA, but allows up to 10^7 transformants.
2. Prepare an overnight saturated culture at 30 °C.
3. Dilute in 100 ml warm YP glucose media to 5×10^6 cells/ml ($A_{600} = 0.2$).
4. Incubate at 30 °C to get 2×10^7 cells/ml ($A_{600} = 0.6$–0.8). Centrifuge and rinse twice with 50 ml of 100 mM LiAc.
5. Centrifuge and resuspend in final volume of 1 ml 100 mM LiAc 100 mM (2×10^9 cells/ml). Use 100 μl/transformation.
6. Prepare a transformation mix of 660 μl of yeast cells and up to 5 μl of DNA (do not exceed 10 μg).
7. Gently mix and immediately heat shock at 42 °C for 1 h.
8. Centrifuge and resuspend in 10 ml of YP glucose.
9. Incubate under shacking at 30 °C for 1–3 h before plating on selective medium.
10. Incubate plates at 30 °C for 3 days (*see* **Notes 13–15**).
11. Meganuclease transformants are plated on 22 × 22 cm plates (QTrays), and single transformants are selected either on MM with glucose, without leucine on YP glucose with G418 (for single transformants expressing homodimers or single chain molecules), or on glucose MM without leucine + G418 for double transformants (strains expressing heterodimers).
12. Target clones are plated either on standard 8 cm dishes or in individual squares of QTrays 48 of glucose-based selective media (MM without uracil). The growth of the clones takes about 3 days at 30 °C.

3.3 In Vivo Cloning in Yeast

1. We use a linearized cloning vector and an insert fragment (generated by PCR or restriction digest) that overlaps with the extremities of the vector by at least 20 pb; however, 100–200 bp of homology with the 2 μm-based replicative vectors are used.

2. Cotransform your yeast strain with both fragment and linearized plasmid and select for the auxotrophic marker of the cloning vector. In our case, to generate Meganuclease's expression vectors, approximately 25 ng of Meganuclease-encoding PCR fragments and either 25 ng of expression vector (either pCLS0542 linearized by digestion with NcoI and EagI or pCLS1107 linearized by digestion with DraIII and NgoMIV) are used to transform the yeast *Saccharomyces cerevisiae* strain FYC2-6A (*MATα*, *trp1Δ63*, *leu2Δ1*, *his3Δ200*) using a high-efficiency LiAc transformation protocol described in Subheading 3.2.
3. Transformants are selected on either synthetic medium lacking leucine (pCLS0542) or rich medium containing G418 (pCLS1107) (*see* **Note 16**).
4. For target cloning, 20 bp of homology with the extremities of the vector are required. 25 ng of linearized vector and 25 ng of PCR fragment are combined to transform *Saccharomyces cerevisiae* strain FYBL2-7B (*MAT a*, *ura3Δ851*, *trp1Δ63*, *leu2Δ1*, *lys2Δ202*) [44]. Transformants are selected on synthetic medium lacking uracil.

3.4 Picking and Rearraying

1. Meganuclease-expressing clones and target strains then have to be picked and transferred into 96-well plates (we use flat bottom plates, but round bottom are performing as well). The picking is handled manually or by using QPix and QPix II XT colony pickers' robots (Genetix).
2. Clones are grown in 200 μl of liquid media (same as selective media used after transformation) for 2–3 days under shaking.
3. After growth, mother plates are either replicated into daughter plates that are grown again for 2 days before entering the screening process or prepared for banking by addition of glycerol storage medium and stored at −80 °C. Once an experiment is designed, plates are taken out and thawed (*see* **Note 17**).
4. If needed, clones are subjected to rearraying into new plates. Sterilize the pins between each 96-well plates to avoid cross contamination. The procedure we implemented uses three washing bathes: water, followed by 50 % ethanol, and then 100 % ethanol.

3.5 Gridding and Mating

1. The screening experiment is entirely performed on a 22 × 22 cm nylon membrane that covers an agar plate (Fig. 3).
2. The first step of the experiment is the mating. Meganuclease-expressing strains are spotted at the desired format (from ~4 to ~20 spots/cm^2) on a filter one after another using a QPix II robot (Genetix). The robot uses a 96-pin head to spot an aliquot

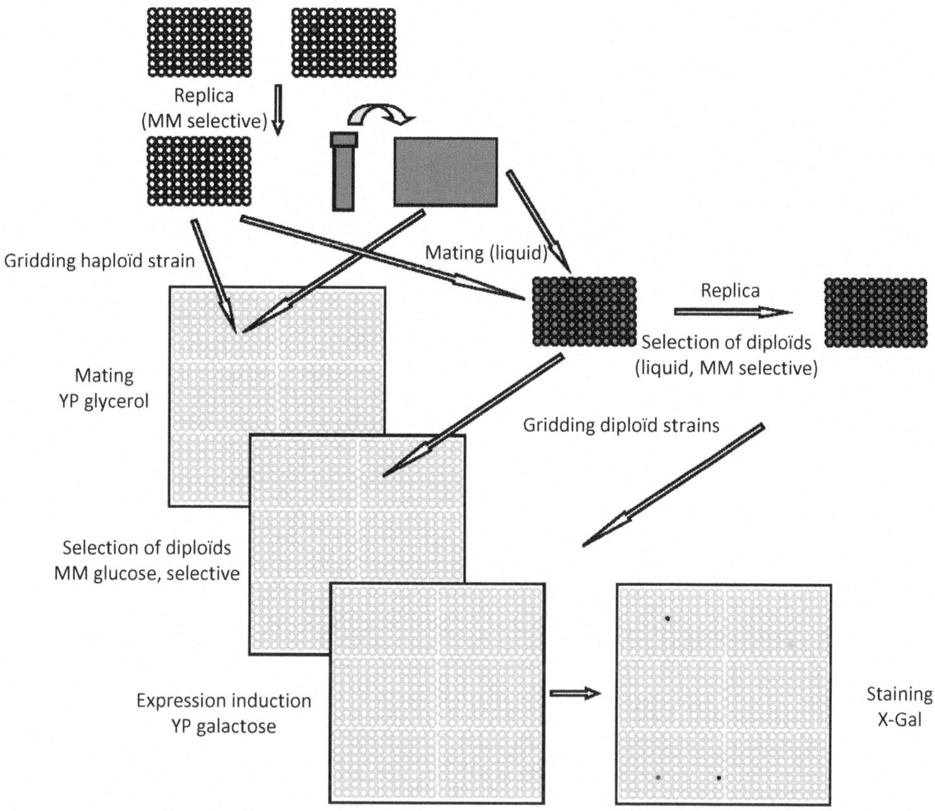

Fig. 3 Steps of the screening assay in yeast. One yeast strain expressing a Meganuclease is stamped on a nylon membrane using a QPix II colony gridder. Another strain transformed by a target reporter vector is stamps right on the previous spot. The mating occurs over night on complete glycerol media. The filter membrane is than manually transferred to a media that allows the selection of cotransformed diploids (minimum synthetic media with glucose lacking the proper amino acids). The filter is again transferred on galactose media for induction of Meganuclease expression. At that step, we can change the incubation temperature and/or the carbon source for modulating the activity of the Meganuclease. Finally, the filter is put on revelation media to measure the beta-galactosidase activity. A picture of the filter is taken and analyzed by a proprietary software that assign an activity value to each dot. Alternative procedures allow mating and selection directly in 96-well plates before gridding

of each well on the filter. Each well can be spotted several times forming a "cluster" (Figs. 3 and 5).

3. A second gridding process is performed on the same filters for spotting of a second layer consisting of reporter harboring yeast strains for each target (*see* **Note 18**).
4. Target strains can be spotted first and Meganuclease clones second.
5. During the gridding, the only important step consists of sterilizing the pins each time that you change a 96-well plate. Use the same procedure in Subheading 3.4.
6. The medium used for mating is a rich medium with glycerol as carbon source. Plates are incubated overnight at 30 °C.

However, an alternative procedure consists of growing, mating, and selecting diploids in liquid media in 96-well plates before spotting (*see* **Note 19**).

3.6 Selection of Diploids

1. The day after mating, the filter is manually transferred onto a second agar plate containing synthetic minimum medium with an amino acids dropout mix (and with G418 for co-expression experiments) for selecting only transformed diploids strains on glucose.
2. Incubation lasts for 2–3 days at 30 °C.
3. The liquid procedure mentioned above can also be continued at that step (*see* **Note 20**).

3.7 Induction of Meganuclease Expression

1. After selection of diploids, the filter is transferred to complete medium with galactose as carbon source to induce the expression of the Meganuclease (Fig. 3; *see* **Notes 21–24**).
2. The incubation temperature is either 30 °C or 37 °C (Table 2, *see* **Note 23**).
3. If you chose the liquid mating/selection procedure, and regardless the "variations" you decided to use (refer to **Notes 21–24**), you need to grid your filters at that step or repeat the selection step detailed above (*see* **Note 26**).

3.8 Colorimetric Readout

1. The filter is placed on an agarose plate for cell disruption and X-Gal staining after incubation at 37 °C (Fig. 3 and Table 2).
2. After 2 days on X-Gal, a picture of the filter is taken with a CCD camera, and proprietary software will acquire the data by measuring the blue color of each spot [36] (*See* **Note 27**).
3. Referring to a scale from 0 (white dot, no cleavage) to 1 (dark-blue dot, maximum intensity detected), a value is assigned to

Table 2
Timelines for the yeast screening assay

	Growth	Mating	Selection	Induction	Staining
Media	MM (selective)	YP glycerol	MM (selective)	YP galactose	X-gal
Time (days)	0.5–3	0.5	2–3	2–3	2 (or 1)
Temperature (°C)	30	30	30	30 and 37	37 (42)

First, the transformed yeast cells are grown in liquid culture (tubes or 96-well plates) with glucose selective media. The mating occurs over night on complete glycerol media. Selection of cotransformed diploids (minimum synthetic media with glucose lacking the proper amino acids) is maintained for 2–3 days to allow growth of the colonies. Then, induction of Meganuclease expression is carried out on complete galactose media for 2 more days (or 3). At that step, modulation of the activity level can be done by changing the incubation temperature and/or the carbon source. Finally, beta-galactosidase activity is measured by X-Gal staining for 2 days

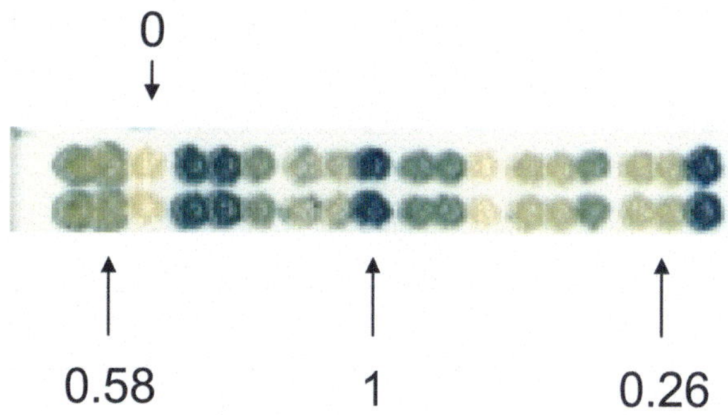

Fig. 4 Read out of the yeast assay. At the end of the assay, a picture of the filter is taken and analyzed by a proprietary software that assign an activity value (from 0 to 1) to each dot. On a filter, a negative control dot gives the background value of 0. The *darkest* positive control *dot* is measured as 1. The scale also includes an intermediate value. A value is then assigned to each dot

each dot and characterizes the activity level of the tested Meganuclease on a particular target (Fig. 4).

4. It is possible to speed up the assay by placing the plate at 42 °C, so that the color develops in about 10 h.
5. The advantage of the liquid process is that you will have very nice spots.

3.9 Workflow

1. It takes almost 10 days from the gridding to the final entry of the results in the database.
2. For the consistency of the data, the process was standardized for minimizing day-to-day variations from a test to another (Fig. 3, Table 2 and *see* **Note 28**).

3.10 Quality Controls

1. Control wells are added at the picking step into each plate, and more control dots are added at the gridding level.
2. On each Meganuclease expression plate, a particular plate design is used. Wells H10-H11-H12 are reserved for intraplate controls, namely, an "empty" strain that does not express any Meganuclease and will give a "white dot," a strain expressing a weak I-SceI variant that will give a "light blue" dot, and a wild-type I-SceI Meganuclease will show a "dark-blue" dot.
3. For target strains, no additional control is added since each target is tested on each cluster.
4. On filters, each cluster of each field, one positive and one negative control, is added.
5. We prefer "loosing" as much as 50 % of the dots on one filter (in the case of a 2 × 2 cluster) to keep a very high quality of the data instead of misinterpreting data.

6. All those controls give the assurance that (1) each target strain can give a blue dot (I-SceI control site allows the use of I-SceI Meganuclease as a positive control to prove that the construct can recombine upon cleavage), (2) the whole filter is valid (each plate responds positively), and if not, give us the opportunity to qualify the data as invalid in our database. They also give plate-to-plate, filter-to-filter, and screening campaign comparisons (*see* **Notes 29** and **30**).

3.11 Screening Formats

1. Depending on the experiment, the screening format can range from ~4 to ~20 spots/cm^2 (Fig. 5 and Table 3).
2. On a filter, up to 9,216 dots can be spotted by the robot (4×4 format). The robot uses 96-pin head that grids 6 fields, where 16 dots can be spotted. Those 16 spots form a "cluster" and can come from 16 different mutant plates tested on 1 target

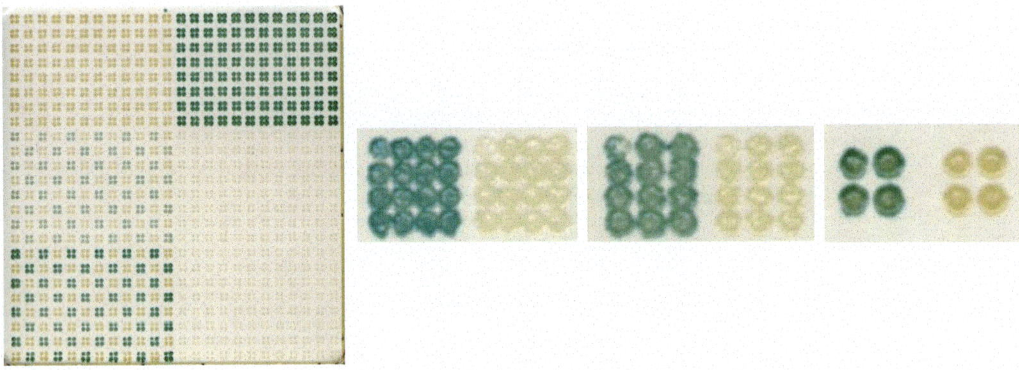

Fig. 5 One filter, different formats. A filter is composed of six fields. On each field, different formats can be used independently. The 2 × 2 format (the four dots form a cluster) is the more quantitative. The 3 × 4 is denser since the 12 dots are closer, but the cluster is not proportionally bigger. The 4 × 4 (16 dots per cluster) is only qualitative but allows higher throughput

Table 3
Screening formats and uses

Cluster format	Nb dots/filter	Quality	Usage
2×2	2,304	Quantitative	II^{ry} and III^{ry} screening
3×2	3,456	Quantitative	II^{ry} and III^{ry} screening
3×3	5,184	Semiquantitative	II^{ry} screening
4×3	6,912	Qualitative	I^{ry} screening
4×4	9,216	Qualitative	I^{ry} screening

The gridding format of the colonies on a 22×22 filters depends upon the type of experiment. The denser format allows high throughput, but only qualitative data. The less dense formats are used for quantitative analysis (and vice versa for profiling experiments or testing target libraries); there can be eight mutant plates, spotted twice on two targets each (*see* **Notes 31** and **32**).

3. The screening process was divided into three consecutive screening steps that are more and more sensitive and give precise information regarding the enzyme characteristics.
4. First, a primary screening (qualitative) is performed to recover positive clones (or negative if needed).
5. Second, the clones with the desired phenotype are consolidated into new plates and are subjected to a secondary screening at a less dense, quantitative format for hit confirmation.
6. Finally, if needed, that DNA can be used to transform *E. coli* for segregation of plasmid population (*see* **Note 33**) and reused for a tertiary "clonal" "tertiary" screening and validation in yeast.

3.12 Yeast Plasmid DNA Extraction

1. Recovering yeast plasmid DNA requires that spheroplasts are formed by digestion of the cell wall with lyticase and then isolated by a DNA extraction protocol.
2. A 96-deep-well plate is grown for 3 days and centrifuged.
3. Keep the pellet and add 50 μl of 33.5 mM KH_2PO_4, pH 7.5, and incubate the deep-well plate under shaking for resuspension.
4. 20 μl of lyticase (2.5 U/μl) are added before a 1 h incubation step at 37 °C.
5. Spheroplasts are "blown out" by adding 10 μl of 20 % SDS.
6. Isolate bacterial DNA using extraction procedure is applied. The amount of plasmid DNA that is recovered is only sufficient for PCR amplification or *E. coli* transformation (*see* **Note 34**). Additional comments can be found in **Notes 35–39**.

4 Notes

1. Autoclave at 121 °C for 15 min. Autoclaving at higher temperatures or for longer periods of time may cause the sugar solution to darken. It will decrease the performance of such media.
2. Galactose must be highly pure and contain less than 0.01 % glucose.
3. Galactose is not stable at high temperature; it has to be sterilized by filtration (0.2 μm) and added to the media at a temperature <55 °C.

4. Even if it is often mentioned that amino acids mixtures must be added after autoclaving, at temperature <55 °C, we have never found problems with adding them prior autoclaving.
5. Powders are pooled together in a dark, sterile plastic container; add 4–5 large "kids" glass beads and shake for mixing. Keep away from humidity. Store at room temperature.
6. DMF is highly toxic; wear gloves.
7. ssssDNA can be long to dissolve: dissolve overnight, boil to denature for 10 min, and aliquot and store at −20 °C.
8. The beta-galactosidase was chosen because its activity can be measured on solid media (X-Gal overlays, X-Gal staining of colonies on filters, etc.) or on liquid cultures (absorbance, light, fluorescence, etc.) [35].
9. We chose in vivo cloning because it allows the generation of large collections of target plasmids at once (using 96-well plates). It is also very powerful to generate large libraries of Meganucleases variants and avoid the generation of such libraries in *E. coli*. However, regular cloning can of course be used. In the case of target plasmid generation, it is important to notice that our construct allows classical blue/white screening in *E. coli*, meaning that the "non-recombined" target vector give raise to a white *E. coli* clone on LB plates containing X-Gal. Transformation leads to spontaneous recombination, and screening the "non-recombined" *E. coli* colonies helps saving time (*see* **Note 11**).
10. The reporter construct was cloned into a low-copy number vector to limit the amount of beta-galactosidase in the yeast cell and minimize the chances of saturation of the assay since the enzyme is known to be very stable in yeast. The LacZ gene is under the control of an ADH1 constitutive promoter [46]. An intervening sequence separates both repeats. It contains a URA3 expression cassette, an I-SceI control cleavage site, and any other cleavage site or DNA fragment (Fig. 2 and Table 1; *see* **Notes 11** and **22**).
11. Since transformation is highly recombinogenic in yeast (and bacteria), each yeast target strain must be tested before entering the screening process: a target plasmid is used to transform the target strain, 4 [URA$^+$] clones are picked, assayed for beta-galactosidase expression, and tested for I-SceI-induced recombination by mating with an I-SceI expression strain. A white [URA$^+$] clone that turns blue after I-SceI testing is validated. A final Sanger sequencing of the target is also done to confirm the sequence of the target.
12. We thought that it would be convenient to have a set of two similar expression vectors with two different selectable markers

to test the co-expression of two different Meganucleases and thus assessing the activity of heterodimers. We also made the choice of a high-copy number expression vector (Fig. 2 and Table 1). The strains used in the laboratory are derived from the S288C yeast strain which has been used for genome sequencing [44].

13. Make a rough calculation of the number of clones you expect, and do not hesitate to plate several dilutions of your transformation to ensure that the clones will not be confluent.
14. If a control transformation shows that the efficiency is too low, change all transformation buffers by fresh ones (ssssDNA, LiAc, TE, etc.).
15. For regular plasmid transformation, the same protocol can be used with 50 μl of competent cells, 100 ng to 5 μg of plasmid DNA, and 300 μl of transformation mix. You can decrease heat shock and recovery time down to 30 min each.
16. Using that scheme, we have been able to assemble 3 DNA fragments into a linearized vector at once (100 pb homologies with the vector and 15–20 bp homologies between fragments).
17. However, mother plates are never directly used for testing. Replicas are always made fresh for testing (but plates taken directly from the freezer can perfectly be used as such).
18. It is possible to either pre-grid a filter with one of the two partners and keep it at 4 °C before gridding the second partner and go for regular steps afterwards. In that case, the only requirement is to work with the very same robot twice to ensure that both spots will really be on top of each other with no physical shift! We recommend performing the second gridding not more than 3 days after the first one.
19. The mating can also take place in liquid media (tube or standard 96-well plate). Saturated cultures of Meganucleases expressing strains and target strains would have to be mixed in equal volume in YP glucose media and incubated for at least 5 h without shaking. Then, 5 μl are taken out and transferred into the appropriate selective media containing glucose. We usually select diploid transformed strains (MM without lysine and without histidine for diploids, and without tryptophan for target vector, and without leucine +/– G418 for expression vector). Grow plates for 3 days at 30 °C under shaking. The drawback of such "liquid processes" is that a very large number of plates are generated.
20. Like mating, the selection of the diploids can be done in liquid culture by replication of the liquid mating into the proper glucose selective media. Grow the 96 shallow well plates for 2–3 days at 30 °C.

21. The yeast system that we developed is very sensitive. If it gives us the opportunity to detect very low levels of cleavage, it is very difficult to assess the "real" activity of enzymes that give a defined level of signal since the saturation of the assay is very fast. At that step, some alternate procedures can be applied to allow a good discrimination among the good cutters. It is possible to play with the copy number of the expression vector in a way to minimize or maximize the amount of enzyme expressed in the yeast cell (*see* **Note 22**). The growing temperature can be controlled since *S. cerevisiae* supports quite a large range (*see* **Note 23**). We directly transfer the filters from glucose media to galactose without any derepression step (*see* **Note 24**). Finally, the colorimetric test can be analyzed at different times points (see bellow and **Note 25**). We also tested few situations where several of those parameters were modified during the same experiment. All those comparisons did lead us to the conclusion that the best screening conditions would be to duplicate each filter and evaluate the Meganuclease's activity under induction condition, but changing the induction temperature from 37 to 30 °C. Those two temperatures give us a good estimation of the activity level of the enzyme providing a high discrimination quality of weak (active only when induction is done at 37 °C) and strong (also active at 30 °C) Meganucleases (Table 2). They also give indication of the activity such enzyme can have in cells or organisms for which the living temperature is around 37 °C (mammalian cells), or 30 °C (yeast), if not less (algae, fungi, plants).
22. We tested the effect of a low-copy (with a centromeric origin of replication) expression vector instead of the 2 μm based classical vector but the effect is not drastic. The option of having two types of vectors has not been kept since the effort would have been doubled for a minimum gain of robustness (not shown).
23. We tested different incubation temperature, either during the whole diploid selection/induction process or only for the galactose induction phase. The best growing temperature for *S. cerevisiae* is 30 °C; however, it can support quite a large spectrum. The range of temperature tested varied between 25 and 37 °C, but we also tested 4 and 42 °C. The later experiments were however too extreme, and the process had to be completely modified to allow either colony to grow large enough (at 4 °C) or to stay alive (at 42 °C) rendering those modifications not suitable for production (not shown).
24. We assayed different carbon source to test variations of the transcriptional activation of the P_{GAL10} promoter and modulate the galactose switch (or diauxic shift) during the assay (from glucose repression to glycerol/lactate or raffinose derepression

[47]) for modulating the activation level of the Meganuclease gene (not shown).

25. The quantification of the Meganuclease activity is directly deduced from the beta-galactosidase activity, namely, the intensity of the blue color of each spot. We showed that the beta-galactosidase activity correlates directly with the activity of the enzyme in vitro [36]. At the same time, we have shown that the spot intensity almost perfectly reflects various level of Meganuclease activity by serial dilution of transformants and subsequent measure of the blueness indicating that the spot intensity is a good indicator of the number of cells with active Meganuclease (i.e., Meganuclease activity).
26. Unlike mating and selection of the diploids, the induction is always done on filter. This is the last step when a growing media is used (with carbon source and nutriments).
27. Time courses were made at the revelation step when the filters are put on the X-Gal medium. Pictures and their analysis were taken 8, 24, and 48 h after the beta-galactosidase tests had started (not shown).
28. Griddings that are performed Tuesday, Wednesday, and Thursday follow a standard process. Friday is a transformation day, so that yeast colonies are fully grown and ready to be picked on Mondays. It is also possible to grow the 96-well plates over the weekend and start gridding on Monday, but the mating is less efficient and the colony growth is not always homogeneous. Monday and Friday can also turn into gridding days since it is possible to either pre-grid a filter with one of the two partners and keep it at 4 °C before gridding the second partner.
29. Always check for filter homogeneity of the controls (cluster controls and plate internal controls). Edge effect can be detected all around the filter. To avoid them, one solution consists of filling the outer wells of each plate with negative controls.
30. The signal-to-noise as well as signal-to-background ratio cannot be easily applied to our assay because of the limited range of our readout (from 0 to 1); however, they can be easily appreciated by the eye. For example, the I-SceI test on each target should saturate (dark blue, value of 1); however, if the positive I-SceI controls are not saturating on a quantitative filter, data are marked as invalid but kept. Nevertheless, if we use all the data stored in our database, the overall Z′ factor associated to those controls may vary from less than 0.1–0.8; however, for secondary screenings, profiling experiments, and particularly our tertiary screening tests, it is consistently above 0.5 (data not shown) [48]. When repeated on different filters, measurements are found to be within 0.1 interval around median activity values in 95 % (gal30) and 90 % (gal37) percent

of the cases (not shown). Finally, our system is robust enough to give a very low false-positive ratio and a very consistent hit confirmation rate [49–52].

31. The denser the format, the smaller are the dots and the assays turns from quantitative to only qualitative. However, for large libraries, a dense primary screening allows the recovery of the positive clones which are later on collected in rearraying plates and reassayed at a less dense format for a secondary screening (Table 3).

32. Since we planned to turn into automation and high throughput, we tested two different SBS plate formats. Internal data favored 96 wells. We had positive data with denser formats (384-well plates); however, the homogeneity of growth all over the plate was not consistent, and we faced shaking issues of the plates to avoid sedimentation. The 96-well-plate format, even if not optimal, allows a decent throughput and is perfectly adapted to our needs (library size, versatility, etc.). Our screening capacity is large enough to reach 8×10^5–1 million dots (one Meganuclease tested on one target) per week (about 100 high density filters), which represents a decent figure given the fact that our test is functional (DSB-induced HR) and in vivo (heterologous expression of the molecule in a living organism) and that a lot of steps are still manually handled (transformation, plating, transferring filters, etc.). It is important to mention that every sample, every aliquot (DNA plasmid, library, target, yeast strain, plates, media, etc.) is barcoded and entered into a Laboratory Information Management System (LIMS) for complete traceability of the data.

33. When individual candidates are tested from a single plasmid construct, the results can be immediately validated. However, when libraries are tested, the yeast transformation can lead to several different plasmids inside a single haploid transformant. We observe this phenomenon if we use either plasmid libraries or linearized plasmid with DNA fragments that are cloned "in vivo" by gene conversion after transformation. The frequency of multiple transformations is approximately the same in both cases. This "polyclonality" is easily observed by sequencing for approximately 1/3 of the transformants, and usually, two or three different sequences can be found, rarely more (unpublished data).

34. The G418 resistance cassette of the pCLS1107 derivatives can efficiently be used to recover kanamycin-resistant *E. coli* colonies.

35. Like any other test, ours has some limitations. However, each data point relies on the expression of the nuclease in a living eukaryotic cell and counter selects toxic molecules that would

kill the host. The molecule needs to be properly folded, active and able to recognize, and bind and cleave its target triggering a homologous recombination event that is monitored and quantified. The expression level of the protein(s) is neither monitored nor controlled. However, the expression vector and the cloning are always strictly identical. We can then assume that the only variation would be due to RNA stability or protein half life variations, but the differences among them being minimum, it is very unlikely.

36. One strength of a screening approach is that each candidate is tested whether it does or does not have the desired cleavage profile. Information can thus be gathered from the nonresponsive clones and enriches the database.

37. Whether homodimers or heterodimers have to be tested, the screening procedures are the same, except the culture media. With standard procedures in place, the decision was also taken to have every experiment done by the screening platform. Development of new enzymes and R&D experiments all enter the process and follow the same procedures. Each data point and each result can be queried and compared to others. Since the clones that express Meganucleases as well as target strains are kept frozen, we always have the possibility to redo an experiment or to add some more tests to enrich a genotype with a broader phenotype. This gives us a large choice among 55,000 engineered frozen enzymes attached to 570,000 cleavage data points to choose candidates for their final desired properties (the total number of data point in our database is about 100 million).

38. Today, the very same process is used in house to test and engineer TALENs. Once again, we use the same mixed tools of molecular biology, yeast genetic, and HTS to learn more about how those molecules work on new and better scaffold for downstream genome surgery applications and produce TALENs with tailored specificity.

39. The major drawback of such an assay relies on the fact that it is an extrachromosomal SSA assay. Most of the final applications rely on mutagenesis (NHEJ) or gene conversion in a living cell, at a wild-type chromosomal locus. First, the biological event is not the same (SSA vs. NHEJ or GC); second, the structure of the target is different. Euchromatin, heterochromatin, or chromatin compaction/structure in general has an impact on the cleavage capacity of LHEs. Cytosine methylation can also impact the accessibility of the target and impair or reduce efficiency of the cleavage by perturbing the binding step [53]. The engineering of nuclease and the choices that are made upon cleavage activity in a heterologous, artificial system like the one

described here must be taken carefully like a hit to lead selection step. In any case, it can prove the final efficacy of the enzyme for its final use. The system is good at giving the best "intrinsic" activity of the protein, but, like an in vitro assay, it remains artificial. It discriminates cutters and non-cutters. It allows characterizing the better possible candidates. However, among those, some might be even better than others, and finally, the chromosomal target might turn out to be "uncleavable" under the conditions of the experiments (cell cycle stage, cell type, differentiation stage, etc.). As powerful as it is, an assay gives the best possible choice, but the final answer lies in "real-life" experiment. A recent work illustrates perfectly the range of activities of engineered Meganuclease at their locus even though they were all chosen as good or very good cutters in yeast. It shows that targeted mutagenesis can be measured between <0.1 and 6 % and homologous gene targeting between <0.1 and 15 % and demonstrate a strong position effect correlated with chromatin structure [53]. The best engineered enzyme will always be a tool sensitive to chromosomal context and epigenetic factors.

References

1. Schleifman EB, Chin JY, Glazer PM (2008) Triplex-mediated gene modification. Methods Mol Biol 435:175–190
2. Ramirez CL, Joung JK (2013) Engineering zinc finger nucleases for targeted genome editing. Top Curr Genet 23:121–146
3. Epinat JC, Silva G, Paques F, Smith J, Duchateau P (2013) Engineering meganucleases for genome engineering purposes. Top Curr Genet 23:147–185
4. Christian M, Cermak T, Doyle EL et al (2010) Targeting DNA double-strand breaks with TAL effector nucleases. Genetics 186:757–761
5. Hafez M, Hausner G (2012) Homing endonucleases: DNA scissors on a mission. Genome 55:553–569
6. Jacquier A, Dujon B (1985) An intron-encoded protein is active in a gene conversion process that spreads an intron into a mitochondrial gene. Cell 41:383–394
7. Kostriken R, Strathern JN, Klar AJ, Hicks JB, Heffron F (1983) A site-specific endonuclease essential for mating-type switching in Saccharomyces cerevisiae. Cell 35:167–174
8. Stoddard BL (2011) Homing endonucleases: from microbial genetic invaders to reagents for targeted DNA modification. Structure 19:7–15
9. Prieto J, Molina R, Montoya G (2012) Molecular scissors for in situ cellular repair. Crit Rev Biochem Mol Biol 47:207–221
10. Taylor GK, Stoddard BL (2012) Structural, functional and evolutionary relationships between homing endonucleases and proteins from their host organisms. Nucleic Acids Res 40:5189–5200
11. Belfort M, Perlman PS (1995) Mechanisms of intron mobility. J Biol Chem 270:30237–30240
12. Belfort M, Roberts RJ (1997) Homing endonucleases: keeping the house in order. Nucleic Acids Res 25:3379–3388
13. Chevalier BS, Stoddard BL (2001) Homing endonucleases: structural and functional insight into the catalysts of intron/intein mobility. Nucleic Acids Res 29:3757–3774
14. Dujon B, Belfort M, Butow RA et al (1989) Mobile introns: definition of terms and recommended nomenclature. Gene 82:115–118
15. Arnould S, Delenda C, Grizot S et al (2011) The I-CreI meganuclease and its engineered derivatives: applications from cell modification to gene therapy. Protein Eng Des Sel 24:27–31
16. Delenda C, Paris S, Arnould S, Balbirnie E, Cabaniols JP (2013) Bio-applications derived from site-directed genome modification technologies. Top Curr Genet 23:353–384
17. Silva G, Poirot L, Galetto R et al (2011) Meganucleases and other tools for targeted genome engineering: perspectives and challenges for gene therapy. Curr Gene Ther 11:11–27

18. Paques F, Duchateau P (2007) Meganucleases and DNA double-strand break-induced recombination: perspectives for gene therapy. Curr Gene Ther 7:49–66
19. Pessach IM, Notarangelo LD (2011) Gene therapy for primary immunodeficiencies: looking ahead, toward gene correction. J Allergy Clin Immunol 127:1344–1350
20. Gimble FS, Moure CM, Posey KL (2003) Assessing the plasticity of DNA target site recognition of the PI-SceI homing endonuclease using a bacterial two-hybrid selection system. J Mol Biol 334:993–1008
21. Molina R, Redondo P, Stella S et al (2012) Non-specific protein-DNA interactions control I-CreI target binding and cleavage. Nucleic Acids Res 40:6936–6945
22. Doyon JB, Pattanayak V, Meyer CB, Liu DR (2006) Directed evolution and substrate specificity profile of homing endonuclease I-SceI. J Am Chem Soc 128:2477–2484
23. Rosen LE, Morrison HA, Masri S et al (2006) Homing endonuclease I-CreI derivatives with novel DNA target specificities. Nucleic Acids Res 34:4791–47800
24. Seligman LM, Chisholm KM, Chevalier BS et al (2002) Mutations altering the cleavage specificity of a homing endonuclease. Nucleic Acids Res 30:3870–3879
25. Sussman D, Chadsey M, Fauce S et al (2004) Isolation and characterization of new homing endonuclease specificities at individual target site positions. J Mol Biol 342:31–41
26. Chames P, Epinat JC, Guillier S, Patin A, Lacroix E, Paques F (2005) In vivo selection of engineered homing endonucleases using double-strand break induced homologous recombination. Nucleic Acids Res 33:e178
27. Gruen M, Chang K, Serbanescu I, Liu DR (2002) An in vivo selection system for homing endonuclease activity. Nucleic Acids Res 30:e29
28. Chen Z, Zhao H (2005) A highly sensitive selection method for directed evolution of homing endonucleases. Nucleic Acids Res 33:e154
29. Chen Z, Wen F, Sun N, Zhao H (2009) Directed evolution of homing endonuclease I-SceI with altered sequence specificity. Protein Eng Des Sel 22:249–256
30. Takeuchi R, Certo M, Caprara MG, Scharenberg AM, Stoddard BL (2009) Optimization of in vivo activity of a bifunctional homing endonuclease and maturase reverses evolutionary degradation. Nucleic Acids Res 37:877–890
31. Prieto J, Epinat JC, Redondo P et al (2008) Generation and analysis of mesophilic variants of the thermostable archaeal I-DmoI homing endonuclease. J Biol Chem 283:4364–4374
32. Volna P, Jarjour J, Baxter S et al (2007) Flow cytometric analysis of DNA binding and cleavage by cell surface-displayed homing endonucleases. Nucleic Acids Res 35:2748–2758
33. Jarjour J, West-Foyle H, Certo MT et al (2009) High-resolution profiling of homing endonuclease binding and catalytic specificity using yeast surface display. Nucleic Acids Res 37:6871–6880
34. Paques F, Haber JE (1999) Multiple pathways of recombination induced by double-strand breaks in Saccharomyces cerevisiae. Microbiol Mol Biol Rev 63:349–404
35. Guthrie C, Fink R (1991) Guide to yeast genetics and molecular and cell biology. Methods Enzymol 194:3–933
36. Arnould S, Chames P, Perez C et al (2006) Engineering of large numbers of highly specific homing endonucleases that induce recombination on novel DNA targets. J Mol Biol 355:443–458
37. Arnould S, Perez C, Cabaniols JP et al (2007) Engineered I-CreI derivatives cleaving sequences from the human XPC gene can induce highly efficient gene correction in mammalian cells. J Mol Biol 371:49–65
38. Grizot S, Duclert A, Thomas S, Duchateau P, Paques F (2011) Context dependence between subdomains in the DNA binding interface of the I-CreI homing endonuclease. Nucleic Acids Res 39:6124–6136
39. Grizot S, Epinat JC, Thomas S et al (2009) Generation of redesigned homing endonucleases comprising DNA-binding domains derived from two different scaffolds. Nucleic Acids Res 38:2006–2018
40. Grizot S, Smith J, Daboussi F et al (2009) Efficient targeting of a SCID gene by an engineered single-chain homing endonuclease. Nucleic Acids Res 37:5405–5419
41. Smith J, Grizot S, Arnould S et al (2006) A combinatorial approach to create artificial homing endonucleases cleaving chosen sequences. Nucleic Acids Res 34:e149
42. Epinat JC, Arnould S, Chames P et al (2003) A novel engineered meganuclease induces homologous recombination in yeast and mammalian cells. Nucleic Acids Res 31:2952–2962
43. Bhargava VO, Rahman S, Newton DW (1989) Stability of galactose in aqueous solutions. Am J Hosp Pharm 46:104–108
44. Clayton RA, White O, Ketchum KA, Venter JC (1997) The first genome from the third domain of life. Nature 387:459–462
45. Gietz RD, Woods RA (2002) Transformation of yeast by lithium acetate/single-stranded carrier DNA/polyethylene glycol method. Methods Enzymol 350:87–96

46. Legrain P, Dokhelar MC, Transy C (1994) Detection of protein–protein interactions using different vectors in the two-hybrid system. Nucleic Acids Res 22:3241–3242
47. Scott A, Timson DJ (2007) Characterization of the Saccharomyces cerevisiae galactose mutarotase/UDP-galactose 4-epimerase protein, Gal10p. FEMS Yeast Res 7:366–371
48. Zhang JH, Chung TD, Oldenburg KR (1999) A simple statistical parameter for use in evaluation and validation of high throughput screening assays. J Biomol Screen 4:67–73
49. Ilouga PE, Hesterkamp T (2012) On the prediction of statistical parameters in high-throughput screening using resampling techniques. J Biomol Screen 17:705–712
50. Brideau C, Gunter B, Pikounis B, Liaw A (2003) Improved statistical methods for hit selection in high-throughput screening. J Biomol Screen 8:634–647
51. Gunter B, Brideau C, Pikounis B, Liaw A (2003) Statistical and graphical methods for quality control determination of high-throughput screening data. J Biomol Screen 8:624–633
52. Sui Y, Wu Z (2007) Alternative statistical parameter for high-throughput screening assay quality assessment. J Biomol Screen 12:229–234
53. Daboussi F, Zaslavskiy M, Poirot L et al (2012) Chromosomal context and epigenetic mechanisms control the efficacy of genome editing by rare-cutting designer endonucleases. Nucleic Acids Res 40:6367–6379

Chapter 10

Rapid Determination of Homing Endonuclease DNA Binding Specificity Profile

Lei Zhao and Barry L. Stoddard

Abstract

Evaluating the binding specificity and identifying the most preferred target sequence for a homing endonuclease often represents a key step during its characterization, engineering, and application for genome engineering. This chapter describes a high-throughput, fluorescence-based, competition-binding assay which can be used to measure the relative binding affinities of the homing endonuclease to a large number of DNA target site variants in a single experiment. The base preference at each position of the target sequence can be quantitated based on this assay, and the overall binding specificity of the enzyme can thereby be determined and compared with its cleavage specificity.

Key words Homing endonuclease, DNA binding protein, Specificity, High throughput, Fluorescence-based, Competition binding assay, Base preference

1 Introduction

The assay described in this chapter can be used to determine the binding specificity profile of a homing endonuclease by simultaneously measuring the binding affinities of the homing endonuclease to a large number of DNA sequences, each of which varies only slightly from the wild-type target sequence. This assay was developed based on previously described methods which use a fluorescence-based, microplate format to facilitate high-throughput analyses that include large numbers of replicate measurements for each DNA target [1, 2]. The main improvement of the assay described in this chapter is the employment of competition-binding strategy, which not only minimizes the cost required for fluorescent labeling of the DNA target but also greatly improves sensitivity and reproducibility [3].

An overview of the assay is illustrated in Fig. 1. Briefly, the assay is performed in 96-well microplates; an independent binding assay for one of the sequence variants (referred to below as a "test") is conducted in each individual well. The recombinant homing

David R. Edgell (ed.), *Homing Endonucleases: Methods and Protocols*, Methods in Molecular Biology, vol. 1123, DOI 10.1007/978-1-62703-968-0_10, © Springer Science+Business Media New York 2014

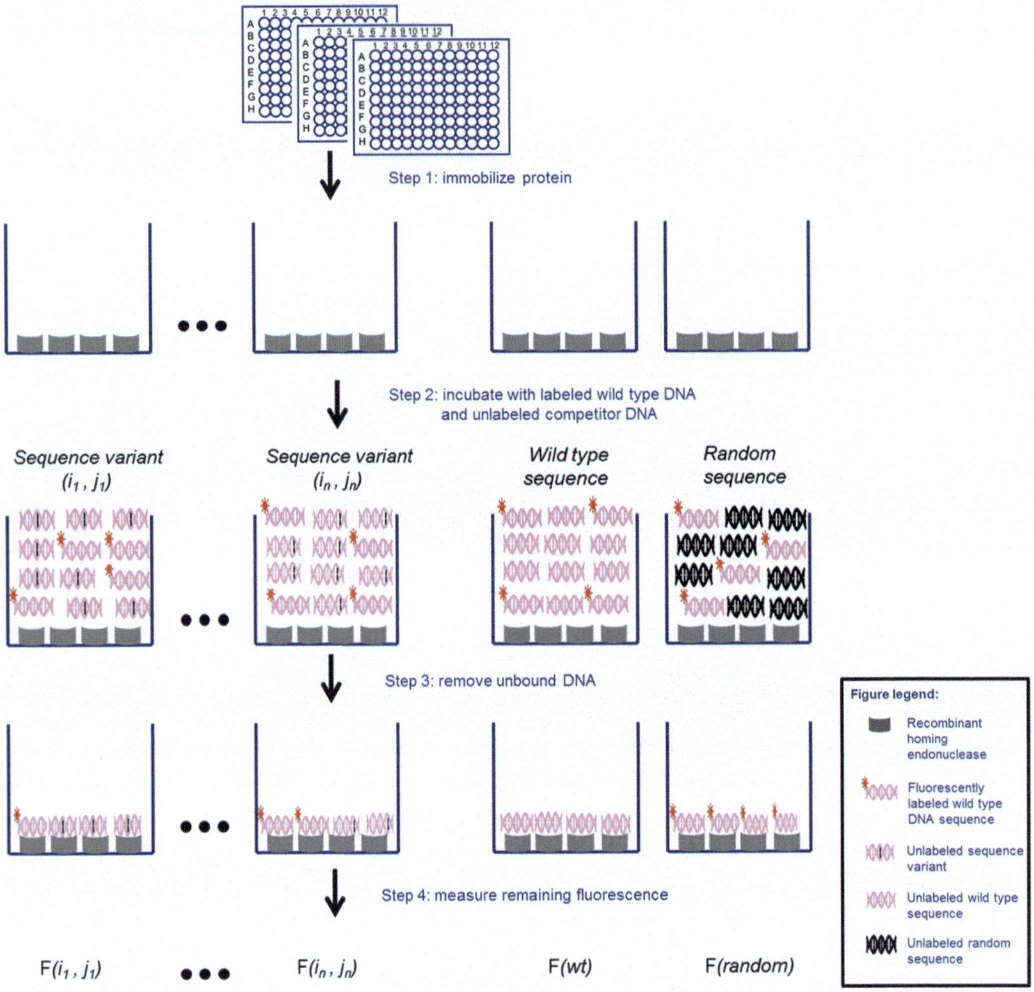

Fig. 1 High-throughput fluorescence competition-binding assay. Histidine-tagged protein is immobilized onto nickel-coated 96-well microplates. A mixture of fluorescently labeled DNA duplex containing the wild-type target site and one of the unlabeled sequence variants is incubated with the protein, allowing determination of the ability of each sequence to compete for binding against the wild-type DNA target. In order to calculate relative affinities, the experiment is also conducted using unlabeled wild-type DNA target and a nonspecific, random DNA sequence

endonuclease with an N- or C-terminal polyhistidine tag is first immobilized onto nickel-coated microplates. The immobilized homing endonuclease is then incubated with a small amount of fluorescence-labeled wild-type DNA duplex, which is presented along with an excess of unlabeled competitor DNA. A different unlabeled DNA sequence variant is added alongside the labeled wild-type DNA target in each well. The relative binding affinities of the homing endonuclease to each individual sequence variant are measured by determining the ability of each DNA sequence variant to compete for binding against the fluorescently labeled wild-type target sequence (*see* **Note 1**). After incubation of the

DNA mixture with the homing endonuclease, the unbound DNA (labeled or unlabeled) is washed off, and the amount of remaining labeled wild-type DNA can be quantified by a fluorescence reader.

The fluorescent signal in each test well is inversely proportional to the binding affinity of the homing endonuclease to the corresponding sequence variant. In order to compute the relative binding affinity, two additional types of tests must be included in the experiment. The first additional test should have wild-type DNA sequence as added unlabeled competitor, and the second additional test should include a random unlabeled DNA sequence (a completely randomized sequence used as a negative control to account for competition by a nonspecific DNA sequence) which is also used to measure binding competition by a nonspecific DNA sequence. The relative binding affinity of each sequence variant to that of the wild-type target can be calculated based on the readings of these three types of tests. The base preference at each position of target DNA of the homing endonuclease can then be derived from the relative binding affinities of a comprehensive matrix of sequence variants.

2 Materials

1. Ni-NTA HisSorb plates, white.
2. Poly dI-dC: dissolve in TBS buffer at 1 μg/μl, aliquot, warm to 45 °C for 5 min, cool, and freeze.
3. Synthesized unlabeled sequence variants (*see* **Note 2**).
4. Synthesized unlabeled wild-type sequence.
5. Synthesized random DNA sequence.
6. 5′Cy3^{TM} labeled wild-type sequence.
7. 6×-His-tagged recombinant homing endonuclease.
8. 5× TBS: 250 mM Tris-HCl, pH 7.5, 750 mM NaCl.
9. TBS/BSA: 1× TBS with 0.2 % BSA.
10. TBS/Tween-20: 1× TBS with 0.05 % Tween-20.
11. Binding buffer: 1× TBS with 0.02 mg/ml poly dI-dC, 10 mM $CaCl_2$.
12. PCR H_2O: milli-Q H_2O filtered by 0.22 μm pore size sterile filters.

3 Methods

3.1 Preparation of DNA Duplexes

1. Dissolve the synthesized fluorescence-labeled oligos into PCR H_2O at 200 mM concentration. Mix equal amount of top strand and bottom strand, which leads to a final concentration

of 100 mM for the annealed duplex. Use a PCR machine to anneal the oligos (Heat to 90 °C for 10 min; slowly cool to RT. Slow cooling is important to avoid hairpin formation; an overnight annealing reaction in a PCR machine is recommended).

2. The synthesized sequence variants can be dissolved in solution at a normalized concentration by the manufacturer upon request. Mix equal amount of top strand and bottom strand in PCR tubes. Anneal the duplex as described in **step 1**.

3.2 Immobilization of the Homing Endonuclease

1. Dilute the his-tagged homing endonuclease sample to 100 nM protein concentration with TBS/BSA buffer, and then load 200 μl protein into each microwell of the HisSorb plates.
2. Incubate at room temperature or in a cold room (depending on the stability of the homing endonuclease under investigation) for at least 2 h. For more efficient binding, incubate on a plate shaker.
3. After incubation, discard the solutions in the plates by quickly inverting the plate and taping on paper towels gently to dry.
4. Wash the plates four times with TBS/Tween-20. Allow soaking for 1 min between washes.
5. Discard the buffer after final wash.

3.3 Binding

Each microwell contains an independent binding assay, in which an unlabeled competitor DNA competes for binding against the fluorescently labeled wild-type DNA. The unlabeled competitor may be one of the following three types: (1) "WT" (the unlabeled wild-type DNA site competing against itself), (2) "random" (a completely randomized sequence used as a negative control to account for competition by a nonspecific DNA sequence), or (3) one of the DNA target site variants. We use a sequence matrix (i,j) to identify individual sequence variant, in which i stands for the position in the target DNA, and j represents the base substitution at that position. All three types of unlabeled DNA competitor are treated independently in all steps.

1. Mix labeled wild-type DNA with unlabeled competitor DNA in binding buffer. The final concentration of the labeled DNA should be 100 nM and that of the unlabeled competitor is 3 μM (*see* **Note 3**).
2. Apply 200 μl of the mixture to each well and incubate for at least 2 h.
3. Wash the plates with TBS for four times. Dry the plate on paper towel as described above.
4. Apply 200 μl TBS to each well after final wash.

3.4 Fluorescence Reading

Measure the retained fluorescence in each well using a SpectraMax® M5/M5[e] microplate reader (excitation: 510 nm, emission: 565 nm, cutoff: 550).

3.5 Data Analysis

The measurements of the retained fluorescent signal (F) for each mismatch sequence variant $[F(i,j)]$ are then converted to relative binding affinities as compared to the wild-type target site using the relationship $rK_a(i,j) = [(F(\text{random}) - F(i,j)) \times F(\text{wt})]/[F(i,j) \times (F(\text{random}) - F(\text{wt}))]$. This formula gives a close approximation of the binding affinity of each sequence variant (Fig. 2).

The relative base pair preference of the homing endonuclease under investigation at each position can then be calculated using the relationship: $\text{BP}(i,j) = K_a(i,j)/\sum_{j=A,C,G,T} K_a(i,j)$. An example of the full calculation of $K_a(i,j)$ and $\text{BP}(i,j)$ from raw fluorescence intensity measurements is provided in the **Note 4**.

3.6 Remarks About This Method and Data

It is agreed that for most if not all homing endonucleases, a considerable fraction of the overall cleavage specificity of the enzyme is derived from the actual catalytic action of the enzyme (i.e., during the course of DNA cutting) rather than during initial recognition and binding of the target site. In other words, the cleavage specificity of most homing endonucleases is substantially higher than is their binding affinity. However, the determination of a homing endonuclease's binding specificity can be quite important, for at least two reasons. First, the mere binding of a homing endonuclease at a large number of sites with significant affinity can have notable effects on genomic fidelity and cell viability, particularly for dividing cell populations. Second, it has become clear that for at least some homing endonucleases, the basis of cleavage specificity across the target site can be unevenly distributed between binding (K_M) and catalysis (k_{cat}) [4]. In either case, a quantitative understanding of binding specificity and affinity can provide a solid foundation for the manipulation, engineering, and application of homing endonucleases for targeted gene modification and genome engineering.

4 Notes

1. The fluorescently labeled sequence used in this assay is often the wild-type target sequence for the homing endonuclease. However, any other sequence with high binding affinity (sub nano-molar range) to the homing endonuclease can be used as the standard sequence. As long as the absolute affinity of the standard sequence is known, the binding affinities of other sequence variants can be inferred following above instructions.
2. We usually add a short flanking random sequence (such as "AAAAAA") to both ends of the DNA substrates to ensure

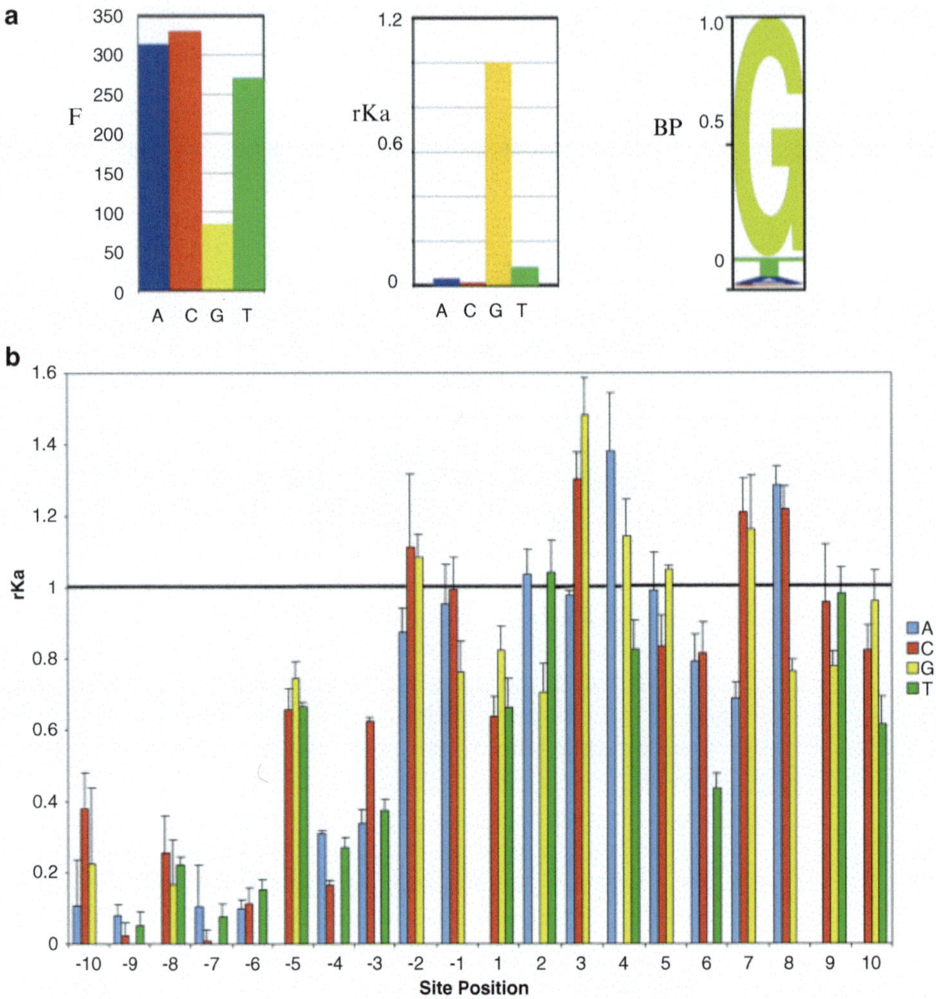

Fig. 2 Panel (**a**) Example data conversion. The raw fluorescent reading, relative K_a (rK_a), and base preference (BP) are plotted in the *left*, *middle*, and *right panel*, respectively. Refer to the text for detail. Panel (**b**) Binding specificity profile for wild-type I-AniI (a LAGLIDADG homing endonuclease). The effect of a single base pair substitution at each position across a 20 base pair sequence corresponding to the endonuclease's wild-type target is shown. The binding affinity of the enzyme to its wild-type target is normalized and denoted as "1"; the relative affinity (rK_a) of the same protein to each of 60 separate DNA targets (each of which contains a single base pair substitution at one position, from base pair −10 to base pair +1) is displayed. The height of the bars denotes the effect of each possible substitution at each position, as described in the text and illustrated in the *middle panel* of (**a**) above. The error bars represent standard deviations for three independent measurements. Note that the enzyme displays considerable binding specificity at positions −10 to −3 and then considerably lower binding specificity at positions +3 to +10. In contrast, the enzyme displays considerable cleavage specificity across the entire target site. *See* ref. 4 for more details

effective binding of the homing endonuclease to its target DNA. This is based on the observation that some homing endonucleases make nonspecific contacts to the phosphate backbone of the flanking sequences from structural studies.

3. To minimize mixing and transferring, a 10× oligonucleotide mix (with 1 μM labeled DNA and 30 μM unlabeled DNA) can be made first. Load 20 μl of this 10× oligo mix into each microwell with immobilized protein and then dilute with 180 μl of binding buffer.
4. Example data conversion (*see* Fig. 2)

$$rK_a(i,j) = [(F(random) - F(i,j) \times F(wt)] / [F(i,j) \times (F(random) - F(wt))],$$

$$BP(i,j) = \frac{K_a(i,j)}{\sum_{j=A,C,G,T} K_a(i,j)}$$

$$(for\ i = position\ at\ the\ target\ DNA, \quad j = \mathrm{A,C,G,T})$$

are used for converting raw data to relative binding affinity (rK_a) and base preference (BP) as described in Subheading 3. Below are details of data conversion at one example base pair substitution at position −5:

(a) Raw fluorescence signals:

$$F(-5,\mathrm{A}) = 313.20, F(-5,\mathrm{C}) = 329.75, F(-5,\mathrm{T}) = 270.00,$$
$$F(-5,\mathrm{G}) = F(wt) = 84.17, F(random) = 342.66$$

(b) Conversion to relative K_a (rK_a) values:

$$rK_a(-5,\mathrm{A}) = \frac{[(F(random) - F(-5,\mathrm{A})) \times F(wt)]}{[F(-5,\mathrm{A}) \times (F(random) - F(wt))]} = \frac{[(342.66 - 313.20) \times 84.17]}{[313.20 \times (342.66 - 84.17)]} = 0.03$$

Using the same equation, $rK_a(-5,\mathrm{C}) = 0.01$, $rK_a(-5,\mathrm{T}) = 0.08$, $rK_a(-5,\mathrm{G}) = rK_a(\mathrm{wt}) = 1$

(c) Conversion to base preference (BP):

$$BP(-5,\mathrm{A}) = K_a(-5,\mathrm{A}) \sum j = \mathrm{A,C,G,T} K_a(-5,\mathrm{G})$$
$$= \frac{K_a(-5,\mathrm{A})}{(K_a(-5,\mathrm{A}) + K_a(-5,\mathrm{C}) + K_a(-5,\mathrm{T}) + K_a(-5,\mathrm{G}))}$$
$$= \frac{0.03}{(0.03 + 0.01 + 0.08 + 1)} = 2.7\%$$

$$BP(-5,\mathrm{C}) = \frac{K_a(-5,\mathrm{C})}{\sum_{j=A,C,G,T} K_a(-5,j)}$$
$$= \frac{K_a(-5,\mathrm{C})}{(K_a(-5,\mathrm{A}) + K_a(-5,\mathrm{C}) + K_a(-5,\mathrm{T}) + K_a(-5,\mathrm{G}))}$$
$$= \frac{0.01}{(0.03 + 0.01 + 0.08 + 1)} = 0.9\%$$

$$BP(-5,\mathrm{T}) = \frac{K_\mathrm{a}(-5,\mathrm{T})}{\sum_{j=\mathrm{A,C,G,T}} K_\mathrm{a}(-5,j)}$$

$$= \frac{K_\mathrm{a}(-5,\mathrm{T})}{(K_\mathrm{a}(-5,\mathrm{A}) + K_\mathrm{a}(-5,\mathrm{C}) + K_\mathrm{a}(-5,\mathrm{T}) + K_\mathrm{a}(-5,\mathrm{G}))}$$

$$= \frac{0.08}{(0.03 + 0.01 + 0.08 + 1)} = 7.2\%$$

$$BP(-5,\mathrm{G}) = \frac{K_\mathrm{a}(-5,\mathrm{G})}{\sum_{j=\mathrm{A,C,G,T}} K_\mathrm{a}(-5,j)}$$

$$= \frac{K_\mathrm{a}(-5,\mathrm{G})}{(K_\mathrm{a}(-5,\mathrm{A}) + K_\mathrm{a}(-5,\mathrm{C}) + K_\mathrm{a}(-5,\mathrm{T}) + K_\mathrm{a}(-5,\mathrm{G}))}$$

$$= \frac{1}{(0.03 + 0.01 + 0.08 + 1)} = 89.3\%$$

References

1. Zhang ZR, Palfrey D, Nagel DA, Lambert PA, Jessop RA, Santos AF, Hine AV (2003) Fluorescent microplate-based analysis of protein-DNA interactions. I: immobilized protein. BioTechniques 35:980–982, 984, 986
2. Hallikas O, Taipale J (2006) High-throughput assay for determining specificity and affinity of protein-DNA binding interactions. Nat Protoc 1:215–222
3. Zhao L, Pellenz S, Stoddard BL (2009) Activity specificity of the bacterial PD-(D/E)XK homing endonuclease I-Ssp6803I. J Mol Biol 385(5): 1498–1510
4. Thyme SB, Jarjour J, Takeuchi R, Havranek JJ, Ashworth J, Scharenberg AM, Stoddard BL, Baker D (2009) Exploitation of binding energy for catalysis and design. Nature 461: 1300–1304

Chapter 11

Quantifying the Information Content of Homing Endonuclease Target Sites by Single Base Pair Profiling

Joshua I. Friedman, Hui Li, and Raymond J. Monnat Jr.

Abstract

Homing endonucleases (HEs) are native proteins that recognize long DNA sequences with high site specificity in vitro and in vivo. The target site specificity of HEs is high, though not absolute. For example, members of the well-characterized LAGLIDADG family of homing endonucleases (the LHEs) recognize target sites of ~20 base pairs, and can tolerate some target site base pair changes without losing site binding or cleavage activity. This modest degree of target site degeneracy is practically useful once defined and can facilitate the engineering of LHE variants with new DNA recognition specificities. In this chapter, we outline general protocols for systematically profiling HE target site base pair positions in order to define their functional importance in vitro and in vivo, and show how information theory can be used to make sense of the resulting data.

Key words Position-specific-scoring/weight matrix (PSSM/PWM), Information theory, Information content, DNA target sequence specificity, Target sequence specificity profiling

1 Introduction

The DNA binding surfaces of homing endonucleases (HE) must be structurally and chemically complimentary to their cognate DNA target sites to specifically recognize and cleave DNA [1, 2]. These surfaces form stabilizing intermolecular interactions with target site DNA that facilitate target site recognition, and stabilize the high-energy transition state leading to catalytic cleavage of the DNA phosphodiester backbone. Efforts to redesign HEs to recognize novel DNA target sites requires knowledge of both the starting specificity as encoded by contacts in the DNA–protein interface, and how these contacts can be modified to alter HE target site recognition specificity [3].

Structural analyses of HEs bound to their target sites have provided many useful insights into HE structure–function relationships [4]. These data have guided efforts to design HEs with altered target sequence specificities, but cannot directly identify

David R. Edgell (ed.), *Homing Endonucleases: Methods and Protocols*, Methods in Molecular Biology, vol. 1123, DOI 10.1007/978-1-62703-968-0_11, © Springer Science+Business Media New York 2014

the changes needed in an existing HE interface to generate a new recognition specificity. The explanation for this is that design alterations to the binding interface induce unanticipated structural rearrangements as residues repack to accommodate a new structural and chemical environment. These structural rearrangements in turn can alter or destroy existing or newly designed contacts to suppress high affinity binding or cleavage of the new target site [5]. Despite these challenges, some positions within HE DNA–protein interfaces have been found to readily accommodate design changes. Thus the systematic identification of base pair positions in the HE-DNA interface that are most conducive to redesign can guide the engineering of HEs with novel DNA recognition specificities.

Information theory is widely used to quantify the contribution of specific positions and base pairs to target site recognition and catalysis [6–9]. By systematically measuring the catalytic activities of HEs against a large set of DNA targets, information theoretic approaches can be used to identify the sequence features that are most important to site recognition and catalysis. These statistical models can accelerate protein engineering and design by screening out prohibitively difficult targets early in the design process, identifying permissive positions and directing subsequent efforts to the regions of the DNA–protein interface that represent design challenges. In this chapter, we provide a brief review of information theory and how it can be used to understand HE target site specificity. We then provide experimental protocols for the generation of HE site specificity profiling data, and suggest several useful ways to interpret and visualize the resulting data to aid HE design and engineering.

1.1 Position-Specific Scoring Matrices

Information theoretic approaches to HE target site modeling are based on large datasets that describe the relative preference of an HE for each DNA base (P_A, P_T, P_G, P_C) at each target site base pair position. These preferences can be concisely represented in the form of a Position-Specific Scoring Matrix (PSSM; also referred to as a Position Weight Matrix, or PWM), in which HE activity (binding and/or catalysis) at a given base or base pair (in rows) is recorded at each of the "N" positions in the target sequence (in columns). These scores are normalized such that each of the columns will sum to 1.

$$PSSM = \begin{pmatrix} P_{1,\mathrm{A}} & \cdots & \cdots & P_{N,\mathrm{A}} \\ P_{1,\mathrm{T}} & \ddots & & P_{N,\mathrm{T}} \\ P_{1,\mathrm{G}} & & \ddots & P_{N,\mathrm{G}} \\ P_{1,\mathrm{C}} & \cdots & \cdots & P_{N,\mathrm{C}} \end{pmatrix}$$

In the case of HEs, probability terms can be derived by simply measuring the relative efficiency with which an HE cleaves a target site that contains a specific base pair substitution in the native DNA target site under single turnover conditions. Fully populating the PSSM/PWM matrix for a given HE requires determining HE activity against all possible "one-off" target sites that contain an A, C, G, or T at each target site base pair position. The number of target site sequences that need to be assayed in this way to fully populate a PSSM matrix is $3N+1$, where N is the target site length in base pairs (*see* Protocols below).

1.2 Definition of Information

Information can be usefully thought of as a reduction in uncertainty about outcomes, or in the parlance of information theory a reduction in "information entropy." For example, when an HE exclusively cleaves target sequences containing only one base at a given position, e.g., only an A (adenine) at position x, position x can be described as having an information entropy of zero: there is no uncertainty as to the identity of a DNA substrate at that position that will be cleaved by the cognate HE.

This relationship between information entropy and a statistical outcome can be further formalized by the Shannon entropy relation, Eq. 1 below, where H_x is the information entropy (a measure of uncertainty) associated with the DNA base at position x, and $P_{x,i}$ is the xth column and ith row of the PSSM matrix [10].

$$H_x = -\sum_{i=\mathrm{A,T,G,C}} P_{x,i} \cdot \log_2 \left| P_{x,i} \right| \qquad (1)$$

If each of the four bases found in DNA is equally likely to occur at a given position, (i.e., if $P_A = P_c = P_G = P_T = 0.25$), then by evaluation of Eq. 1, the informational entropy of that base position would be 2 bits. The explanation for this value is that with four possible DNA bases at a position, two binary digits or bits are needed to uniquely specify the four possible bases at that position (e.g., in one possible encoding scheme A = (0 0), T = (0 1), G = (1 0), C = (1 1)). Site-specific DNA proteins by definition do not recognize all possible DNA bases with equal probability (*thus* $P_A \neq P_T \neq P_G \neq P_C$), and thus the informational entropy of specifically recognized DNA positions (following Eq. 1) will always be ≤2 bits of uncertainty. By extension, the information content of a single base, commonly written as R_{info}, is given by $R_{info} = 2 - H_x$, where 2 is the information entropy of a randomly selected base and H_x is the information entropy of all the possible cognate bases at that position.

1.3 Information Content of a HE DNA Target Site

One way to calculate the information content of HE DNA target site would be to simply sum the R_{info} values across all target site base pair positions. This approach assumes that base pair recognition is independent of sequence context, but this is known not to be the case: specific base pair recognition often involves additional binding avidity contributions from adjacent base pairs.

This influence of additional positions X on the informational entropy of recognition at sequence position Y is specified in Eq. 2 below, a conditional entropy equation.

$$H(X \mid Y) = -\sum_{i=A,T,G,C}\sum_{j=A,T,G,C} P(X_i \mid Y_j) \cdot log_2 \left| \frac{P(X_i \mid Y_j)}{P(Y_j)} \right| \quad (2)$$

Fully accounting for all the interdependencies in a target sequence using Eq. 2 would require evaluating an exponentially increasing number of terms $P(X_i| Y_{1j}, Y_{2j}, \ldots, Y_{nj})$ for each additional base in the target site. Experimentally evaluating these interdependent probabilities for long HE target sites is prohibitively difficult, even using new high-throughput approaches. Thus cases more complicated than the simplest assumption of Eq. 1 are rarely, if ever, considered.

A work-around that is still highly informative and practically useful is to experimentally determine "one-off" dependences, in which informational entropy is measured within a fixed sequence context where Eq. 1 remains phenomenologically valid. These experimentally derived measures of information content are target site sequence-dependent, and as a result capture a portion of the information contained in and contributed by adjacent or nearby base pairs. In the protocols below we describe how to perform target site single base pair scans, and show how information theory and visualization can be used to make sense of the resulting data.

2 Materials

2.1 Cloning HE Target Sites into Plasmid DNA

1. The pDR-GFP-universal plasmid harbors the target sites and is used for combined in vitro/in vivo cleavage analyses (*see* map and details at: http://depts.washington.edu/monnatws/plasmids/pDR-GFP%20univ.pdf).
2. Oligonucleotides need to be synthesized for all single base pair target site variants on a 25 nmol scale. The complementary pairs should be designed so that when annealed they generate a dsDNA target site insert with XhoI/SacI sticky ends to facilitate directional cloning into the pDR-GFP-universal vector. The number of oligonucleotides that need to be synthesized for a systematic scan of a target site that is N base pairs long is $2 \times (3N+1)$.
3. DR-GFP target site sequencing primer, 5′-GGGGAGGGC CTTCGTGCGTCGC-3′. Primers should be synthesized on a 25 nmol scale and resuspended in oligo storage buffer (10 mM Tris-Cl, pH 8.5) to generate a 100 μM stock. Suspended oligos can be stored at −20 °C and thawed as needed.

4. Luria Broth: 10 g tryptone, 5 g yeast extract, 10 g NaCl in 1 L of water. Autoclave and store at room temperature or 4 °C.
5. *E. coli* DH5α chemically competent host cells.
6. 10× T4 polynucleotide kinase (PNK) reaction buffer: 0.7 M Tris–HCl pH 7.6 (at 25 °C), 0.1 M $MgCl_2$, 50 mM dithiothreitol, and 10 mM rATP.
7. T4 polynucleotide kinase (PNK): 10,000 U/mL.
8. DNA annealing buffer: 50 mM Tris–HCl pH 7.6, 0.5 M NaCl.
9. T4 DNA ligase: 400,000 U/mL.
10. PCR Cleanup Kit.

2.2 In Vitro "Barcode" Cleavage Assay

1. Reaction buffer: 10 mM $MgCl_2$, 20 mM Tris–HCl pH 8.0 (*see* **Note 1**).
2. Stop buffer (3×): 300 mM EDTA, 0.3 % SDS (w/v), 3.9 % Ficoll 400 (w/v).
3. 1× TBE buffer: mix 10.8 g Tris base, 5.5 g boric Acid, and 20 mL of 0.5 M EDTA and add water to 1 L.
4. Taq thermophilic DNA polymerase: 5,000 U/mL.
5. Taq PCR buffer (10×): 500 mM KCl, 15 mM $MgCl_2$, 100 mM Tris–HCl pH 8.3.
6. dNTP stock: equimolar mix of 10 mM dATP, 10 mM dTTP, 10 mM dGTP and 10 mM dCTP.
7. Betaine: 4 or 5 M stock solution in H_2O.
8. PCR Cleanup Kit.
9. Purified homing endonuclease protein.
10. Primer pairs for amplification of target sites from pDR-GFP-universal: for 1.3 kb substrate fragments forward primer 5′-GG GGAGGGCCTTCGTGC GTCGC-3′ and reverse primer 5′-GTGGTATGGCTGATTATGATCTAGA GTCGC-3′; for 1.6 kb substrate fragments forward primer for 1.6 kb fragment 5′-TTTATGGTAATCGTGCGAGAGGGCGCAGGG-3′ and reverse primer 5′-TTGTGATGCTATTGCTTTATTTGTAAC CATTATAAGCTGC-3′; for 1.9 kb substrate fragments forward primer 5′-GCGGGG CTCGCCGTGCC-3′ and reverse primer 5′-CCTCTGTTCCACATACACTT CATTCTCAGT ATTGTTTTGCC-3′; and for 2.2 kb substrate fragments forward primer 5′-GGGCTGCGAGGGGAACAAAGGCTGCGT GCGGGG-3′ and reverse primer 5′-CCAAATTAAGGGCCA GCTCATTCCTCCCAC TCATG-3′. Primers are synthesized on a 25 nmol scale, and resuspended in nuclease-free water to generate 100 μM stocks that are store at −20 °C until use.
11. Electrophoresis gel image quantification software (e.g., ImageQuant and Image J).

2.3 In Vivo Cleavage Assay in Human Cells

All of the following reagents should be purchased or prepared sterile in order to ensure the success of the in vivo cleavage profiling protocol outlined below.

1. Complete growth medium: Dulbecco-modified Eagle's medium supplemented with 10 % (v/v) fetal bovine serum and 1 % penicillin/streptomycin.
2. Human HEK 293 T cells.
3. Sterile 0.25 mM $CaCl_2$.
4. Sterile 2× BBS buffer.
5. 1× phosphate-buffered saline.
6. 0.25 % trypsin–EDTA.
7. Flow cytometry analysis software (e.g., FlowJo).
8. Plasmids: pEGFP-C1 plasmid (Clontech), a transfection efficiency and flow cytometry positive control plasmid; expression plasmids for the homing endonuclease being profiled; pDR-GFP-universal reporter target site plasmids with single base pair variant target sites cloned into the XhoI/SacI cloning site.

3 Methods

An overview of the experimental protocols outlined below for in vitro and in vivo cleavage profiling assays is shown in Fig. 1.

3.1 Cloning HE Target Sites into Plasmid DNA

This protocol details how to generate a library of all possible single base pair variant HE target sites in a common plasmid vector backbone that can then be used for both in vitro and in vivo cleavage profiling assays.

1. The synthetically prepared top and bottom strands of each target site should be designed to form XhoI and SacI sticky-ends when annealed to facilitate directional cloning into the pDR-GFP-universal plasmid. Resuspend individual top and bottom strand oligonucleotides at 100 μM in nuclease-free sterile water.
2. Target site oligonucleotides need to be phosphorylated prior to annealing to facilitate cloning. In separate PCR tubes combine 5 μL of each DNA oligonucleotide with 1 μL of 10× T4 polynucleotide kinase (PNK) reaction buffer and 4 μL of nuclease-free H_2O. Mix and then add 1 μL of 10,000 U/mL T4 PNK, mix again gently by pipetting up and down, then incubate at 37 °C for 1 h.
3. Mix pairs of complementary, phosphorylated oligonucleotides in a single PCR tube to anneal the top and bottom strands of each test target site. Dilute to a final concentration of ~50 nM dsDNA by adding 180 μL of DNA annealing buffer. Heat to 95 °C for 5 min followed by a slow cooling to 25 °C (at −1 °C/min).

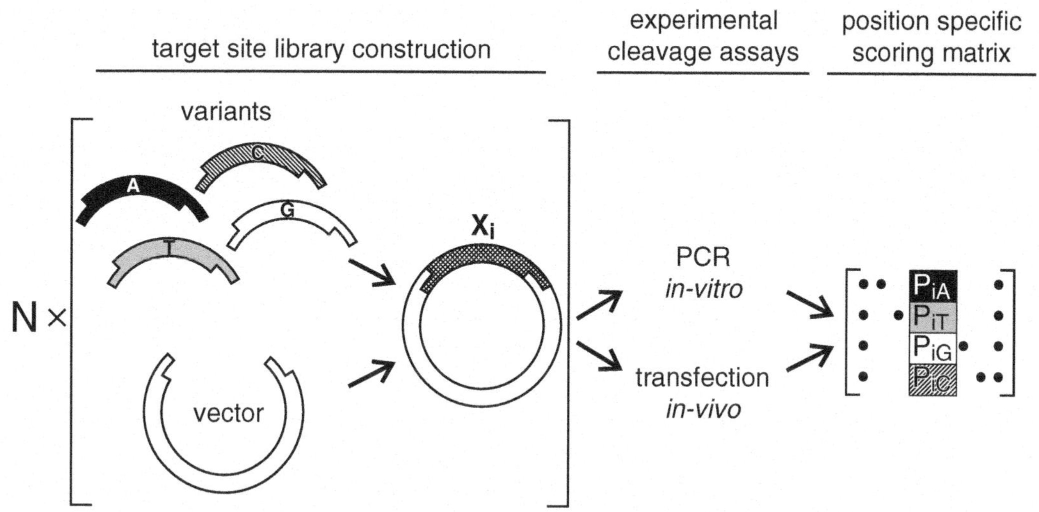

Fig. 1 General outline of profiling protocols. A library of plasmids harboring all target site single base pair variants of a target site *N* bases long is first assembled. HE specificity is assessed by in vitro cleavage of substrate DNAs PCRed from the plasmid site library, or by the in vivo, cleavage-dependent generation of GFP+ cells. The cleavage activity in both assays can be used to generate HE-specific target site cleavage matrices that form the basis for assessing the information content of an HE target site. These data can also be used as a statistical description of HE target site specificity and activity

4. Prepare pDR-GFP universal target site plasmid vector by double-digest 50 ng of plasmid (or ~3 μg for a library of 61 sites) with the restriction enzymes XhoI and SacI using manufacturer's recommended double digest conditions. Stop the reaction by heating at 80 °C for 20 min, then purify cleaved plasmid DNA using a PCR cleanup kit. Adjust the concentration of cleaved plasmid stock with nuclease-free water to a final concentration of ~25 ng/μL (or A_{260} ~0.5). A small aliquot can be run on an agarose gel to check the completeness of digestion if desired.
5. Ligate target sites into plasmid DNA by combining 1 μL of cleaved pDR-GFP-universal plasmid DNA from **step 4** in a fresh tube with 1 μL of phosphorylated, annealed target site insert from **step 3**. Add 6 μL nuclease-free water and 1 μL 10× T4 DNA ligase buffer, then mix gently. Add 1 μL (10 U) of T4 DNA ligase and mix gently, then incubate at room temperature for ≥1 h. To facilitate subsequent steps it is desirable to do all of the ligations needed to generate a library in parallel in a 96-well PCR plate (*see* **Note 2**).
6. Following ligation, chill samples on ice for 15 min, then add ~30 μL of chemically competent DH5α *E. coli* host cells/well. Heat shock by placing plates in a thermocycler pre-equilibrated at 42 °C. After a 45 s return the plate to an ice bucket for 1 min. Add 500 μL of sterile LB media to each well, then transfer well contents into individual deep-well 96-well plate

wells and cover with a gas-permeable top prior to shaking gently to recover at 37 °C for 1 h.

7. Plate 200 μL of each transformation onto LB-agar plates containing 100 μg/mL ampicillin. Grow plates overnight at 37 °C, or until well-delineated individual colonies are visible.
8. Use sterile toothpicks to transfer well isolated bacterial colonies into a 96-well PCR plate-containing 20 μL of ultrapure water/well. Mix gently to break up colony cell clumps.
9. Transfer 10 μL of each resuspended colony into a fresh, deep-well, 96-well plate containing 500 μL of LB or TB media/well supplemented with 50 μg/mL ampicillin. Grow overnight in a 37 °C plate shaker.
10. Isolate target site plasmids from the overnight cultures using your preferred plasmid DNA isolation method.
11. Combine 10 μL of purified plasmid DNA with 5 μL of a 1:100 dilution of sequencing primer and submit for DNA sequencing to verify all the targeted sequences (*see* **Note 3**).

3.2 In Vitro "Barcode" Cleavage Profiling of HE Target Sites

This protocol uses pooled sets of oligonucleotide substrates generated by PCR amplification of target sites cloned into pDR-GFP-universal in competitive cleavage reactions. In each reaction the cleavage sensitivity of all four base pair variants at a target site base pair position are directly compared in the same single tube digest. Full "one off" target site libraries can be easily profiled using this "barcode" cleavage protocol, and the resulting cleavage reactions displayed on a single agarose gel, as shown in outline in Fig. 2a.

1. Four forward and four reverse primer pairs need to be designed and synthesized, to amplify individual target site variants cloned into pDR-GRP-universal in Subheading 3.1 above. Primer sets should be designed to generate DNA fragments that are different enough in size to be easily resolved on a 1.0–1.2 % agarose gel, and in which the DNA fragment size is coded to be directly informative of the base pair variant present at base pair positions in that fragment. For example, our primer sets generate PCR products of approximately 1.3, 1.6, 1.9 and 2.2 kb, with the HE cleavage site located at the center of the fragment and in which all 2.2 kb fragments contain A's (adenines), 1.9 kb fragments C's, 1.6 kb fragments G's and 1.3 kb fragments T's as the variable base pair in amplified target sites across all base pair positions. This design allows four PCR fragments containing all four base pairs at each target site base pair position to be combined, digested and displayed in a single tube-1 lane agarose gel assay to generate target site position- and base pair-specific cleavage "barcodes" (*see* **Note 4**).
2. Adjust the volume of each site primer to a final concentration of 10 μM, and each target site plasmid to ~50 ng/μL (A_{260} = 1.0).

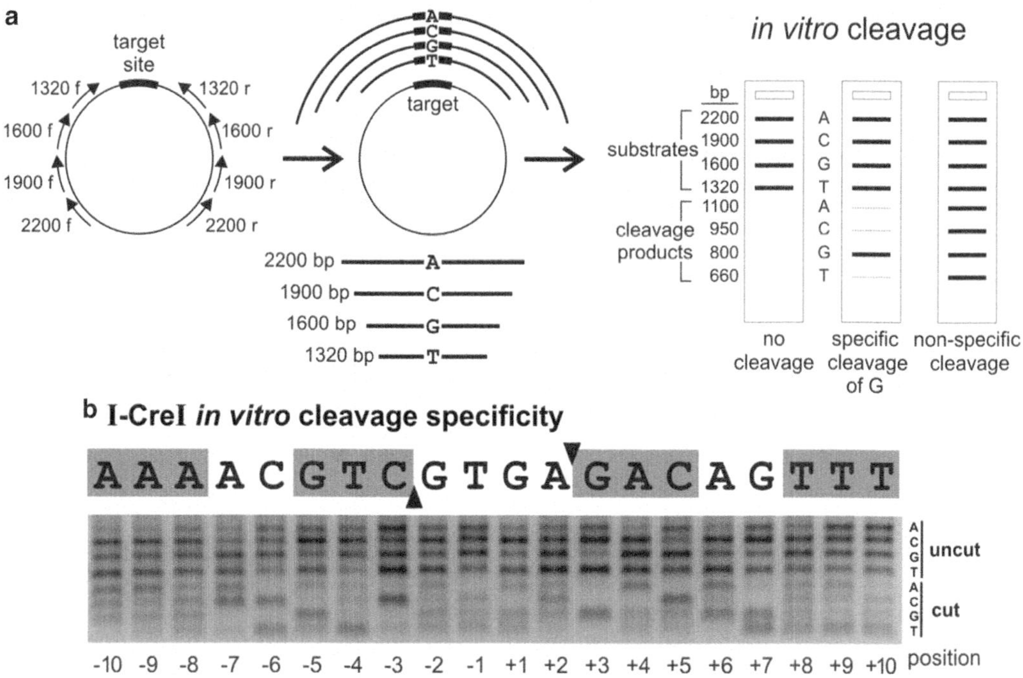

Fig. 2 In vitro "barcode" cleavage profiling of HE target sites. (**a**) Substrate DNA fragments are generated by PCR amplifying individual HE target site variants from the pDR-GFP-universal plasmid backbone using primer pairs that generate different sized substrate molecules in which fragment size encodes the variant base pair identity. Pools of the four PCR fragments covering each target site position and nucleotide possibility are then used in a 1-tube competitive cleavage assay. This approach allows all target site single base pair variants and their cleavage products to be assayed and quantified in a single experiment on an agarose gel. (**b**) Example of barcode cleavage profiling of the I-CreI HE cleavage site using the monomerized version of I-CreI (i.e., mCreI; [ref]) as the cognate HE. (Panels (**a**) and (**b**) are taken from Li H, Ulge UY, Hovde BT, Doyle LA, Monnat RJ Jr. (2012) Comprehensive homing endonuclease target site specificity profiling reveals evolutionary constraints and enables genome engineering applications. *Nucleic Acids Res.* 40(6):2587–2598. Epub 2011 Nov 25. PMID:22121229)

3. Set up PCR amplification reactions for each target site variant. For each target site plasmid mix 2 μL target site plasmid DNA, 4 μL base pair-specific forward primer, 4 μL base pair-specific reverse primer, 1 μL dNTP mix, 5 μL 10× Taq Buffer, 12.5 μL 4 M betaine, 2 μL formamide, 24 μL nuclease-free water, and 1 μL of Taq DNA polymerase (*see* **Note 4**).
4. PCR amplify target sites using 30 cycles of 95 °C for 30 s, 60 °C for 30 s, and 68 °C for 120 s followed by a single final incubation at 68 °C for 5 min.
5. Clean up PCR reactions using a PCR cleanup kit and elute DNA in ~30 μL nuclease-free water. Calculate the amount of DNA and its concentration based on A_{260} using the equation $[C] = A_{260} \times (50/\lambda \times 650)$, where [C] is concentration in μM and λ is the length of the PCR product in kb.

6. For each target site base pair position, mix in a single tube an equimolar ratio of the four PCR products (10 nM each) corresponding to each of the base pair variants (A, G, C, and T) at that position in a final volume of 10 μL (*see* **Note 5**).
7. Add homing endonuclease protein to each pooled template tube to a final concentration of 40 nM (this corresponds to 1:1 HE molecule/substrate DNA molecule in the example here). Adjust volume to 20 μL using Reaction Buffer and mix gently. Incubate the reaction mixture at 37 °C for 15 min, then stop digestions with 4 μL/tube of 6× Stop Buffer (*see* **Note 6**).
8. Load digests into single lanes of a 1.0–1.2 % agarose/1× TBE gel and run at 90 V for 2 h (*see* Fig. 2b for an example). Stain the gel in 1 μg/mL ethidium bromide gel buffer for 40 min with gentle shaking, then destain in water for 10 min prior to visualizing bands under 302 nm UV illumination (wear proper eye protection!). Take care not to oversaturate the detector in any part of the gel (*see* **Note** 7).
9. Integrate the intensities of each band on the gel using image quantification software. The signal intensity of each cleaved band (the two cleaved products in the substrates designed as described above run as a double-intensity, unresolved doublet; *see* Fig. 2b and **Note 4**) should be divided by the signal intensity of its corresponding un-cleaved substrate band plus the cleaved product band(s) to find the fractional cleavage of each substrate. The relative cleavage efficiency of target sites with single base pair changes is calculated by dividing the cleavage efficiency of target sites with single base pair changes by the cleavage efficiency of native target site base in the same lane. These results can be used directly to populate different displays of the results (*see* Subheading 4 below, **Note 8**).

3.3 In Vivo Cleavage Profiling of HE Target Sites

The same pDR-GFP-universal target site library used in Subheading 3.2 above can be used directly to profile the cleavage sensitivity of all single base pair variant HE target sites in human cells. This is done by co-transfecting each target site plasmid together with an HE coding plasmid into cells, then using flow cytometry to detect and quantify cleavage events that promote recombination and the generation of GFP+ cells. An example of this assay is shown in Fig. 3.

This in vivo cleavage assay takes advantage of having target sites cloned into the 5′ copy of the GFP (green fluorescence protein) genes contained in the pDR-GFP-universal plasmid. Target site insertion inactivates the 5′ GFP gene, which can be repaired after cleavage off the downstream, inactive 3′ GFP copy to restore the GFP open reading frame (Fig. 3). In this system the efficiency of in vivo cleavage of pDR-GFP-universal target site plasmids can be estimated from the frequency of GFP+ cells.

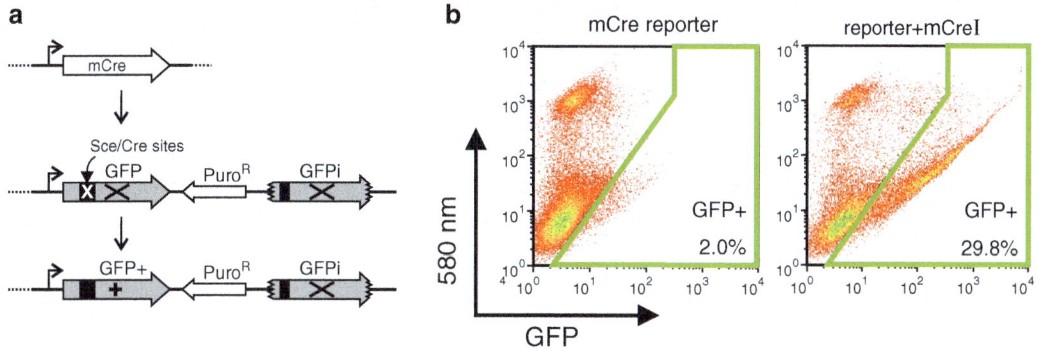

Fig. 3 In vivo cleavage profiling of HE target sites in human cells. (**a**) The pDR-GFP-universal target site plasmid facilitates in vivo cleavage profiling of HE target site variants that are cloned into the 5′ copy of GFP (green fluorescence protein gene) contained in pDR-GFP-universal. The HE target site insertion together with stop codons inactivate the 5′ GFP gene, whereas the downstream duplicated 3′ GFP gene is inactivated by open reading frame truncations. (**b**) Cleavage of the 5′ HE target site in cells by a co-transfected and co-expressed HE stimulates homology-dependent repair of the 5′ GFP copy off the downstream, inactive 3′ GFP copy to restore the GFP open reading frame. GFP+ cells can then be detected and quantified by flow cytometry. GFP+ cell generation closely parallels in vivo cleavage of the pDR-GFP substrate, and thus can be used to profile HE cleavage activity in vivo on target site variants. (Figure from Li H, Ulge UY, Hovde BT, Doyle LA, Monnat RJ Jr. (2012) Comprehensive homing endonuclease target site specificity profiling reveals evolutionary constraints and enables genome engineering applications. *Nucleic Acids Res.* 40(6):2587–2598. Epub 2011 Nov 25. PMID:22121229)

1. Seed 1.5×10^5 HEK 293 T cells in 500 μL of complete growth medium into each well of a 24-well plate, then incubate in a 37 °C humidified, 5 % CO_2 incubator for 24 h. Set up at least duplicate wells for each target site sample you plan to assay.
2. Prepare a transfection mix for each target site plasmid in a 1.5 mL microfuge tube that consists of 1.5 μg of plasmid DNAs in 10 μL H_2O to which 40 μL 0.25 M $CaCl_2$ and 40 μL 2× BBS buffer are added. The plasmid DNA amount for 24-well format transfections should be a total of 1.5 μg/well with a 3:1 M ratio mix of HE expression plasmid to pDR-GFP-universal target site plasmid. For control transfections, pUC19 plasmid DNA can be used to make up the difference in plasmid DNA amount to ensure a consistent 1.5 μg for transfections.
3. Mix transfection reactions by finger flicking tubes, incubate for 15 min at room temperature (~22 °C), then add drop by drop into plate wells. Place cells in a humidified, 37 °C, 3 % CO_2 incubator for 24 h to allow precipitate to form and transfection to occur (*see* **Notes 9** and **10**).
4. Refeed transfections by gently aspirating medium from transfected cells, and replacing it with 500 μL fresh complete growth medium before moving the plates to a humidified, 37 °C 5 % CO_2 incubator for an additional 24 h.
5. Harvest cells for flow cytometry: 48 h after transfection, aspirate the growth medium from each well and gently wash cells

once with pre-warmed PBS. Add 100 μL of trypsin–EDTA per well, and incubate the plate in 37 °C incubator for a couple of minutes to allow cells to detach. Add 400 μL fresh medium per well and pipet cells up and down gently to generate a single cell suspension. Transfer cells in media to a flow cytometry tube (a sterile 5 or 7 mL snap cap polystyrene or polypropylene tube), then transport on ice to the cytometer.

6. Flow cytometry analysis: use positive- and negative-control cell/plasmid combinations to determine gating to reliably identify GFP+ cells, then count 50,000 events for each transfected sample. Quantify the fraction of events that are GFP+ over total events.
7. Calculate target site cleavage efficiency: divide the number of GFP+ positive cells in co-transfected samples by the number of GFP+ positive cells observed in reporter-only transfections. The relative cleavage efficiency of a given target site variant can be determined by dividing the GFP+ frequency of that target site by the GFP+ frequency of the native target site determined in the same assay. This transient transfection assay is simple to multiplex and can detect even low levels of target site-specific cleavage despite a relatively high background that results from reporter plasmid DNA breakage upon transfection.

3.4 Data Analysis and Visualization

The protocols above yield information on the relative preferences of HEs for different target sequences. Thus a first step in visualizing these data is to appropriately normalize these ratios. There are two ways these relative data can be normalized: (1) by assigning the native DNA target site a relative activity value of 1.00, or (2) alternatively, normalizing the relative activities across all four target site base pair possibilities at a position such that each column of the PSSM sums to 1.00. This second normalization is required to calculate information entropy via Eq. 1. As this chapter is focused on the information content in DNA target sites, we typically use normalization 2, where the probability terms can be thought of as representing the fraction of HEs that would bind and cleave a specific substrate when in the presence of the other three competing target sites. Once appropriately normalized, PSSMs can be displayed as a matrix (Fig. 4 top) or transformed into a graphical representation of the data.

Sequence logos are a common way to visualize the data in PSSMs: the height of each base letter in a sequence logo is proportional to its corresponding value in the PSSM. To generate a sequence logo from a PSSM, first translate it into TRANSFAC motif format [11]. Briefly, TRANSFAC format is an ASCII text file that begins with a header line "P0 A C G T," and lists the contents of the PSSM as "xx $P_{xx,A}$, $P_{xx,C}$ $P_{xx,G}$ $P_{xx,T}$" in separate lines where xx

PSSM

	A	A	A	A	C	G	T	C	G	T	G	A	G	A	C	A	G	T	T	T
	-10	-9	-8	-7	-6	-5	-4	-3	-2	-1	1	2	3	4	5	6	7	8	9	10
A	0.61	0.65	0.41	0.44	0.00	0.26	0.21	0.00	0.34	0.00	0.43	0.36	0.00	0.66	0.00	0.49	0.17	0.39	0.27	0.00
C	0.22	0.13	0.24	0.46	0.49	0.12	0.11	0.85	0.19	0.28	0.00	0.36	0.00	0.00	0.67	0.12	0.00	0.09	0.12	0.06
G	0.17	0.16	0.23	0.00	0.00	0.38	0.00	0.00	0.28	0.25	0.36	0.28	1.00	0.19	0.19	0.39	0.42	0.11	0.01	0.23
T	0.00	0.06	0.12	0.10	0.51	0.23	0.68	0.15	0.19	0.47	0.21	0.00	0.00	0.15	0.14	0.00	0.41	0.41	0.61	0.71

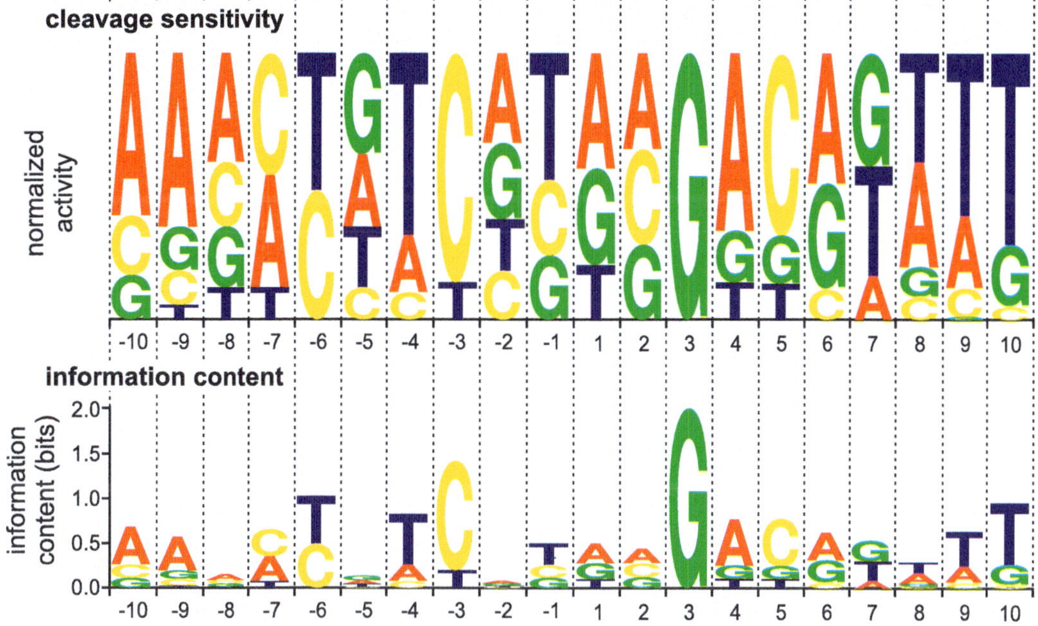

Fig. 4 Common methods for the visualization of target site specificity information. (**a**) Position Specific Scoring Matrix (PSSM) for the HE I-CreI where the native base is colored at each position; (**b**) the PSSM depicted as a sequence iconograph where the height of each letter is proportional to the preference of the HE for target sites containing that base; and (**c**) a sequence iconograph where the height of each letter is proportional to the information in bits that base provides to the HE I-CreI upon binding

is the target sequence position number beginning with "01" and the $P_{xx,i}$ terms are the appropriate entries from the PSSM. This file can be fed directly into WebLogo [12] (*see*: http://weblogo.berkeley.edu), or into many other freely available logo generators that will produce plots representing sequence preference or information content.

Sequence logos are also a useful way to visualize the information content of each position in a target site sequence, and can again be automatically generated by WebLogo using the same TRANSFAC input file described above. In contrast to direct representations of the PSSM, representations of information content emphasize DNA target site positions most important to recognition by a site-specific binding protein. The reader is referred to http://weblogo.berkeley.edu for detailed instructions on how to generate and customize these target site representations.

4 Notes

1. Optimal cleavage reaction conditions vary between HEs, and should be determined and optimized in advance.
2. The ligation can be extended at 16 °C overnight to improve ligation efficiency. In addition, inactivation of ligase by heating the ligation reaction at 70 °C for 20 min before transformation can also improve ligation efficiency.
3. Many commercial DNA sequencing services will now perform colony sequencing and PCR cleanup for an additional fee.
4. To optimize signal in barcode cleavage assays, the target sites in amplified fragments should be located in the center of the amplicon. This ensures the cleaved products will appear as a single band of double intensity on agarose gels. Added betaine is essential for efficient amplification of target sites from pDR-GFP-universal plasmid DNA.
5. Accurate quantification of enzymatic activity requires equal molar ratios of each DNA substrate in the substrate mixture.
6. The cleavage conditions used here were chosen to favor 50 % native target site cleavage to provide the best dynamic range for assessing target site cleavage sensitivities. Cleavage conditions and sampling of the cleavage time course need to be determined and optimized for each HE of interest.
7. Using 1× TBE buffer typically results in better separation of DNA fragments in agarose gels than the use of 1× TAE buffer. Electrophoresis conditions should again be optimized in advance in order to achieve the best separations to facilitate easy quantification of substrate and product bands.
8. As the PCR "barcode" substrate fragments are of different lengths, it is important to note the band intensity is not directly proportional to concentration. Each substrate and corresponding reaction product band should be self-normalized before comparing the different substrates. These complications can be avoided if the fragments are instead end-labeled.
9. We have had good luck using CMV promoter plasmids to drive HE expression in several different human cell types. A wide range of mammalian expression vectors can be used for this purpose once verified.
10. pEGFP C1 (Clontech) is transfected as positive control to monitor transfection efficiency. Sterile water or DNA buffer can be used as negative control. pDR-GFP-universal plasmid containing a native HE target site is also included.

Acknowledgements

This work was supported by US National Institutes of Health Training Grant award to H.L. (5RL9HL092555); by a US National Institutes of Health U54 Interdisciplinary Research Roadmap award (1RL1 CA133831) to R.J.M. Jr; and by a Bill and Melinda Gates Foundation/Foundation for the National Institutes of Health Grand Challenges in Global Health award to R.J.M. Jr.

References

1. Choo Y, Klug A (1997) Physical basis of a protein-DNA recognition code. Curr Opin Struct Biol 7(1):117–125
2. Garvie CW, Wolberger C (2001) Recognition of specific DNA sequences. Mol Cell 8(5): 937–946
3. Chevalier BS, Kortemme T, Chadsey MS, Baker D, Monnat RJ Jr, Stoddard BL (2002) Design, activity, and structure of a highly specific artificial endonuclease. Mol Cell 10(4): 895–905
4. Stoddard BL (2005) Homing endonuclease structure and function. Q Rev Biophys 38(1):49
5. Thyme SB, Baker D, Bradley P (2012) Improved modeling of side-Chain-Base interactions and plasticity in Protein-DNA interface design. J Mol Biol 419:255–274
6. Chevalier B, Turmel M, Lemieux C, Monnat RJ, Stoddard BL (2003) Flexible DNA target site recognition by divergent homing endonuclease isoschizomers I-CreI and I-MsoI. J Mol Biol 329(2):253–270
7. Rosen LE, Morrison HA, Masri S, Brown MJ, Springstubb B, Sussman D et al (2006) Homing endonuclease I-CreI derivatives with novel DNA target specificities. Nucleic Acids Res 34(17):4791–4800
8. Ulge UY, Baker DA, Monnat RJ (2011) Comprehensive computational design of mCreI homing endonuclease cleavage specificity for genome engineering. Nucleic Acids Res 39(10):4330–4339
9. Li H, Ulge UY, Hovde BT, Doyle LA, Monnat RJ (2012) Comprehensive homing endonuclease target site specificity profiling reveals evolutionary constraints and enables genome engineering applications. Nucleic Acids Res 40(6):2587–2598
10. Schneider TD, Stormo GD, Gold L, Ehrenfeucht A (1986) Information content of binding sites on nucleotide sequences. J Mol Biol 188(3):415–431
11. Wingender E (1988) Compilation of transcription regulating proteins. Nucleic Acids Res 16(5 Pt B):1879
12. Crooks GE, Hon G, Chandonia JM, Brenner SE (2004) WebLogo: a sequence logo generator. Genome Res 14(6):1188–1190

Chapter 12

Homing Endonuclease Target Site Specificity Defined by Sequential Enrichment and Next-Generation Sequencing of Highly Complex Target Site Libraries

Hui Li and Raymond J. Monnat Jr.

Abstract

Homing endonucleases (HEs) are DNA sequence-specific enzymes that recognize and cleave long target sites (14–40 bp) to generate double-strand breaks (DSBs). Their high site recognition specificity and tight coupling of binding and cleavage make HEs attractive reagents for targeted genome manipulation. In order to delineate the target site specificity of HEs and facilitate HE engineering, we have developed a method for comprehensive target site profiling of HEs cleavage specificity using partially randomized target site libraries and high-throughput DNA sequencing.

Key words Homing endonuclease, Cleavage specificity, Sequential enrichment, Next-generation sequencing, Targeted genome engineering

1 Introduction

Highly site specific endonucleases are powerful tools for genome engineering that can enable targeted gene manipulation by generating DSBs in specific target genomic loci [1]. These DSBs stimulate cellular DNA repair to generate different outcomes depending on repair pathway activity and the presence or absence of a repair template. For example, DBSs in the absence of a repair template can promote non-homologous end joining (NHEJ) of DSBs to generate small deletions or insertions at or near the DSB site. These targeted modifications may alter or disrupt gene function depending on their size and the target gene reading frame [2]. Alternatively, homology-dependent recombinational repair (HDR/HR) of DSBs in the presence of a homologous repair template can be used to introduce specific genetic modifications [3].

Three different endonuclease protein scaffolds are now being used to generate highly site specific DSBs in vivo to facilitate genome engineering: zinc-finger nucleases (ZFNs), TAL effector nucleases (TALENs, TALNs, or TALs), and LAGLIDADG

David R. Edgell (ed.), *Homing Endonucleases: Methods and Protocols*, Methods in Molecular Biology, vol. 1123, DOI 10.1007/978-1-62703-968-0_12,

homing endonucleases (LHEs, also known as "meganucleases"). ZFNs and TALENs are novel endonuclease scaffolds that are generated by combining modular DNA binding motifs with a non-sequence-specific nuclease domain derived from the Type II restriction endonuclease FokI [4, 5]. The LHE nucleases, in contrast, are small, naturally occurring proteins in which DNA recognition and catalytic motifs are tightly integrated. Native LHE proteins are found throughout all kingdoms of life, most often as open reading frames in mobile introns or inteins [6, 7]. Despite their small size (<50 kDa), the LHEs recognize long DNA sequences (~20 base pairs) and cleave these sites in vivo to promote lateral gene transfer or "homing" of their respective mobile intron or intein. LHE target sites are cleaved with high though not absolute specificity: we and others have shown that LHEs can tolerate some target site base pair changes without losing their site binding or cleavage activities [8]. This modest degree of LHE site degeneracy is practically useful, as it can enable the engineering of new DNA recognition specificities [9].

In order to better delineate the cleavage specificity of LHEs, we have previously developed a sequential enrichment protocol to identify cleavage-sensitive HE target sites contained in highly complex target site libraries [10]. This method is very robust and has been used successfully to profile the target site sequence degeneracy of several other LHEs [11, 12]. In this chapter we describe the use of this sequential enrichment protocol together with next-generation sequencing (NGS) to identify and characterize cleavage-sensitive target sites contained in highly complex target site libraries. These results have begun to provide insight into practically important questions such as the cleavage co-dependence of target site base pair positions and base pair combinations, and the cleavage sensitivity of different "central 4" base pair combinations. The "central 4" are the four contiguous base pairs that reside at the center of LHE target sites between the scissile phosphates on the target site top and bottom strands. These four base pairs are known to strongly influence target site cleavage, despite the paucity of DNA–protein contacts to these base pairs. We provide an example of how sequential enrichment together with NGS was used to rank-order the cleavage sensitivity of all 256 central 4 base pair combinations for the canonical LHE I-CreI. Our results using enrichment and sequencing are in close agreement with prior results from our lab and independently reported in vivo cleavage assay data [13].

Many additional LHE proteins with different target site specificities are being identified by genome sequencing [14, 15]. The protocol outlined below can be used to rapidly define the target site specificity of these LHEs, or of other highly site-specific nucleases to facilitate targeted genome engineering applications.

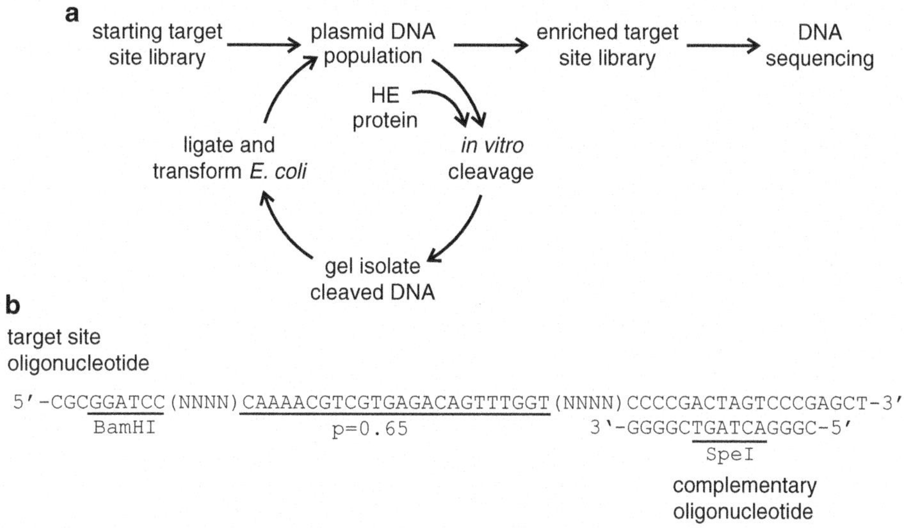

Fig. 1 Protocol overview and target site design. (**a**) Schematic overview of sequential enrichment protocol beginning (*upper left*) with construction of a complex target site library in a plasmid vector backbone. (**b**) Design of target site insert for library construction, using the I-CreI/mCreI target site as an example. The target site sequence (*underlined*) is partially randomized during synthesis, whereas flanking ("N") nucleotides are fully randomized to serve as simple barcodes to allow different versions of the same target site sequence to be distinguished. The complementary primer oligonucleotide shown right anneals to a constant region of the target site oligonucleotide, and primes the DNA synthesis needed to convert single-stranded, partially randomized target site oligonucleotides into dsDNA for plasmid capture (see text for additional detail). This oligonucleotide also contains a unique, target site-specific restriction site to allow the easy assessment of the fraction of a starting plasmid library that contains a target site insert

2 Materials

Figure 1 provides an overview of our sequential enrichment protocol, together with an example of target site oligonucleotide design to generate a highly complex target site library for sequential enrichment and NGS characterization. We have used the canonical homodimeric LHE I-CreI and variants such as monomerized I-CreI (referred to as mCreI; [16]) as an example in the following protocols.

2.1 Generation of Partially Degenerate Target Site Library

1. A single-stranded target site DNA oligonucleotide harboring the partially degenerate I-CreI LHE target site and constant flaking DNA sequences:

   5′-CGCGGATCCNNNNCAAAACGTCGTGAGACAGTTTGGTNNNNC CCCGACTAGTCCCGAGCT-3′ in which the LHE target site is underlined. This template oligonucleotide is synthesized using defined ratios of the four DNA bases to ensure the resulting target site library will be both complex and diverse. The example shown was synthesized with 65 % wild type base, and 11.6 % of each of the three other bases at each position. The four "N" bases flanking the target site are fully randomized during synthesis to provide barcodes that help identify individual target sites (*see* Fig. 1 and **Note 1**).

2. Complementary primer oligonucleotide: 5′-CGGGACTAGT CGGGG-3′.
3. Klenow DNA polymerase, T4 DNA ligase, and restriction enzymes.
4. 25 mM dNTP mix: generate this by mixing equal volumes of 100 mM dNTP stocks. Store frozen at −20 °C until use.
5. Phenol–chloroform–isoamyl alcohol mix (25:24:1): made by mixing equal volumes of water-saturated phenol and a 24:1 mix of chloroform–isoamyl alcohol. Store at −20 °C until use.
6. DNA gel extraction kit.
7. 20 mg/mL glycogen.
8. pBluescript SK(+) plasmid (New England Biolabs).
9. DH5α-E electroporation-competent *E. coli* host cells.
10. SOC medium: 0.5 % yeast extract, 2 % tryptone, 10 mM NaCl, 2.5 mM KCl, 10 mM $MgCl_2$, 10 mM $MgSO_4$, and 20 mM glucose.
11. Plasmid midiprep kit.

2.2 Sequential Enrichment of Cleavage-Sensitive Target Sites

1. 5 μM purified I-CreI or mCreI homing endonuclease or other enzyme (*see* **Note 2**).
2. 10× reaction buffer: 100 mM $MgCl_2$, 200 mM Tris–HCl pH 8.0 (*see* **Note 3**).
3. 6× stop buffer: 300 mM EDTA, 0.3 % SDS (w/v), 3.9 % Ficoll 400 (w/v).
4. 1× TAE buffer: 40 mM Tris, 20 mM acetic acid, and 1 mM EDTA.
5. NuSieve GTG Agarose.
6. DNA gel extraction kit.
7. T4 DNA ligase.
8. DH5α-E electroporation-competent *E. coli* host cells.
9. SOC medium: 0.5 % yeast extract, 2 % tryptone, 10 mM NaCl, 2.5 mM KCl, 10 mM $MgCl_2$, 10 mM $MgSO_4$, and 20 mM glucose.
10. Plasmid midiprep kit.
11. Gel image analysis software (e.g., ImageQuant, ImageJ, etc.).

2.3 Sample Preparation for Illumina Sequencing

1. Oligonucleotides for target site PCR amplification: forward primer:
5′-AATGATACGGCGACCACCGAGTAAAACGACGGCC AGTG-3′ and reverse primer: 5′-CAAGCAGAAGACGGCAT ACGAGGAAAC AGCTATGACCATG-3′.
2. Oligonucleotide primer for target site sequencing: 5′-GAAT TCCTGCAGCCCGGGGGATCC-3′.
3. A high fidelity DNA polymerase (e.g., Phusion polymerase from New England Biolabs).
4. DNA gel extraction kit.
5. 1× TAE buffer: 40 mM Tris, 20 mM acetic acid, and 1 mM EDTA.
6. NuSieve GTG Agarose.

3 Methods

3.1 Generation of Partially Degenerate Target Site Library

1. Dissolve the degenerate target site oligonucleotide and the complementary primer oligonucleotide in H_2O to generate 100 μM working stocks of each. Mix equal molar amount of both oligonucleotides to a final concentration of 5 μM in 1× Klenow polymerase buffer, and incubate at 95 °C for 10 min, then slowly cool to room temperature.
2. Add 6 units of Klenow DNA polymerase and 2 μL of the 25 mM dNTP mix, then incubate at 37 °C for 10 min to generate a double-stranded version of the target site oligonucleotide (dsDNA) (*see* **Note 4**).
3. Extract the polymerase extension reaction with an equal volume of phenol–chloroform–isoamyl alcohol (25:24:1) by vortexing for 1 min followed by centrifugation at maximum speed in a tabletop centrifuge.
4. Carefully extract the aqueous phase with a pipettor and stand on ice of 5 min. If the aqueous phase is still hazy or milky, respin and extract residual phenol from the bottom of the tube with a pipettor.
5. Add 2 volumes of ethanol and 1 μL of 20 mg/mL glycogen, mix well, then precipitate overnight at −20 °C.
6. Spin down the precipitated dsDNA target site for 5 min at top speed in a tabletop centrifuge, and carefully wash once with 70 % ethanol. Air-dry or SpeedVac-dry, then dissolve in 20 μL H_2O.
7. Digest the dsDNA with 5 units of restriction enzyme/μg DNA, followed by a second round of extraction and precipitation. BamHI is shown in the example in Fig. 1b (**steps 3–6** above; *see* **Note 4**).

8. Digest pBluescript SK(+) plasmid DNA using the same restriction enzyme(s) and digest conditions, and purify the digested linear plasmid DNA using a DNA gel extraction kit.
9. Mix cleaved target site dsDNA with digested pBluescript SK(+) plasmid vector DNA at 3:1 molar ratio of vector to insert, add 10 units of T4 DNA ligase, and incubate overnight at 16 °C.
10. Heat the ligation reaction at 65 °C for 10 min, then purify the DNA by repeating **steps 3–6** above and resuspend in 20 μL sterile H_2O.
11. Electroporate the purified and resuspended ligation products into 50 μL DH5α-E electroporation-competent cells, then add 1 mL SOC medium and incubate for 1 h at 37 °C with gentle shaking.
12. Dilute 1 μL of transformed cells in 1 mL SOC or sterile H_2O, and plate 50 μL of the diluted cells on a 10 cm LB Agar plate containing 100 μg/mL carbenicillin. Incubate overnight inverted at 37 °C. Add the rest of the transformed cells to 50 mL LB medium with 100 μg/mL carbenicillin and grow overnight at 37 °C in a shaking incubator.
13. Count colonies on the LB Agar plate to calculate the library size by multiplying the colony number by the dilution factor.
14. Purify target site library DNA from the 50 mL overnight culture using a plasmid midiprep kit.
15. Randomly pick at least 12 colonies from the LB Agar plate, and inoculate each into 5 mL fresh LB medium with 100 μg/mL carbenicillin. Grow cultures overnight in a 37 °C shaking incubator.
16. Extract plasmid DNA and sequence the target site region of each colony plasmid to estimate the likely complexity of the target site library.

3.2 Sequential Enrichment of HE Cleavage-Sensitive Target Sites

1. Mix reaction components in a 1.5 mL microcentrifuge tube as follows: 10 μL of 10× reaction buffer, 10 μg of pBluescript target site library DNA (final concentration: ~10 nM), 5 μL of 5 μm purified I-CreI or mCreI, and H_2O to 100 μL (*see* **Note 5**).
2. Incubate at 37 °C for 1 h, then add 20 μL of 6× stop buffer and incubate at room temperature for 30 min (*see* Fig. 2 and **Note 5**).
3. Electrophorese the cleavage reaction products through a 1 % TAE-buffered NuSieve GTG agarose at 1.4 V/cm overnight at room temperature.
4. Stain the resulting gel with 100 ng/mL ethidium bromide and visualize products on a UV light box, taking care to wear proper UV eye protection.

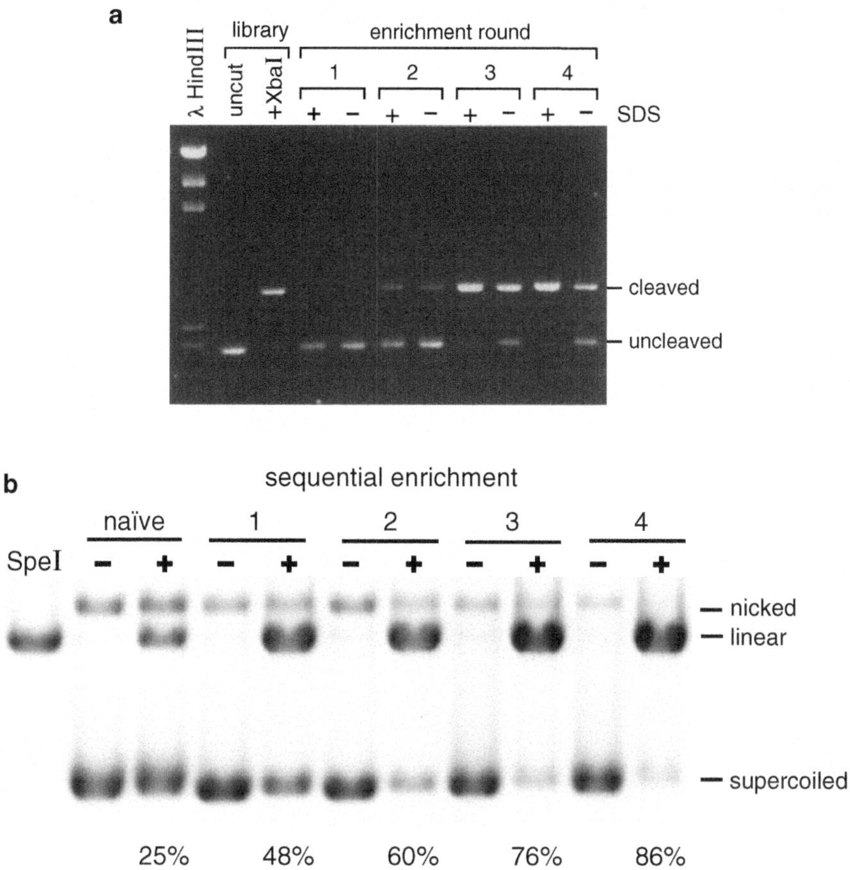

Fig. 2 Sequential enrichment examples. The two gels shown demonstrate sequential enrichment of cleavage-sensitive I-CreI/mCreI LHE target sites using low (**a**) and high (**b**) enzyme:substrate ratios of cleavage-sensitive sites present in the starting library over four successive rounds of cleavage-based recovery and amplification. Enrichment cycles in (**a**) were performed with or without the addition of SDS after completion of the cleavage reactions to facilitate product release [10]. *Numbers below* (**b**) represent the proportion of linearized, cleavage-sensitive plasmid DNA molecules after each round of enrichment. Panel (**a**) has previously been published as Fig. 3a in ref. 10

5. Excise the gel regions containing linear plasmid DNA (which contains cleavage-sensitive target sites) and supercoiled plasmid DNA (which contains cleavage-resistant target sites).
6. Extract DNA using a gel extraction kit and elute DNAs in sterile H_2O.
7. Mix the purified linear DNA plasmid with 800 units of T4 DNA ligase, 10 μL of 10× ligase buffer, and sterile water to 100 μL, then incubate at 16 °C overnight.
8. Repeat Subheading 3.1 **steps 11–14**.
9. Repeat **steps 1–6** for the supercoiled plasmid DNA fraction if you are interested in characterizing cleavage-resistant target sites in addition to cleavage-sensitive target sites.

10. Repeat **steps 1–8** above three times, or until the enrichment of cleavage-sensitive target sites reaches a plateau as assessed by the fraction of cleaved, linear DNA molecules (*see* Figs. 1 and 2).

3.3 Sample Preparation for NGS/Illumina Sequencing

Note: In the following protocol we are using Illumina 36 base pair single end reads as a NGS sequencing platform for target site library analyses. The same general protocol can be easily adapted to different read lengths and Illumina sequencers, and to other NGS sequencing platforms.

1. Set up PCR reaction: mix 20 μL of 5× polymerase buffer, 0.8 μL of 25 mM dNTPs, 0.5 μL of 100 μM forward primer, 0.5 μL of 100 μM reverse primer, 200 ng plasmid DNA, and 1 μL of high fidelity DNA polymerase (at 2 units/μL) with sterile H_2O to 100 μL.
2. Run PCR using the following conditions: initial denaturation: 98 °C for 30 s; 30 elongation cycles: 98 °C for 10 s, 55 °C for 30 s and 72 °C for 30 s; and final elongation step: 72 °C for 5 min.
3. Electrophorese PCR products through a 4 % TAE-buffered NuSieve GTG agarose at 6 V/cm for 2 h at room temperature. Be sure to include an appropriate size standard!
4. Stain the resulting gel with 100 ng/mL ethidium bromide and visualize products on a UV light box taking care to wear proper UV eye protection.
5. Excise the region containing PCR products of the desired size range, and extract the DNA using a gel extraction kit. Elute the resulting DNA in 20 μL of sterile H_2O.
6. Determine the quality and quantity of the purified PCR product by checking 1 μL of the isolated DNA on a 4 % TAE agarose gel (*see* **Note 6**).
7. Send three samples (naïve library, cleavage-sensitive plasmids and, if isolated, cleavage-resistant plasmids) for Illumina single end 36 bp sequencing. Each sample should consist of 10 μL of 200 nM DNA and 20 μL of 100 μM sequencing primer (*see* **Note 7**).

3.4 Data Analysis of Illumina Sequencing Results (*see* Note 8)

1. Search across all positions of each read, and eliminate any reads that have read positions with quality scores = "B".
2. Among these high quality reads identify the target read subset that has the expected end sequence for the target site oligonucleotide shown in Fig. 1b (CCCC, from target site base pair positions 33 to 36; Fig. 1b).

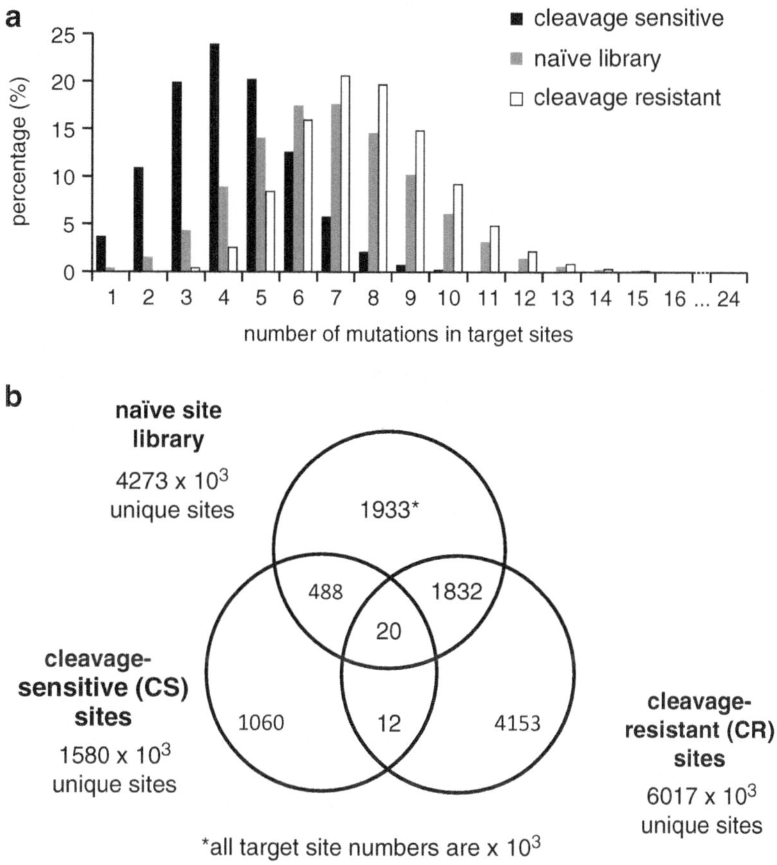

Fig. 3 Starting site library and enriched fractions. (**a**) Distribution of number of target site mutations across the starting/naïve library and cleavage-sensitive and cleavage-resistant library fractions. The distribution and type(s) of mutations in the starting library serve as a quality control measure for library synthesis. The most common oligonucleotide synthesis errors are single base deletions and insertions. The *left* shift to lower numbers of mutation in cleavage-sensitive sites, and the *right* shift in cleavage-resistant sites provide two additional indications of successful functional enrichment for these functional site classes. (**b**) Venn diagram showing the distribution of unique target sites and sites found in more than one site library (naïve, cleavage sensitive (CS) and/or cleavage resistant (CR) libraries) after four rounds of enrichment

3. Identify the subset of high quality target site sequences with the correct ends that are present in both the naïve and cleavage-sensitive (and/or -resistant) target site library fractions (*see* Fig. 3b).
4. Sum the number of reads for each unique target site in each library. Identical target sites with different barcodes are considered as different target sites, as they originated from different library templates (*see* Table 1).
5. The frequency of each target site is calculated by dividing the read number of each target site by the total read number of the sample.

Table 1
Target site enrichment sequencing results[a]

Samples	Total reads[b]	Total sites[c]	Unique sites[d]
Naïve library	13,089,102	10,925,984	6,303,279
Cleavage-sensitive	12,524,976	11,275,812	1,399,651
Cleavage-resistant	24,200,450	20,344,154	8,595,486

[a]This experiment followed closely the protocol and used the target site library design outlined in Fig. 1. The read numbers differ from the experimental analysis shown in Fig. 3b

[b]Number of sequencing reads with quality score above "B" at each position (**step 1** of Subheading 3.4)

[c]Number of sequencing reads with the expected target site oligonucleotide end sequence ("CCCC" from position 33 to 36) (Fig. 1b; **step 2** of Subheading 3.4)

[d]Number of unique sites in each library with different target site sequences or the same target site sequence but different barcodes (**step 3** of Subheading 3.4)

6. The "fold enrichment" of each target site is calculated by dividing the frequency of one target site in cleavage-sensitive sample by its frequency in naïve library sample (Fig. 4).

4 Notes

1. A 35 % randomization ratio of the target site was chosen so that a target site library for I-CreI/mCreI with a site length of 22 bp and a typical *E. coli* plasmid-based library size (10^6–10^7 independent colonies/transformants) would contain several independent copies of the native/wild type I-CreI/mCreI target site in the starting library.
2. Make sure your purified HE sample is free of contaminating DNases. This can be done by incubating serial dilutions of your purified HE, with supercoiled and linear fractions of a target site plasmid that *does not* contain a cognate target site, under optimal reaction conditions. Even trace amounts of nuclease contamination will lead to plasmid nicking or linearization (endonucleolytic activity/contamination) or linear DNA degradation (endo- and/or exonucleolytic activity/contamination). Either contaminating activity will reduce your recovery of desired target site-specific plasmid molecules.
3. Optimal cleavage reaction conditions vary between HEs and need to be determined in advance to optimize the recovery or cleavage-sensitive or -resistant target site fractions.
4. dsDNA product generation and digestion can be easily monitored by checking 1 μL of the reaction mix on a 4 % TAE agarose gel.
5. The cleavage conditions used here were chosen to favor near-complete target site cleavage. Cleavage conditions and sampling of the cleavage time course can be used to isolate target sites with differing cleavage sensitivities (*see* Fig. 2 for an example).

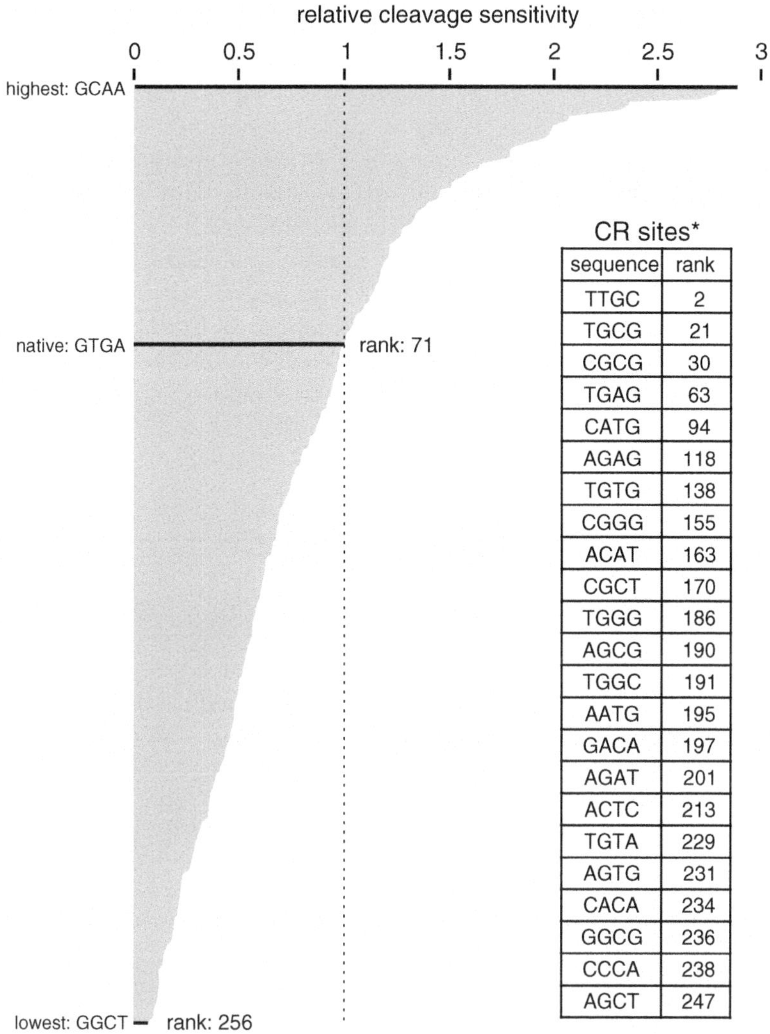

sequence	rank
TTGC	2
TGCG	21
CGCG	30
TGAG	63
CATG	94
AGAG	118
TGTG	138
CGGG	155
ACAT	163
CGCT	170
TGGG	186
AGCG	190
TGGC	191
AATG	195
GACA	197
AGAT	201
ACTC	213
TGTA	229
AGTG	231
CACA	234
GGCG	236
CCCA	238
AGCT	247

Fig. 4 Use of sequential enrichment and sequencing to define central 4 bp cleavage preferences. The main panel shows the rank ordering of relative cleavage sensitivity of 256 different, "central 4" base pair possibilities in the I-CreI/mCreI target site. Cleavage sensitivity is scaled to a native target site central 4 bp GTGA cleavage sensitivity of 1.0, versus the most-sensitive (GCAA) and least sensitive (GGCT) central 4 bp sequence in the context of different flanking sequences (Hui Li, Zafer Aydin, William S. Noble, and R.J.M., Jr., unpublished results). *Inset*: an independent determination of the central 4 bp sequences most resistant to cleavage by I-CreI using a reduced complexity target site library in which only the central 4 bp positions of the I-CreI/mCreI target site were fully randomized (Thai Akasaki, Megan S. Chadsey, and R.J.M., Jr., unpublished results)

6. A Bioanalyzer (Agilent) can be used to characterize DNAs going to NGS, and is a better alternative if available as it will give provide DNA quality and sizing data that make it easier to assess sample adequacy for NGS.

7. A typical Illumina sequencing run on a GAII should generate more than ten million reads with reasonable quality scores.
8. Data analysis of NGS sequencing results is a *very* rapidly evolving area of bioinformatics. The general outline we give can be used to implement a sequence analysis pipeline and generate and analyze results with any of a growing number of open source or commercial NGS analysis software. Some useful online resources include Galaxy tools, especially the NGS Toolbox and SAMTools (https://main.g2.bx.psu.edu/), the Phred, Phrap, and Consed suite developed by Phil Green (University of Washington Department of Genome Sciences: http://www.phrap.org/), and The Broad Institute's Software Tools resource (http://www.broadinstitute.org/scientific-community/software).

Acknowledgments

This work was supported by US National Institutes of Health Training Grant award to H.L. (5RL9HL092555); by a US National Institutes of Health U54 Interdisciplinary Research Roadmap award (1RL1 CA133831) to R.J.M. Jr; and by a grant from the Foundation for the National Institutes of Health funded by the Bill and Melinda Gates Foundation as a Grand Challenges in Global Health award to R.J.M. Jr.

References

1. Scharenberg AM, Duchateau P, Smith J (2013) Genome engineering with TAL-effector nucleases and alternative modular nuclease technologies. Curr Gene Ther 13:291–303
2. Certo MT, Gwiazda KS, Kuhar R, Sather B, Curinga G et al (2012) Coupling endonucleases with DNA end-processing enzymes to drive gene disruption. Nat Methods 9:973–975
3. Yusa K, Rashid ST, Strick-Marchand H, Varela I, Liu P-Q et al (2011) Targeted gene correction of a1-antitrypsin deficiency in induced pluripotent stem cells. Nature 478:391–394
4. Porteus MH, Baltimore D (2003) Chimeric nucleases stimulate gene targeting in human cells. Science 300:763
5. Miller JC, Tan S, Qiao G, Barlow KA, Wang J et al (2011) A TALE nuclease architecture for efficient genome editing. Nat Biotechnol 29: 143–148
6. Belfort M (2005) Back to basics: structure, function, evolution and application of homing endonucleases and inteins. In: Belfort M, Stoddard BL, Wood DW, Derbyshire V (eds) Homing endonucleases and inteins. Springer, Berlin, pp 1–10
7. Stoddard BL (2011) Homing endonucleases: from microbial genetic invaders to reagents for targeted DNA modification. Structure 19:7–15
8. Li H, Ulge UY, Hovde BT, Doyle LA, Monnat RJ (2012) Comprehensive homing endonuclease target site specificity profiling reveals evolutionary constraints and enables genome engineering applications. Nucleic Acids Res 40:2587–2598
9. Ashworth J, Taylor GK, Havranek JJ, Quadri SA, Stoddard BL et al (2010) Computational reprogramming of homing endonuclease specificity at multiple adjacent base pairs. Nucleic Acids Res 38:5601–5608
10. Argast GM, Stephens KM, Emond MJ, Monnat RJ Jr (1998) I-PpoI and I-CreI homing site sequence degeneracy determined by random mutagenesis and sequential in vitro enrichment. J Mol Biol 280:345–353
11. Doyon JB, Pattanayak V, Meyer CB, Liu DR (2006) Directed evolution and substrate

specificity profile of homing endonuclease I-SceI. J Am Chem Soc 128:2477–2484

12. Scalley-Kim M, Connell-Smith A, Stoddard BL (2007) Coevolution of a homing endonuclease and its host target sequence. J Mol Biol 372: 1305–1319

13. Molina R, Redondo P, Stella S, Marenchino M, D'Abramo M et al (2012) Non-specific protein–DNA interactions control I-CreI target binding and cleavage. Nucleic Acids Res 40:6936–6945

14. Takeuchi R, Lambert AR, Mak AN-S, Jacoby K, Dickson RJ et al (2011) Tapping natural reservoirs of homing endonucleases for targeted gene modification. Proc Natl Acad Sci U S A 108:13077–13082

15. Szeto MD, Boissel SJS, Baker D, Thyme SB (2011) Mining endonuclease cleavage determinants in genomic sequence data. J Biol Chem 286:32617–32627

16. Li H, Pellenz S, Ulge U, Stoddard BL, Monnat RJ Jr (2009) Generation of single-chain LAGLIDADG homing endonucleases from native homodimeric precursor proteins. Nucleic Acids Res 37:1650–1662

Chapter 13

Homing Endonuclease Target Determination Using SELEX Adapted for Yeast Surface Display

Kyle Jacoby and Andrew M. Scharenberg

Abstract

Knowing the target sequence of a DNA-binding protein is vital in obtaining fundamental characteristics of the protein and evaluating properties of the protein–DNA interaction. For example, novel homing endonucleases cannot be proven to be functional until a predicted target site is tested. Unfortunately, target site prediction is not always easy, or even possible, depending on the amount of sequence data available. Here we describe a modification of SELEX using yeast surface display that can quickly and inexpensively resolve DNA-binding targets in high throughput for proteins without any prior assumptions or knowledge regarding the target site. This protocol is easily integrated into the yeast surface display pipeline and is leveraged by the expansive number of existing tools for both SELEX and yeast surface display.

Key words SELEX, Yeast surface display, Homing endonuclease, Meganuclease, Flow cytometry, Binding affinity, Protein–DNA interaction, In vitro selection

1 Introduction

Homing endonucleases (HEs) are a class of exquisitely specific DNA-cleaving proteins. Their potential use as genome-modifying or targeted gene therapy reagents has attracted intense investigation into the alteration of their specificities [1–7]. Although design attempts have been met with some success, limits in our current ability to reengineer HE specificity has prompted a search for more enzymes with varied targets [8, 9]; when engineering can begin at any of a variety of starting points, the number of modifications necessary to reach a given specificity is reduced. Fortunately, the HE family is quite large and diverse [9], and the nature of their existence helps reveal their native target specificity, aiding our endeavors. Homing endonucleases are self-propagating genetic elements which reside within a host gene, typically contained within an intron, intein, or precise fusion such that the inserted HE coding sequence does not interrupt the functionality of its host gene. Careful examination of host intron–exon borders or

David R. Edgell (ed.), *Homing Endonucleases: Methods and Protocols*, Methods in Molecular Biology, vol. 1123, DOI 10.1007/978-1-62703-968-0_13, © Springer Science+Business Media New York 2014

comparison with an insert-less allele can reveal the HE's native target site [10, 11]. The insert-free allele must possess the site that the HE originally cut, or "homed" to, when creating the initial double-strand break that lead to its introduction into the host allele [12]. While this method of target determination works with undeniable success, it explicitly requires precise intron–exon borders or sequence of the insert-less allele: something frequently unobtainable. Consequently, although we may predict the presence of a novel homing endonuclease—or more generally, any DNA-binding protein—the frequent lack of a putative DNA target precludes analysis of the protein's fundamental characteristics.

To address this issue, and thereby better exploit the vast amount of genomic data being generated, we present a method for determining a DNA-binding protein's target specificity a priori. This approach is a combination of two existing technologies: SELEX and yeast surface display [13, 14]—the appropriate sections on yeast surface display have been included in this chapter [15] (*see* Subheadings 3.2 and 3.3). In essence, the protein of interest is expressed on the surface of yeast, which acts as a solid support for the binding of an initially randomized library of potential DNA targets (under non-cleaving conditions); the yeast are washed, and bound targets are released, amplified, and subjected to further selection in a cyclical fashion known as SELEX (*see* Fig. 1). The output of successful SELEX experiment is the collection of target oligos which the protein binds best.

Synergy of the two parent technologies yields a powerful new tool that can be easily integrated into existing yeast surface display platforms. Vectors made for yeast surface display can be used to characterize proteins in high throughput, and subsequently shunted into SELEX or directed evolution pipelines and vice versa [8, 16]. For example, a panel of putative HEs can be assayed for proper folding, its targets determined by SELEX, its binding and cleavage properties interrogated in detail, all in a multi-well, high-throughput manner [16, 17]; enzymes specificities can then be altered by directed evolution using the same platform. Our approach to SELEX also benefits from the ability to test oligo binding in high throughput using yeast surface display and flow cytometry. SELEX conditions can be optimized and tuned to the investigators precise needs (*see* Subheading 3.4), and can serve as an indication of the protocols success (*see* Subheading 3.6). Selection conditions can subsequently be modulated to yield more diverse or narrow target pools depending on whether the investigator desires a binding profile, or simply a few best-bound targets [18]. Lastly, SELEX using yeast surface display is an improvement upon traditional SELEX insofar as it allows quick, easy, and inexpensive expression of protein that does not need further purification or modification. It also removes the explicit need for expensive consumables such as magnetic beads or antibodies.

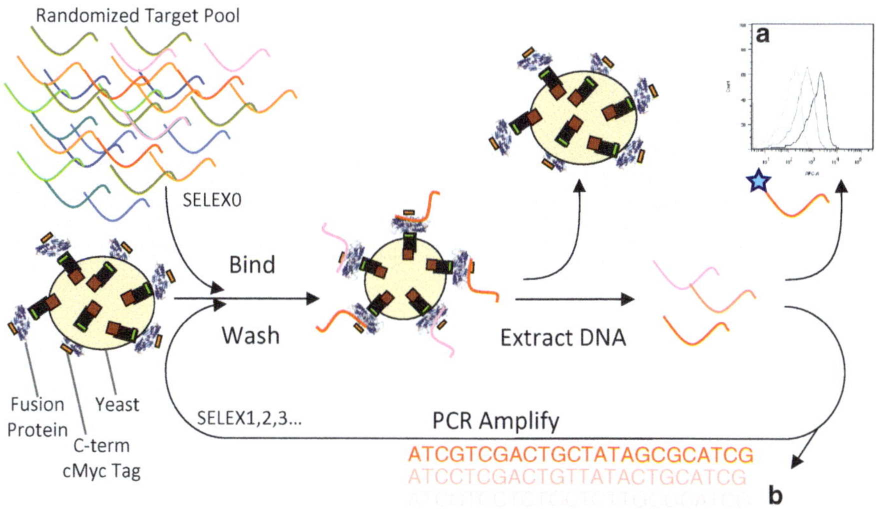

Fig. 1 Schematic representation of SELEX using yeast surface displayed protein. The homing endonuclease of interest is C-terminally cMyc-tagged and fused with the yeast surface-expressed protein, Aga2P. Aga2P forms a disulfide linkage with Aga1P, which is co-induced in the EBY100 strain upon Galactose induction. The yeast is used as a solid support for the protein to allow binding of randomized oligos and subsequent wash steps. The best-bound sequences are extracted, amplified, and subjected to iterative rounds of selection. The initial optimization and later, the success of the experiment are assayed using fluorescently tagged oligo and flow cytometry (**a**). After sufficient enrichment for high-affinity targets, the oligos are sequenced and aligned (**b**)

This method should be regarded as a unique alternative to traditional SELEX, as well as an additional tool for the yeast surface display toolbox.

SELEX can be used in a variety of scenarios. To name a few, it can be used to obtain targets of DNA-binding proteins where (1) no prediction methodology exists, (2) engineering methods yield unpredictable specificity, and (3) their target predictions rely on missing, partial, or improperly annotated sequence data. SELEX can also be modified to use genomic sequence as the input pool [19], providing a set of best-bound sequences *in that genome* for the protein (and any off-targets, which may be of keen interest [20]). That being said, certain things should be kept in mind when considering SELEX. First, this protocol will provide the investigator with a set of sequences best bound by a given protein. Although binding discrimination is integral to achieving cleavage specificity, binding and cleavage specificity are not synonymous [21]; that said we have had outstanding success finding cleavable substrates and recapitulating known cleavage data by using this method with I-OnuI family enzymes. Furthermore, it may also be possible to add a cleavage selection step to help rectify any binding/cleavage discrepancy. Second, it should be noted that multiple or alternative nucleic acid binding domains may obfuscate SELEX results, and this protocol should be *carefully* applied to such proteins.

Ideally, any such domains should be separated or eliminated before testing. Though we have described a large number of potential applications and uses for yeast surface display SELEX, we will provide the reader only with the methods for basic a SELEX experiment with the intention on determining the binding site of a given DNA-binding protein.

2 Materials

Prepare all solutions using ultrapure RNAse- and DNAse-free water (0.22 μm filtered, deionized water) and analytical grade reagents. Prepare and store all reagents at 4 °C unless otherwise indicated. Cultures and reagents may be prepared at the bench, but care should be taken to use sterile components and aseptic technique.

2.1 Randomized Pool (SELEX0) Generation from ssDNA Template

1. SELEX single stranded randomized oligo pool template with flanking SELEX primer sites, standard desalting, resuspended to 100 μM (*see* **Notes 1–4**) (Table 1).
2. SELEX reverse primer and 5′-A647-labeled SELEX reverse primer, standard desalting, resuspended to 100 μM each (*see* **Note 5**) (Table 1).
3. PCR kit (*see* **Note 6**).
4. 10 mM dNTPs.
5. 0.2 mL thin wall PCR strip tubes.
6. Thermal cycler.

2.2 Polyacrylamide Gel

1. 30 % acrylamide–bis acrylamide (19:1) solution.
2. 10× Tris–borate–EDTA (TBE) buffer (*see* **Note 7**).
3. *N,N,N′,N′*-tetramethylethylenediamine (TEMED).

Table 1
Example oligo sequences

Oligo	Sequence
Forward primer	CAGGGATCCATGCACTGTACG
Reverse primer	(A647)*-AGGATCCGCAGTCAAGTGG
30N SELEX0	<u>CAGGGATCCATGCACTGTACG</u>TTT(N30)AAA<u>CCACTTGACTGCGGATCCT</u>

This primer and template pair has been tested and used with the given PCR conditions, cycle numbers, and SELEX parameters described in this chapter. *The reverse primer should be ordered in its 5′-A647 labeled and unlabeled forms. Note the three thymidine bases preceding, and three adenine bases following the 30N randomized region of the SELEX0 template not part of the primer-binding region (*underlined*); they are included to discourage high-affinity protein interactions with the constant regions, which would otherwise complicate analysis. *See* **Note 17** for high-throughput sequencing primer details

4. Ammonium persulfate: 10 % w/v solution in water.
5. Plastic gel cassette, 1.0 mm.
6. Vertical gel electrophoresis apparatus.
7. 6× DNA loading buffer: 18 % (w/v) Ficoll-400 in 6× TBE, with no added dyes (*see* **Note 8**).

2.3 Yeast Transformation

1. EBY100 yeast (Invitrogen).
2. 2× YPAD non-selective yeast media: 20 g Bacto yeast extract, 40 g Bacto peptone, 100 mg Adenine hemisulfate, 50 g Glucose, water to 1 L, pH to 6.0. Filter-sterilize or autoclave and store at 4 °C (*see* **Note 9**).
3. Salmon sperm DNA, 2 mg/mL solution in 1× TE.
4. 1 M Lithium acetate solution in water.
5. 50 % w/v Polyethylene glycol in water, MW 3350 (PEG 3350).
6. Plasmid DNA encoding endonuclease (or protein of interest) cloned into the pETCON yeast surface display vector (*see* **Note 10**).
7. 42 °C water bath.
8. Yeast selective growth media "SC–Ura–Trp": 6.7 g Yeast nitrogen base without amino acids, 1.4 g yeast synthetic drop-out media supplement without Trp, Ura, His, Leu, 76 mg Histidine, 380 mg Leucine, 4.34 g MES, and water to 900 mL. Adjust pH to 5.25 with HCl. Sterilize by autoclaving 20 min. Prior to use, add penicillin (100 i.u./mL), streptomycin (100 μg/mL), and kanamycin (25 μg/mL). Store at 4 °C.
9. 20 % w/v glucose solution, filter-sterilized. Store at 4 °C.
10. Selective growth media agar plates: add 20 g of bacteriological agar to 900 mL of SC–Ura–Trp selective growth media and autoclave for 20 min. Add 100 mL pre-warmed (55 °C) 20 % w/v glucose and penicillin (100 i.u./mL), streptomycin (100 μg/mL), and kanamycin (25 μg/mL). Pour into petri dishes and let solidify at room temperature. Store plates at 4 °C.
11. Water-jacketed incubator.

2.4 Yeast Growth and Induction

1. EBY100 yeast transformed with surface-expression vector containing homing endonuclease of interest.
2. SC–Ura–Trp selective growth media (for Recipe, *see* Subheading 2.3).
3. 20 % w/v glucose solution, filter sterilized, store at 4 °C.
4. 20 % w/v d-(+)-raffinose pentahydrate + 0.1 % w/v glucose solution, filter sterilized, store at room temperature.
5. 20 % w/v d-(+)-galactose solution, filter sterilized, store at 4 °C.
6. Disposable 15-mL culture tubes or deep-well 96-well V-bottom plate and breathable sealing film.

7. Shaking incubator.
8. Spectrophotometer.

2.5 Binding Selection

1. Induced EBY100 yeast with surface expressed homing endonuclease (*see* **Note 11**).
2. 10× Base Bind and Wash Buffer: 0.1 M NaCl, 0.1 M HEPES, 0.05 M K-Glu (l-Glutamic Acid Potassium Salt Monohydrate), 0.5 % BSA, adjusted to pH 7.5 with KOH (*see* **Notes 12** and **13**). Filter-sterilize solution and store at 4 °C in a light-protected or foil-wrapped container.
3. Supplementary binding-mitigation reagent(s): 1 M KCl. Filter sterilize solution and store at room temperature (*see* **Note 14**).
4. Supplementary binding cofactor: 1 M $CaCl_2$ solution. Filter-sterilize and store at room temperature (*see* **Note 15**).
5. Costar 96-well V-bottom plate.
6. Plate sealing tape (clear or aluminum).
7. Plastic multichannel pipette basins.

2.6 Binding Selection Condition Optimization or SELEX Analysis by Flow Cytometry

1. Materials from Subheading 2.5.
2. Fluorescent ds SELEX0 (for binding condition optimization, described in Subheading 3.1 and made from materials under Subheading 2.1) or Labeled ds DNA pools from each round (for SELEX analysis, described in Subheading 3.6 and made from materials under Subheading 2.8).
3. 10× Yeast Staining Buffer (YSB): 1.8 M KCl, 0.1 M NaCl, 0.1 M HEPES, 2 % BSA, 1 % w/v d-(+)-Galactose, adjust pH to 7.5 with KOH (*see* **Note 12**). Filter sterilize and store at 4 °C in a light-protected or foil-wrapped container.
4. FITC-conjugated chicken anti-cMyc antibody (*see* **Note 16**).
5. BD FACScalibur or LSRIITM cytometer (BD Biosciences with high-throughput sampler (HTS) attachment or other cytometer with equivalent optics).
6. FloJo software (Tree Star Inc.).

2.7 SELEX Protocol

1. Materials from Subheading 2.5.
2. Non-labeled ds SELEX0 pool.
3. SELEX forward primer and (non-labeled) SELEX reverse primer, standard desalting, resuspended to 100 μM each (Table 1).
4. PCR kit (*see* **Note 6**).
5. 0.2 mL thin wall PCR strip tubes or 96-well plates.
6. Thermal cycler.

2.8 Labeled Double-Stranded DNA Pools for SELEX Analysis by Flow Cytometry

1. Thermostable DNA Polymerase with PCR buffer and 50 mM $MgSO_4$ (*see* **Note 17**).
2. SELEX pool templates (from each sample and round), with flanking SELEX primer sites.
3. SELEX forward primer (*see* **Note 1**) (Table 1).
4. 5′-A647-labeled SELEX reverse primer (*see* **Note 1**) (Table 1).
5. 10 mM dNTPs.
6. 0.2 mL PCR strip tubes and 96-well PCR plates.
7. Thermal cycler.
8. Exonuclease I.
9. 96-well filter plate (*see* **Note 18**).
10. illustra Sephadex G-100 (GE Healthcare): Make sephadex solution at 1 g/20 mL in water; allow at least 24 h for bead hydration; store at room temperature for a maximum of 4 months (*see* **Note 19**).
11. Odyssey infrared imaging system (Li-Cor Biosciences) or UV transilluminator.
12. Microvolume spectrophotometer (e.g., NanoDrop or similar instrument).

2.9 Sequencing and Analysis

1. TA or Blunt end cloning kit (e.g., CloneJET) and associated required *E. coli* transformation and growth materials (see specific kit for details).
2. Sanger sequencing services (see facility requirements for additional materials) (*see* **Note 20**).
3. Sequence trimming software (e.g., Geneious or custom scripts).
4. High-throughput sequencing services (*see* **Note 21**) and associated preparatory reagents.
5. High-throughput, barcoded sequencing forward primers in 96-well plate format, standard desalting, resuspended to 10 μM each (*see* **Note 22**).
6. High-throughput sequencing reverse primer, standard desalting, resuspended to 100 μM.
7. PCR kit (*see* **Note 6**).
8. 0.2 mL thin wall PCR strip tubes or 96-well plates.
9. Thermal cycler.
10. Barcode binning software (e.g., Geneious or custom scripts).
11. Sequence alignment software (e.g., MEME) (*see* **Note 23**).

Table 2
Example reaction volumes for a SELEXO dsDNA pool preparation

Stock []		Final []		1× Volumes	8× Volumes	Unit, reagent
10	×	1	×	10.0	80.0	μL, PCR buffer
50	mM	1.5	mM	3.0	24.0	μL, $MgSO_4$
10	mM	1.2	mM	12.0	96.0	μL, dNTPs
100	μM	0	μM	0.0	0.0	μL, Forward primer
100	μM	30	μM	30.0	240.0	μL, Reverse primer
100	μM	10	μM	10.0	80.0	μL, ssDNA pool
5	U/μL	0.05	U/μL	1.0	8.0	μL, Polymerase
				34.0	272.0	μL, ddH_2O
				100.0	800.0	μL, Total

Sample volumes are given for 1 and 8 PCR reactions

3 Methods

Prepare PCR reactions on ice and carry out all other procedures at room temperature unless otherwise specified.

3.1 Randomized Pool (SELEXO) Generation from ssDNA Template

1. Prepare a 1× PCR master mix to make a number of small aliquots of the double stranded (ds) randomized oligos (*see* **Note 24**). Prepare a separate, small amount of the PCR mix A647-labeled reverse primer (in place of unlabeled reverse primer) to make fluorescent dsDNA used in selection condition optimization (*see* Subheading 3.4).
2. *See* Table 2 for PCR reagent concentrations and sample volumes (*see* **Note 25**).
3. Divide the PCR mix into 100 μL aliquots in multiple thin-wall PCR tubes (or plate), cap (or seal) them, put them in a thermal cycler, and incubate them as follows: 95 °C for 5 min, 59 °C for 10 min, 72 °C for 10 min, 4 °C hold.
4. Keep one sample at 4 °C for immediate use and store the remaining dsDNA randomized pool at −20 °C.
5. Run target oligos on a 10 % polyacrylamide gel. For a 7.5-mL gel, combine 0.75 mL 10× TBE, 2.5 mL 30 % acrylamide–bis acrylamide (19:1), and 4.15 mL water. Add 100 μL 10 % APS and 10 μL TEMED, then immediately mix and pipette into the prepared gel cassette (*see* **Note 26**). Once the gel is set, load 0.5–1.0 μL ssDNA, dsDNA, and labeled dsDNA, each diluted with 1.0 μL 6× loading buffer and 4 μL water (*see* **Note 8**). Use a short length ladder (~50–500 bp) as a standard. Run the gel for 90 min at 120 V.

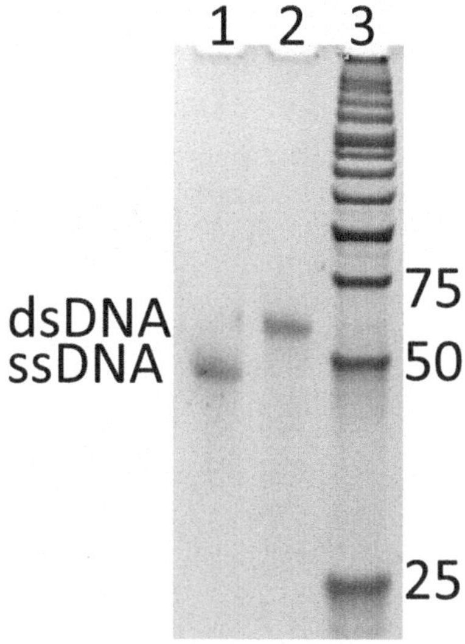

Fig. 2 Gel of ssDNA and dsDNA. Using a low molecular weight marker (*lane 3*), we verify conversion of the single-stranded randomized oligo (ssDNA, *lane 1*) to the double-stranded oligo used to seed SELEX experiments (dsDNA, *lane 2*). The ssDNA template should be fully converted to, and run higher than the final dsDNA library

6. Stain the gel with 1× SybrGold in 1× TBE for 20 min to allow visualization on a Licor Odyssey infrared imager (using the 700 nM laser) or UV transilluminator (*see* **Note 27**). After washing once in 1× TBE, the gel should show a prominent single PCR product which runs higher than the ssDNA template (Fig. 2).

3.2 Yeast Transformation

1. Thaw and spin down a frozen aliquot of EBY100 competent yeast cells (*see* **Note 28**).
2. Resuspend the pellet in the following transformation mixture: 50 μL denatured 2 mg/mL salmon sperm DNA (heat at 95 °C for 5 min, then transfer immediately to ice), 36 μL 1 M LiAc, 260 μL 50 % PEG 3350, and 14 μL water plus plasmid DNA (up to 1 μg) (*see* **Notes 29** and **30**).
3. Incubate the yeast and transformation mixture at 42 °C for 40 min (*see* **Note 31**).
4. Fill the tube with SC–Ura–Trp + 2 % glucose media and spin down the cells. Remove supernatant.
5. Resuspend the yeast pellet in 1 mL SC–Ura–Trp + 2 % glucose media.
6. Plate 1–10 μL transformed yeast on selective growth media agar plates (SC–Ura–Trp + 2 % glucose) and incubate in a 30 °C water-jacketed incubator. Colonies of an appropriate size for picking should appear by 48 h

3.3 Yeast Growth and Induction

1. Transfer a single colony of transformed yeast into 1.5 mL SC–Ura–Trp + 2 % raffinose + 0.1 % glucose media (*see* **Notes 32** and **33**).
2. Incubate overnight in a 15-mL culture tube at 30 °C with 250 RPM shaking until the cells reach a density of 90–120 million/mL and place on ice for up to 24 h (*see* **Notes 34** and **35**).
3. Determine the density of yeast of induced yeast. This can be done using a hemocytometer or by spectrophotometry (*see* **Note 36**).
4. Wash 30 million cells twice with water and transfer to 1.5 mL of SC–Ura–Trp + 2 % galactose media (*see* **Notes 37–39**).
5. Incubate the galactose culture on the benchtop (room temperature with no shaking) for 16–18 h for optimal induction.

3.4 Binding Selection Condition Optimization

1. Prepare 1× base bind and wash buffer with binding cofactor (2 mM $CaCl_2$) and varying amounts of KCl to create a final bind and wash buffer (BWB) (*see* **Notes 11** and **40**). Keep these buffers at room temperature (or appropriate selection temperature) for the duration of the experiment. 1× bind and wash buffers can be store at 4 °C for up to 1 month.
2. Determine the density of induced yeast as in Subheading 3.3, **step 3**.
3. Transfer enough yeast to test multiple selection conditions for each protein you wish to assay—three million yeast per sample per condition—to a V-bottom plate. Each protein should occupy 1 well, and will be split after staining.
4. Wash cells twice with 200 μL 1× YSB, centrifuging the V-bottom plate at 2,000 × *g* for 1 min and discarding the supernatant.
5. Gently resuspend cells in 50 μL in 1× YSB with 1:100 dilution of anti-cMyc-FITC antibody (i.e., consider the antibody to be a 100× stock; adjust the total volume to stay close to 100 million yeast/mL). Incubate at 4 °C for 30–60 min, mixing gently every 10–15 min. Be sure to include an unstained control for subsequent gating analysis.
6. Wash twice with 200 μL BWB with an intermediate amount of KCl (e.g., 150 mM). Resuspend in a small amount of the same buffer to a final concentration of 500,000 yeast/μL.
7. Perform one round of selection against the A647-labeled SELEX0 dsDNA with each protein in each buffer: mix 5 μL SELEX0 dsDNA with 89 μL of BWB. Distribute 6 μL (three million) yeast bearing each protein to each well in the series of buffers (using a multichannel pipette for multiple proteins). Seal with tape and incubate for 30 min at room temperature with agitation. Wash six times in 150 μL corresponding BWB; remove supernatant by spinning plates with a swinging rotor centrifuge at 2,000 × *g* for 1 min, inverting forcefully over a

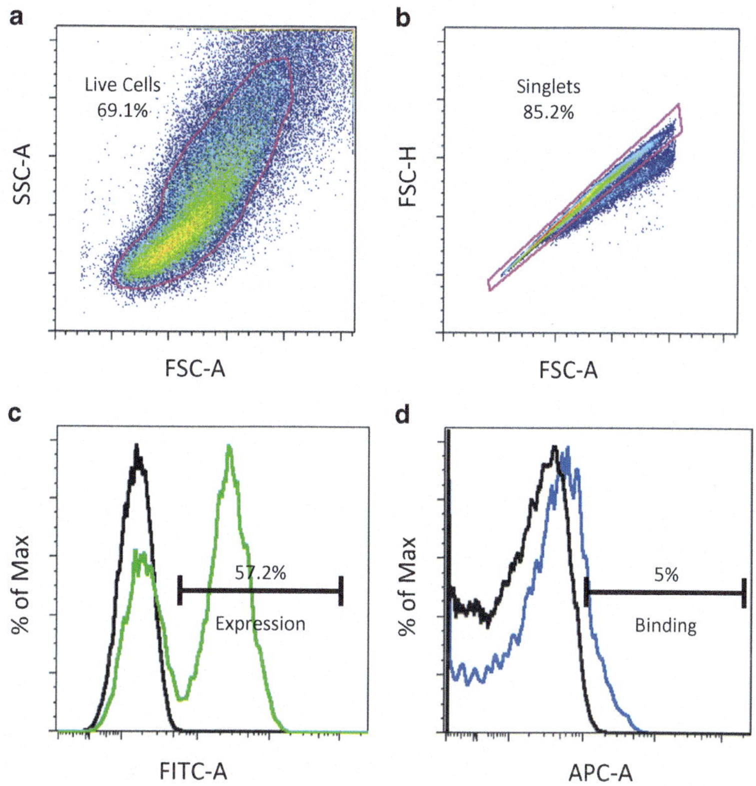

Fig. 3 The gating strategy for flow cytometric analysis of SELEX binding. Live cells are gated from the bulk population on FSC-A and SSC-A (**a**). From these, singlets are identified on their FSC-H and FSC-A (**b**). From this population, expressing cells (*green*) are gated on FITC-A by comparison to a negative (*black*) control (**c**). The expressing live singlets are viewed on an APC histogram to determine how much A647-labeled DNA is bound to the protein (**d**). The goal during initial buffer optimization is to identify a condition where approximately 5 % of expressing cells are marked (*blue*) above a negative control (*black*)

liquid-waste receptacle, and blotting forcefully on a stack of paper towels.

8. Resuspend the yeast in 100 μL BWB and acquire data on a BD Biosciences LSRII with HTS. Record FSC-A, FSC-H, SSC-A, SSC-H, APC, and FITC.
9. Analyze the flow cytometry data using FloJo. Gate live cells (FSC-A by SSC-A), then singlets (FSC-H by FSC-A), then cells staining for FITC (representing full length expression of the C′ terminus) compared to a negative control. Visualize this final subset as a histogram of APC fluorescence (binding) (Fig. 3).
10. For each protein, buffer that allowed a small, but detectable amount of binding (~5 %) is the buffer that allows the desired binding, which will be used in the next section; prepare a large volume (~500 mL) of this buffer.

Table 3
Example reaction volumes for a (*n*)SELEX PCR amplification step

Stock []		Final []		1× Volumes	8× Volumes	Unit, reagent
10	×	1	×	2.50	20.0	μL, PCR buffer
50	mM	1.5	mM	0.75	6.0	μL, MgSO4
10	mM	0.2	mM	0.50	4.0	μL, dNTPs
100	μM	0.67	μM	0.17	1.3	μL, Forward Primer
100	μM	0.67	μM	0.17	1.3	μL, Reverse Primer
				×	×	μL, Template DNA
5	U/μL	0.05	U/μL	0.25	2.0	μL, Polymerase
				QS to 25	QS to 200	μL, ddH$_2$O
				25.0	200.0	μL, Total

The amount of template DNA, and therefore amount of water added, is varied depending on the step's requirements

3.5 SELEX Protocol

1. Determine the density of induced yeast as in Subheading 3.3, **step 3**.
2. Transfer enough yeast to perform multiple rounds of SELEX for each protein you wish to assay—three million yeast per protein per round—to a V-bottom plate (*see* **Note 33**).
3. Wash twice with 200 μL BWB (with an amount of KCl determined in Subheading 3.4), and resuspend in a small amount BWB to a final concentration of 500,000 yeast/μL (*see* **Note 41**).
4. Perform the binding selection as in Subheading 3.4, **step 7**, using non-labeled DNA (using the same BWB as in **step 3**).
5. After the wash step, resuspend each sample in 40 μL 10 % buffer EB and seal the plate securely with sealing tape.
6. Release the DNA by heating the protein past its melting temperature to 70 °C for 10 min (*see* **Notes 42** and **43**). *See* **step 8** during the incubation.
7. *Immediately* spin the plate to pellet the yeast at 2,000 × *g* for 1 min, remove the tape, tilt the plate, and quickly transfer the supernatants (containing the selected DNA) to a 96-well plate for storage using a multichannel pipette. Seal and store the plate at −20 ° C when not in use.
8. Setup a PCR master mix, into which you will be adding 8.5 μL of the selected DNA (template) per sample (*see* Table 3).
9. Aliquot 16.5 μL of the PCR master mix to each well and add 8.5 μL of the selected DNA template to each well using a multichannel pipette.

Table 4
(n)SELEX PCR program

95 °C	5 min	
95 °C 59 °C 68 °C	10 s 15 s 15 s	Repeat (n − 1) times
68 °C	3 min	
4 °C	Hold	

The melting, annealing, and extension steps (items 2, 3, and 4) are repeated (n − 1) times, where n is varied throughout the protocol depending on the step's requirements

10. Seal the plate and run the (20×)SELEX program in a thermal cycler (*see* Table 4). When complete, transfer the product to empty wells in the storage plate (or a new plate) from **step 7**.
11. Setup a secondary PCR master mix, into which you will be adding 6.25 μL of the primary PCR (template) from the preceding step per sample (*see* Table 3) (*see* **Note 44**).
12. Aliquot 18.75 μL of the secondary PCR master mix to each well and add 6.25 μL of each primary PCR DNA template to each well using a multichannel pipette.
13. Seal the plate and run the (2×)SELEX program in a thermal cycler (*see* Table 4).
14. Repeat the steps in this section as necessary, starting at **step 4**, using 2 μL of the amplified DNA pool from the previous round in place of the 5 μL of SELEX0 pool (*see* **Notes 45** and **46**).

3.6 SELEX Analysis by Flow Cytometry

1. Setup a PCR master mix using 5′-A647 labeled reverse primer, into which you will be adding 0.5 μL of each DNA (template). Make enough PCR mix to make product for each round of SELEX for each sample (e.g., 8 samples × 5 rounds of SELEX = 40) (*see* **Note 47**) (*see* Table 3).
2. Aliquot 24.5 μL of the PCR master mix to each well and add 0.5 μL of the (20×)SELEX product of each DNA template to each well using a multichannel pipette (*see* **Note 48**).
3. Seal the plate and run the (6×)SELEX program in a thermal cycler (*see* Table 4) (*see* **Note 49**).
4. Setup a secondary PCR master mix using labeled reverse primer, into which you will be adding 6.25 μL of the amplified DNA (template) from the preceding step per sample (*see* Table 3) (*see* **Note 39**).
5. Aliquot 18.75 μL of the secondary PCR master mix to each well of a new 96-well PCR plate and add 6.25 μL of the primary PCR DNA template to each well using a multichannel pipette.

6. Seal the plate and run the (2×)SELEX program in a thermal cycler (*see* Table 4).
7. Digest excess single-stranded DNA with Exonuclease I: Add 2 units of ExoI to each 25 μL PCR reaction in 2 μL total volume of water (*see* **Notes 50** and **51**). Digest 2–18 h at 37 °C. This product can be stored at 4 °C overnight or at −20 °C for extended periods.
8. Load the hydrated sephadex G-100 suspension into the filter plate. For each 25 μL PCR reaction to be purified, add 500 μL total volume of suspension to a filter plate well. This is best accomplished by loading 320 μL sephadex suspension (using wide bore tips) into each necessary well of the filter plate, centrifuging briefly up to a speed of 500 × *g*, discarding water, and adding the remaining 180 μL sephadex (*see* **Note 52**). The plate should then be dehydrated by centrifugation at 2,000 × *g* for 7 min (*see* **Note 53**).
9. Load the 22 μL PCR+ExoI reaction directly to the center of each sephadex column. Secure a 96-well PCR plate below the filter plate to catch the purified flow-through, using tape if necessary. Centrifuge for 5 min at 2,000 × *g*. Approximately 15 μL of flow-through should be present in each recipient well.
10. Verify the presence of purified target site substrates using a NanoDrop spectrophotometer. Target concentration should be approximately 25 ng/μL.
11. Run target oligos on a 15 % polyacrylamide gel. For a 7.5-mL gel, combine 0.75 mL 10× TBE, 3.75 mL 30 % acrylamide-bis (19:1), and 2.9 mL water. Add 100 μL 10 % APS and 10 μL TEMED, then immediately mix and pipette into the prepared gel cassette. Once the gel is set, load 0.5–1.0 μL purified target, diluted with 1.0 μL 6× Ficoll loading buffer and 4 μL water (*see* **Note 8**). Use labeled SELEX0 dsDNA a size standard. Run the gel for 60 min at 120 V.
12. Visualize the gel on a Licor Odyssey infrared imager, using the 700 nM laser. The gel should show a prominent single PCR product and minimal contamination by other bands or leftover primers. Alternatively, the gel can be stained with 1× SybrGold in 1× TBE for 20 min, washed in 1× TBE or water, and visualized on a UV transilluminator (*see* **Note 54**).
13. Perform the staining and single round of binding, washing, and analysis as described in Subheading 3.4, using only the bind and wash buffer used during the SELEX rounds (you are testing each protein's binding of different SELEXn pools rather than each protein's binding of SELEX0 in different buffers). You will need to stain enough yeast expressing each protein to test binding fluorescent oligo from each round of SELEX for each protein (*see* **Note 55**).

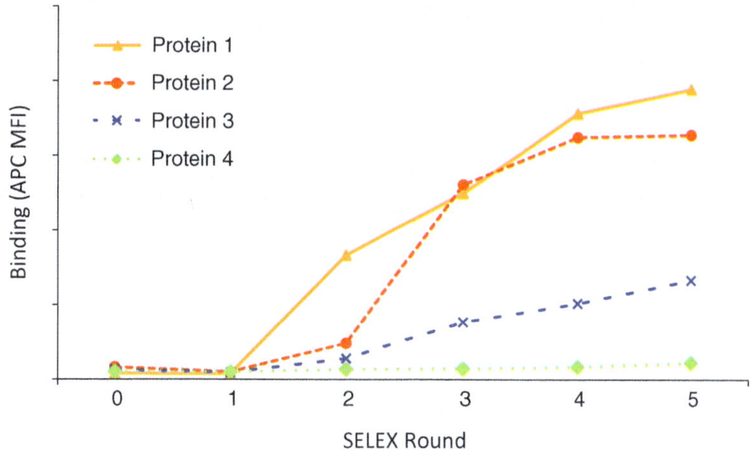

Fig. 4 Example of flow cytometric analysis of a SELEX experiment. This plot was obtained by measuring the median APC of expressing yeast stained with fluorescent oligo made from each SELEXn step. A successful SELEX experiment should yield increasingly higher affinity oligos as progress is made from one round to the next, as seen with Protein 1, 2, and 3. The lower levels of binding for Protein 3 may not be a problem; different proteins may have lower overall binding affinities. Only an absence of binding represents a failed experiment, as with Protein 4

14. Export median fluorescence intensity of the APC channel of expressing [FITC+] cells and plot these versus SELEX round (Fig. 4) (*see* **Note 56**).

3.7 Sequencing and Analysis

1. Clone (2×)SELEX PCR product from the desired round(s) into a vector using your kit of choice (*see* **Note 57**). Transform and plate *E. coli* on selective media according to your kit's instructions.
2. Prepare sequencing reactions for 24 colonies per sample for sequencing according to your service provider's instructions (*see* **Notes 58** and **59**).
3. Trim out constant regions of the SELEX sequences using your software of choice, so that only the randomized regions are analyzed (*see* **Note 60**).
4. Align trimmed sequences using MEME (Fig. 5) (*see* **Note 61**).
5. Using the binding and pilot sequencing data, select proper SELEX*n* samples to be analyzed in depth by high-throughput sequencing (e.g., SELEX3–5 for each protein) (*see* **Note 16**).
6. Prepare a corresponding volume of PCR master mix containing everything except the 1.67 μL forward (barcoded, unique) primers and 0.5 μL (20×)SELEX template. Distribute master mix to each well of a thin-wall 96-well PCR plate (*see* Table 3).

a

Name	Strand	Start	*p*-value	Sites
9T5D0:00793:02235	+	6	5.00e-10	AATTA TCCTCATATACCTTATC GGCTTACC
9T5D0:00462:02164	-	6	5.00e-10	AACCGCTC TCCTCATATACCTTATC TGTAC
9T5D0:02346:00937	-	12	5.00e-10	TG TCCTCACATACCTTATC ACGTTCTTCG
9T5D0:03385:02018	-	13	5.00e-10	A TCCTCATATACCTTATC ATCCCAAGTA
9T5D0:03241:02275	+	2	6.29e-10	A TCCTCATAAGCCTTATC GCGTCGGCAA
9T5D0:00858:01924	+	8	6.29e-10	AGGAGGC TCCTCACAAGCCTTATC CGTTGC
9T5D0:03231:02265	+	8	6.29e-10	AGGAGGC TCCTCACAAGCCTTATC CGTTGC
9T5D0:00125:01076	-	3	8.45e-10	CACAGAATCG TCCTCATACGCCTTATC AC
9T5D0:01572:01914	+	8	8.45e-10	AAGGTTA TCCTCACACGCCTTATC GCGCGG
9T5D0:03304:02633	+	13	9.73e-10	ACTTACCTTA TCCTCATAGACCTTATC CC
9T5D0:01350:03189	-	2	1.16e-09	CCGGGGATGA TCCTCACAATCCTTATC T
9T5D0:00636:02759	-	9	1.56e-09	CCGCCC TCCTCATACTCCTTATC ATCGTATA

b

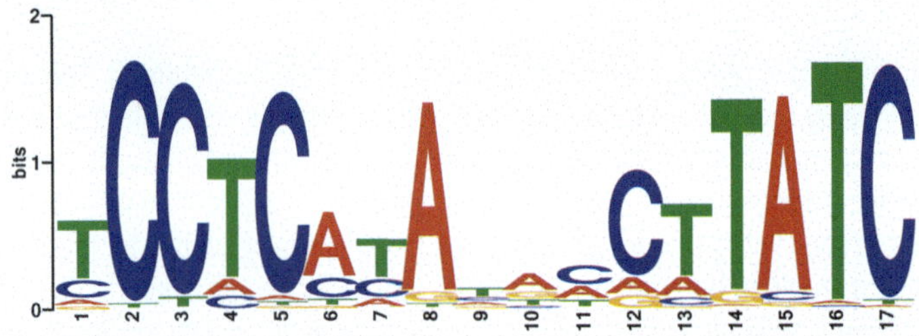

c

-11	-10	-9	-8	-7	-6	-5	-4	-3	-2	-1	1	2	3	4	5	6	7	8	9	10	11
G	C	T	C	C	T	C	A	T	A	A	T	C	C	T	T	A	T	C	A	A	G

Fig. 5 SELEX sequence output. Trimmed sequence alignments (**a**), and position weight matrices and corresponding sequence logos (**b**) are produced by MEME. For comparison, the known native target site for a representative homing endonuclease is shown alongside the sequence logo generated from a SELEX experiment (**c**)

7. Use a multichannel pipette to add 0.5 μL of the template DNA and 1.67 μL of the unique forward primer to each well (*see* **Note 43**).
8. Seal the plate and run the (6×)SELEX program in a thermal cycler (*see* Table 4) (*see* **Note 44**).
9. Setup a secondary PCR master mix, into which you will be adding 6.25 μL of the primary PCR (template) from the preceding step per sample (*see* Table 3) (*see* **Note 39**).
10. Aliquot 18.75 μL of the PCR master mix to each well and add 6.25 μL of the selected DNA template to each well using a multichannel pipette.
11. Seal the plate and run the (2×)SELEX program in a thermal cycler, excluding the 59 °C annealing step (*see* **Note 62**) (*see* Table 4).
12. Pool the PCR products and prepare them for sequencing according to your service provider's instructions, and submit them for sequencing (*see* **Note 63**).
13. Trim and align sequences as above.
14. Once consensus sequences are found from one or both of the sequencing methods outlined above, validate the results by binding or cleavage (*see* **Note 64**).

4 Notes

1. If different primer sequences than those in Table 1 need to be used, carefully select new ones with optimal PCR properties and re-optimize the PCR cycles and conditions [22–24].
2. We have found good quality and randomization in the SELEX0 pool when using Integrated DNA Technology (IDT) oligos with the randomized bases hand-mixed at 25 % each.
3. Do not use other types of purification for the randomized oligo. Purification other than desalting can bias the pool.
4. Keep the number of randomized bases higher than the expected size of the target (by ~50 %); it allows more than one recognition site to be present on each oligo and reduces the number of oligos required for complete coverage (or effectively increases the pool diversity). Furthermore, you may find unexpected, extended sequence specificities.
5. It is imperative that the reverse—and not forward—primer be used in second-strand synthesis. Use of the incorrect primer will fail to generate a double-stranded SELEX0 pool.
6. Either a 2× PCR master mix or a multicomponent PCR kit can be used (e.g., PCR Master from Roche or Platinum Taq DNA Polymerase High Fidelity from Invitrogen), but the corresponding extension temperature suit the chosen polymerase.
7. 10× Tris–acetate–EDTA (TAE) buffer can be used in place of 10× TBE buffer. If this substitution is made, be sure to use 1× TAE as the gel running buffer in place of 1× TBE.
8. While it is more difficult to load samples without dye in the loading buffer, colored dyes commonly fluoresce under the UV or Licor excitation and confound the image.
9. If autoclaving, add the glucose *after* autoclaving. To increase the shelf life of this media, make a 50× stock of the adenine hemisulfate and add it at 1× concentration just prior to use. Store the 50× stock at −20 °C (upon thawing, there will be a small amount of precipitation which will not go back into solution).
10. pETCON vector for LHE surface display: Gal1-10, galactose inducible promoter; HA, hemagglutinin epitope; 3×G4S, thrice repeated gly–gly–gly–gly–ser linker sequence; NdeI restriction site; protein coding sequence (insert); XhoI restriction site; gly–gly–gly–ser linker sequence; Myc, c-Myc epitope. Ampicillin resistance (AmpR) is included for selection in *E. coli*, TRP1 marker used for auxotrophic selection in *S cerevisiae*. pCT-CON2 vector can also be used in place of pETCON. For complete sequence and a vector map, and information on obtaining the vector, see Addgene.org, plasmid #41522 (empty vector)

and plasmid #41523 (I-LtrI homing endonuclease insert for positive control).

11. Induction from a transformed colony or frozen stock requires 48–72 h, depending on whether yeast is cultured in rich media before growth in raffinose and subsequent and galactose induction. Keep this in mind when planning your experiments.
12. Optimal selection buffer components and temperature will be dependent on the family of proteins. A buffer which closely mimics its host's intracellular conditions will likely be a good starting point, such as the buffer provided here which has a final KCl concentration of 150 mM. Some groups have used PBS or DPBS (without $MgCl_2$ or $CaCl_2$) as a base, supplementing with BSA and required binding co-factors [25]. A reasonable range of added KCl to test might be between 75 and 225 mM.
13. Solution should be adjusted to pH 7.5 using KOH, thereby limiting the introduction of additional sodium ions.
14. Some SELEX protocols will call for the use of nonspecific DNA, or DNA analogues such as poly(dI-dC) [25], but our lab has had the best luck by titrating KCl to reduce nonspecific binding (for homing endonucleases). What works best will likely be dependent on the family of proteins at hand.
15. LAGLIDADG homing endonucleases will bind but not cleave their DNA target in the presence of Ca^{++}, as opposed to Mg^{++} which allows both binding and cleavage [26]. A cofactor that has similar properties for your protein of interest is necessary, as a cleavage event will result in the loss of the proper sequences from the pool.
16. Immunology Consultants Laboratory Inc. provides an antibody which has been validated by our lab. Not all αMyc-FITC antibodies provide robust staining in this assay, and any alternates should be tested before use.
17. Invitrogen provides Platinum® Taq DNA Polymerase High Fidelity, which has been validated by our lab for dsOligo synthesis. Depending on the template, some other polymerases (e.g., standard Taq) can produce elevated levels of background PCR product. Alternative polymerases should be validated before use.
18. Our lab uses filter plates with hydrophilic PVDF membranes compatible with nucleic acid purification. The pore size should be no larger than 25 % of the size of the Sephadex beads (roughly 5 μm).
19. Our lab uses GE Healthcare's illustra Sephadex G-100 DNA Grade SF. If an alternative product is chosen, be sure to determine the optimal speed and duration of centrifugation required to obtain optimal purification for the desired dsOligo size.

20. We recommend validating the success of the SELEX experiment by sequencing and aligning a small number of oligos from the pool using Sanger sequencing before subjecting the entire pool to (relatively expensive) high-throughput sequencing. When dealing with large numbers of samples, however, it becomes very valuable to barcode, pool, and sequence the samples in parallel using high-throughput sequencing (after initial validation of a few samples).
21. Obtaining sequences via high-throughput sequencing may not be necessary when assaying only one or a few proteins, or if only a few sequences are needed to obtain a (limited) dataset.
22. If high-throughput sequencing primers are needed, they should include the primer sequences shown in Table 1 with any barcodes appended to the 5′ end of the SELEX forward primer, and provider-specific primer sequences appended to the extreme 5′ ends of both the forward and reverse SELEX primers.
23. MEME provides simple graphical interface, Web access for small data sets, and an easy way to search for consensus motifs in a set of sequences [27]. We use the following settings with good results: "-nostatus -time 7200 -maxsize 60000 -mod zoops -nmotifs 1 -minw 18 -maxw 22 -minsites 5 -revcomp," where nmotifs is the number of motifs to find in one data set, minw/maxw is the minimum/maximum width of the consensus sequence (for homing endonucleases this is a good starting point), and minsites is the minimum number of sequences to align to obtain the consensus; this number should be approximately half the total number of sequences processed per sample.
24. We recommend creating a number of small (100 μL) aliquots of the SELEX0 pool, such that many experiments can be performed using the same batch of oligo. Creating eight such aliquots would provide enough starting material to perform 160 single SELEX experiments.
25. Note the substantially higher amount of template, primer, and dNTPs used in this PCR. Normal PCR master mixes will not contain this concentration of dNTPs, and will need to be supplemented to obtain the proper final concentration.
26. You may want to make a number of gels at once so that product from various stages throughout the protocol can be analyzed. Extra gels can be wrapped in wet paper towels and stored in a zip lock bag at 4 °C for many weeks.
27. An ethidium bromide staining solution will also allow visualization of the gel on a UV transilluminator, but the fluorescence of the DNA bands will not be as intense.
28. Frozen competent cells are prepared according to the published protocol by Gietz and Schiestl [28], and will require

approximately 48 h. Add 2.5 × 109 EBY100 cells from an overnight 2×YPAD culture to 500 mL fresh 2×YPAD media. Grow at 30 °C to a density of at least 20 million/mL. Pellet the cells and wash with sterile water. Resuspend the washed cell pellet in 5 mL of 5 % v/v glycerol + 10 % v/v DMSO in water. Aliquot 50-μL volumes to microcentrifuge tubes. Pack the tubes into a styrofoam rack with lid (or similar form of insulation) and place at −80 °C. (The insulation allows for gradual freezing of the cells.)

29. When transforming a library of variant homing endonucleases, increase the number of yeast and volume of the transformation mixture according to Gietz and Schiestl [29].
30. When transforming high quality plasmid DNA, a single frozen aliquot of competent yeast cells in the described volume of transformation mixture can be used for multiple reactions. In this case, divide the resuspended cells (*prior* to addition of DNA) into up to 15 equal volumes and add up to 1 μL total volume of plasmid DNA to each aliquot. Proceed to the incubation step.
31. We have found that an incubation time of 40–42 min at 42 °C provides the highest transformation efficiency with lowest cell death. Longer incubation times can lead to significant cell death. If using a high-quality plasmid, a shorter incubation time of 20 min will suffice for the generation of transformed clones.
32. Raffinose cultures can be started using a single colony from a selective media + glucose plate. Alternatively, we have found that an initial overnight incubation in YPAD media (at 30 °C with 250 RPM shaking) can substantially increase induction efficiency, and the absence of selective media at this stage does not result in significant plasmid loss.
33. Cultures can also be grown in deep-well 96-well plates to will also facilitate storage and re-inoculation if you are working with many samples at once. To setup this format, first inoculate 80 μL YPAD per well in costar 96-well V-bottom plates and allow to grow for 2 days at 30 ° C in a moist environment (e.g., in an breathable, lidded container with moistened paper towels) until the yeast have reached a maximum density; this serves to normalize inoculation densities for new cultures. From this stock, start 500 μL cultures in deep-well 96-well V-bottom plates.
34. After growing (and not inducing) cultures in YPAD or SC media, yeast can be frozen down for long-term storage in 15 % glycerol/media v/v at −80 °C. Subsequently start cultures by inoculating a new "normalizing YPAD culture plate" (*see* **Note 28**) or culture tube by scraping frozen culture with a pipette

tip (or multichannel pipette, or plastic multi-well inoculator) and submerging in media.

35. When using vertical tube racks inside a shaking incubator, position the 15-mL culture tubes at a slant to allow for maximum aeration, and do not use more than 1.5 mL of media. If using deep-well plates, increase shaking speed to 300–320 RPM.
36. On our spectrophotometer, the density of a yeast culture can be estimated by mixing a 1:10 dilution of yeast in water and measuring the resulting OD_{600}. A simple calculation of $OD_{600} \times 300$ provides an estimated value for density of the culture in millions of cells per mL. The validity of this estimate should be checked when using a different instrument.
37. Care should be taken to wash yeast from the raffinose culture at least twice before transferring to the galactose media. This limits carry-over of raffinose and/or glucose.
38. Depending on your protein and your desired final pool diversity, greater than three rounds of SELEX may be needed to narrow the randomized pool sufficiently. If yeast are grown in plate-format, growing yeast of the same sample in multiple wells may be necessary to obtain enough yeast to complete all rounds of the experiment (3 million yeast/round × 5 rounds requires 15 million yeast; with 10 million per well, two wells of each sample would be needed to have enough yeast).
39. A typical day will allow enough time for about three rounds of SELEX. Many enzymes are stable enough on the surface of yeast to be stored at 4 ° C overnight, some for several days. If this is not the case for your protein, be sure to induce yeast in a staggered fashion so fresh yeast are available for each day of selection or analysis. Induced cultures should be kept on ice or at 4 °C following the galactose induction in their induction media.
40. If possible, include a protein of known functionality in the same family as the protein in question. This will help approximate binding conditions and validate the success of a SELEX experiment.
41. After three rounds of SELEX it may be beneficial to increase the selective pressure by slightly increasing the salt concentration (or other binding-mitigation reagent). Keeping binding sites on the yeast surface unsaturated is pivotal in achieving maximum selective pressure, similar to our goal when finding optimal binding conditions at the offset. The level of saturation should be apparent after analysis of the SELEX pool's fluorescent oligo (*see* Subheading 3.6). Keep in mind, however, that increasing selective pressure also diminishes the ability of a "correct" sequence from binding, too; aim to allow higher binding in this stage than during the SELEX0 pool optimization.

Our lab frequently performs round 4 and 5 in 175 mM KCl (+25 mM compared to round 1–3), depending on the stringency and subsequent final diversity desired. This is a similar concept employed in fixed-stringency SELEX [14].

42. Most LAGLIDADG homing endonucleases melt below this temperature. Chemical release of the entire Aga2P fusion protein from the surface of the yeast may be necessary for proteins with high melting temperatures. This can be accomplished by incubating for 30 min at 37 °C in a 5 mM dithiothreitol BWB solution, and purifying the DNA from the protein by standard methods [16].
43. Sealed costar plates fit conveniently atop the wells of most heated-lid thermal cyclers, while still allowing the lid to close.
44. The secondary (2×)SELEX PCR (after the primary (20×) SELEX amplification PCR) is absolutely required in order to ensure that every oligo is double-stranded and not mismatched—as might be caused after a PCR melting step if no extension resources remain. Single-stranded and mismatched oligos cannot be properly bound in subsequent steps and will cause the experiment to fail.
45. In our experience, three rounds of SELEX are sufficient to obtain a diverse collection of medium to medium-high-affinity binding targets. Up to two additional rounds is typically enough to reduce this pool to just a few high-affinity targets. Greater or fewer rounds may be required for your protein of interest, and after five rounds analysis of SELEX progress is recommended using protocols described in Subheading 3.6.
46. Run a polyacrylamide gel of the (20×)SELEX products from each round after the first two to three rounds of SELEX to ensure that you are getting proper amplification and seeding for the next pool. There should be a prominent band the same size as the dsSELEX0 pool (and sometimes a single band a few base pairs larger whose origin remains speculative, but has no negative impact on the experiment), and no accumulation of other products.
47. Maintain 20–25 μL PCR reaction volumes; larger volumes will overload the sephadex in the purification step. Purified fluorescent oligo can be stored at −80 °C in a light-protected container.
48. (20×)SELEX PCR product is used as the template because it does not contain leftover non-labeled SELEX primers present at the end of a (2×)SELEX program. These primers would otherwise be incorporated in place of the proper primers and contaminate the product.
49. Given the high concentration of starting template (PCR product), only a small number of cycles are required to use

all of the starting material and generate sufficient product. Simultaneously, 6 cycles is enough to sufficiently dilute out the starting template 64-fold.

50. ExoI is diluted with water to a total volume of 2 μL per sample for ease and accuracy of transfer to the PCR reaction. Do not add any of the supplied ExoI buffer.

51. ExoI digest and subsequent sephadex purification is not strictly required if the only when using the product to assay binding; purification will reduce background caused by nonspecific binding of labeled, unincorporated primer. However, these steps *are* required if the oligo is to be used in any flow-based cleavage assays [8, 16]. If subjecting a diverse pool of targets (such as one generated by SELEX) to cleavage analysis, a positive signal can be regarded as the signature of a successful SELEX experiment. Conversely, a negative result may not be indicative of a failed selection: cleavable substrate may not be abundant enough to detect, even though alignment of the sequences may provide the proper, cleavable consensus sequence (*see* **Note 57**).

52. Special care should be taken to avoid any bubbles in the sephadex suspension when mixing or aliquoting to the filter plate. Bubbles will lead to cracks within the final, centrifuged sephadex columns, and cracked columns should be discarded. We find that careful pipetting, using wide bore tips or standard p200 tips cut at approximately the 50 μL gradation, reduces frequency of cracking. We also find that allowing 30 min between pipetting and centrifugation can significantly reduce column cracking.

53. Duration of spin should be modified if your oligo length differs from the one provided here. Longer oligos may not need to be centrifuged for as long, and vice versa. Determine a spin time for your oligo which eliminates smaller products, but allows most of your product to be recovered.

54. Imaging the gel using the infrared imager (exciting the fluorophore) allows a more sensitive readout of leftover primers compared to a UV transilluminator.

55. The volume of the binding analysis can be scaled down by a factor of 4 compared to the selection step; only a sampling of the pool is necessary. Use a total volume of 25 μL, 0.5 μL dsDNA, and 750,000 yeast. This experiment can also be performed in a 384-well plate given the reduced volumes.

56. Successful selection and enrichment of high-affinity binders is noted by an upward trend in binding (median fluorescence intensity of the APC channel of expressing [FITC+] cells) as progress is made from one SELEX round to the next.

57. When working with a large number of samples, pick a few representative samples for pilot analysis. Applying the SAGE adaptation and/or barcoding at this point can increase the throughput of Sanger sequencing [18].
58. We find that as few as 8 good sequences are often enough to find a consensus motif, depending on the diversity of the pool; sequences from pools as diverse as SELEX3 have demonstrated behavior. Sequences from pools prior to SELEX3 are difficult to align, as pools at this point have rarely converged enough.
59. Direct "colony sequencing" can increase throughput if the copy number of your vector is high enough and your sequencing facility allows it (compared to preparing and sequencing purified vector). If not, preparing sequencing template by colony PCR and Exo/SAP treatment (instead of vector purification), can also be used to increase throughput. Colony PCR or sequencing use a bacterial colony picked into 6 μL of ddH_2O and boiled in lieu of purified template.
60. Any constant region (including the flanking AAA or TTT not found in the primer, but present in the template) that is included in an alignment will tend to "lock in" the alignment of those constant regions and taint the analysis.
61. Beware of proteins whose motifs overlap with the constant regions flanking the randomized bases of the SELEX oligos. Such proteins will likely bind to the constant region and overlap only partially with the randomized region. Alignments will therefore be missing part of the motif and include an unknown length of the constant region since the constant regions are removed during analysis. A telltale sign of this situation will be a bias in the location of the motif within the randomized region toward one end (i.e., not located at random internal positions). In this case you may need to select different constant regions for your SELEX oligos and repeat the procedure, depending on your application.
62. The entire high-throughput-sequencing primer *and* SELEX primer regions will bind to the amplified template, removing the need for a low-temperature annealing step; Its exclusion will yield higher specificity in the secondary PCR.
63. Preparation of your pool will likely involve purification of the PCR product using gel purification and a cleanup kit (e.g., Qiaex II). Given the diversity of the pool, the band may appear to have a slight smear on an agarose gel. Be generous with your excision, taking a large swath above the bright band, as exclusion of the smear in its entirety can bias your results.

64. Depending on the protein, searching for the consensus sequence(s) in the protein's host genome by using MAST can validate your SELEX results. MAST is part of the MEME package and searches for the motif identified by MEME in a given (set of) sequence(s). In the case of homing endonucleases, finding the consensus sequence within the homologous host gene(s) without the insert can serve to validate the results [30]. Clearly this will not always be an option, but may help validate some results nonetheless, especially for some other families of proteins. It will also be useful to synthesize the consensus sequence and perform binding or cleavage experiments [26].

References

1. Thermes V, Grabher C, Ristoratore F et al (2002) I-SceI meganuclease mediates highly efficient transgenesis in fish. Mech Dev 118:91–98
2. Gouble A, Smith J, Bruneau S et al (2006) Efficient in toto targeted recombination in mouse liver by meganuclease-induced double-strand break. J Gene Med 8:616–622
3. Arnould S, Perez C, Cabaniols J-P et al (2007) Engineered I-CreI derivatives cleaving sequences from the human XPC gene can induce highly efficient gene correction in mammalian cells. J Mol Biol 371:49–65
4. Gao H, Smith J, Yang M et al (2010) Heritable targeted mutagenesis in maize using a designed endonuclease. Plant J 61:176–187
5. Windbichler N, Papathanos PA, Catteruccia F et al (2007) Homing endonuclease mediated gene targeting in Anopheles gambiae cells and embryos. Nucleic Acids Res 35:5922–5933
6. Smih F, Rouet P, Romanienko PJ et al (1995) Double-strand breaks at the target locus stimulate gene targeting in embryonic stem cells. Nucleic Acids Res 23:5012–5019
7. Smith J, Grizot S, Arnould S et al (2006) A combinatorial approach to create artificial homing endonucleases cleaving chosen sequences. Nucleic Acids Res 34:e149
8. Jacoby K, Metzger M, Shen BW et al (2012) Expanding LAGLIDADG endonuclease scaffold diversity by rapidly surveying evolutionary sequence space. Nucleic Acids Res 40:4954–4964
9. Takeuchi R, Lambert AR, Mak AN-S et al (2011) Tapping natural reservoirs of homing endonucleases for targeted gene modification. Proc Natl Acad Sci U S A 108:13077–13082
10. Heath PJ, Stephens KM, Monnat RJ et al (1997) The structure of I CreI, a group I intron-encoded homing endonuclease. Nat Struct Biol 4:468–476
11. Duan X, Gimble FS, Quiocho FA (1997) Crystal structure of PI-SceI, a homing endonuclease with protein splicing activity. Cell 89: 555–564
12. Jurica MS, Stoddard BL (1999) Homing endonucleases: structure, function and evolution. Cell Mol Life Sci 55:1304–1326
13. Pepper LR, Cho YK, Boder ET et al (2008) A decade of yeast surface display technology: where are we now? Comb Chem High Throughput Screen 11:127–134
14. Djordjevic M (2007) SELEX experiments: New prospects, applications and data analysis in inferring regulatory pathways. Biomol Eng 24: 179–189
15. Baxter SK, Lambert AR, Scharenberg AM et al (2013) Flow cytometric assays for interrogating LAGLIDADG homing endonuclease DNA binding and cleavage properties, Methods Mol Biol(Clifton, NJ) 978:45–61
16. Jarjour J, West-Foyle H, Certo MT et al (2009) High-resolution profiling of homing endonuclease binding and catalytic specificity using yeast surface display. Nucleic Acids Res 37: 6871–6880
17. Jolma A, Kivioja T, Toivonen J et al (2010) Multiplexed massively parallel SELEX for characterization of human transcription factor binding specificities. Genome Res 20:861–873
18. Roulet E, Busso S, Camargo AA et al (2002) High-throughput SELEX-SAGE method for quantitative modeling of transcription-factor binding sites. Nat Biotech 20:831–835
19. Lorenz C, von Pelchrzim F, Schroeder R (2006) Genomic systematic evolution of ligands by exponential enrichment (Genomic SELEX) for the identification of protein-binding RNAs independent of their expression levels. Nat Protoc 1:2204–2212

20. Petek LM, Russell DW, Miller DG (2010) Frequent endonuclease cleavage at off-target locations in vivo. Mol Ther 18:983–986
21. Thyme SB, Jarjour J, Takeuchi R et al (2009) Exploitation of binding energy for catalysis and design. Nature 461:1300–1304
22. Piasecki SK, Hall B, Ellington AD (2009) Nucleic acid pool preparation and characterization. Methods Mol Biol (Clifton, NJ) 535:3–18
23. Hall B, Micheletti JM, Satya P et al (2009) Design, synthesis, and amplification of DNA pools for in vitro selection. In: Frederick M, Ausubel et al (eds). Current protocols in molecular biology. Chapter 24, Unit 24.2
24. Untergasser A, Cutcutache I, Koressaar T et al (2012) Primer3 – new capabilities and interfaces. Nucleic Acids Res 40:e115
25. Miller JC, Tan S, Qiao G et al (2011) A TALE nuclease architecture for efficient genome editing. Nat Biotechnol 29:143–148
26. Volná P, Jarjour J, Baxter S et al (2007) Flow cytometric analysis of DNA binding and cleavage by cell surface-displayed homing endonucleases. Nucleic Acids Res 35:2748–2758
27. Bailey TL, Boden M, Buske FA et al (2009) MEME SUITE: tools for motif discovery and searching. Nucleic Acids Res 37: W202–W208
28. Gietz RD, Schiestl RH (2007) Frozen competent yeast cells that can be transformed with high efficiency using the LiAc/SS carrier DNA/PEG method. Nat Protoc 2:1–4
29. Gietz RD, Schiestl RH (2007) Large-scale high-efficiency yeast transformation using the LiAc/SS carrier DNA/PEG method. Nat Protoc 2:38–41
30. Gogarten JP, Hilario E (2006) Inteins, introns, and homing endonucleases: recent revelations about the life cycle of parasitic genetic elements. BMC Evol Biol 6:94

Chapter 14

Engineering and Flow-Cytometric Analysis of Chimeric LAGLIDADG Homing Endonucleases from Homologous I-OnuI-Family Enzymes

Sarah K. Baxter, Andrew M. Scharenberg, and Abigail R. Lambert

Abstract

LAGLIDADG homing endonucleases (LHEs) are valuable tools for genome engineering, and our ability to alter LHE target site specificity is rapidly evolving. However, widespread use of these enzymes is limited due to the small number of available engineering scaffolds, each requiring extensive redesign to target widely varying DNA sequences. Here, we describe a technique for the chimerization of homologous I-OnuI family LHEs. Chimerization greatly expands the pool of unique starting scaffolds, thereby enabling more effective and efficient LHE redesign. I-OnuI family enzymes are divided into N- and C-terminal halves based on sequence alignments, and then combinatorially rejoined with a hybrid linker. The resulting chimeric enzymes are expressed on the surface of yeast where stability, DNA binding affinity, and cleavage activity can be assayed by flow cytometry.

Key words Homing endonuclease, Meganuclease, Chimera, Chimerization, Protein engineering, Yeast surface display, Flow cytometry, LAGLIDADG, Assembly PCR

1 Introduction

Significant advances have been made in engineering LAGLIDADG homing endonucleases (LHEs) to cleave novel DNA target sequences, but large-scale redesign remains a challenging task. Fortunately, the need for extensive engineering can be avoided when an existing native enzyme closely matches the final DNA target sequence desired. Expansion of the pool of available LHE design scaffolds may be the fastest and simplest means to overcome difficult engineering tasks. The recent identification of new enzymes from the monomeric I-OnuI family [1] provides a number

David R. Edgell (ed.), *Homing Endonucleases: Methods and Protocols*, Methods in Molecular Biology, vol. 1123, DOI 10.1007/978-1-62703-968-0_14, © Springer Science+Business Media New York 2014

of unique engineering scaffolds, but the list of target sites recognized by native enzymes is still dwarfed by the total number of target sites possible.

The dimeric/pseudodimeric architecture of LHEs provides an opportunity for expansion of our existing scaffold collection through the creation of chimeric enzymes (increasing the number of available scaffolds from N to N^2). The feasibility of mixing the N- and C-terminal domains of various native LHEs was initially verified through the successful generation of an active chimera from the monomeric LHE I-DmoI and the homodimeric I-CreI [2, 3]. The structural dissimilarity of these two parent enzymes made necessary both computational modeling and in vitro selection to repack the chimeric interface. Based on this work, it was hypothesized and confirmed that the generation of chimeras from within a family of highly homologous monomeric LHEs could produce a large number of active enzymes with minimal engineering [4].

Here we describe an efficient and structure-independent method for the generation and analysis of chimeric LHEs from I-OnuI family homologues (Fig. 1). Break points defining the N- and C-terminal halves of each parent enzyme are determined through sequence alignment to the well-characterized I-OnuI endonuclease [1]. These N- and C-terminal halves are then recombined and joined with an artificial linker. In our experience, the choice of protein linker can have a significant effect on the stability and activity of the resulting chimera, so for the sake of simplicity and overall compatibility, we suggest using a "half-and-half" hybrid linker [4]. This strategy incorporates residues from both native parents to maintain important interactions with each respective domain, and introduces a flexible and minimally invasive tri-residue bridge at the point of fusion. The resulting chimeras are expressed on the surface of yeast under a galactose-inducible promoter, and flow cytometry is used to assess a variety of enzyme properties [5, 6]. Antibody staining of a C-terminal Myc epitope tag reports the presence of stable, full-length enzyme on the surface of yeast

Fig. 1 (continued) assembly PCR (Subheading 3.2). The pETCON yeast surface expression vector is digested to prepare open vector for cloning (Subheading 3.3), and the parent endonuclease sequences are digested into N- and C-terminal halves with the help of a restriction site in the SGT linker (Subheading 3.4). The endonuclease domains are then ligated into the open pETCON vector in the desired combinations (Subheading 3.5), and transformed into yeast for expression on the yeast surface (Subheadings 3.8 and 3.9). Finally, the stability and DNA binding/cleavage activities of the surface-expressed chimeras are assessed using flow cytometry (Subheadings 3.10 and 3.11)

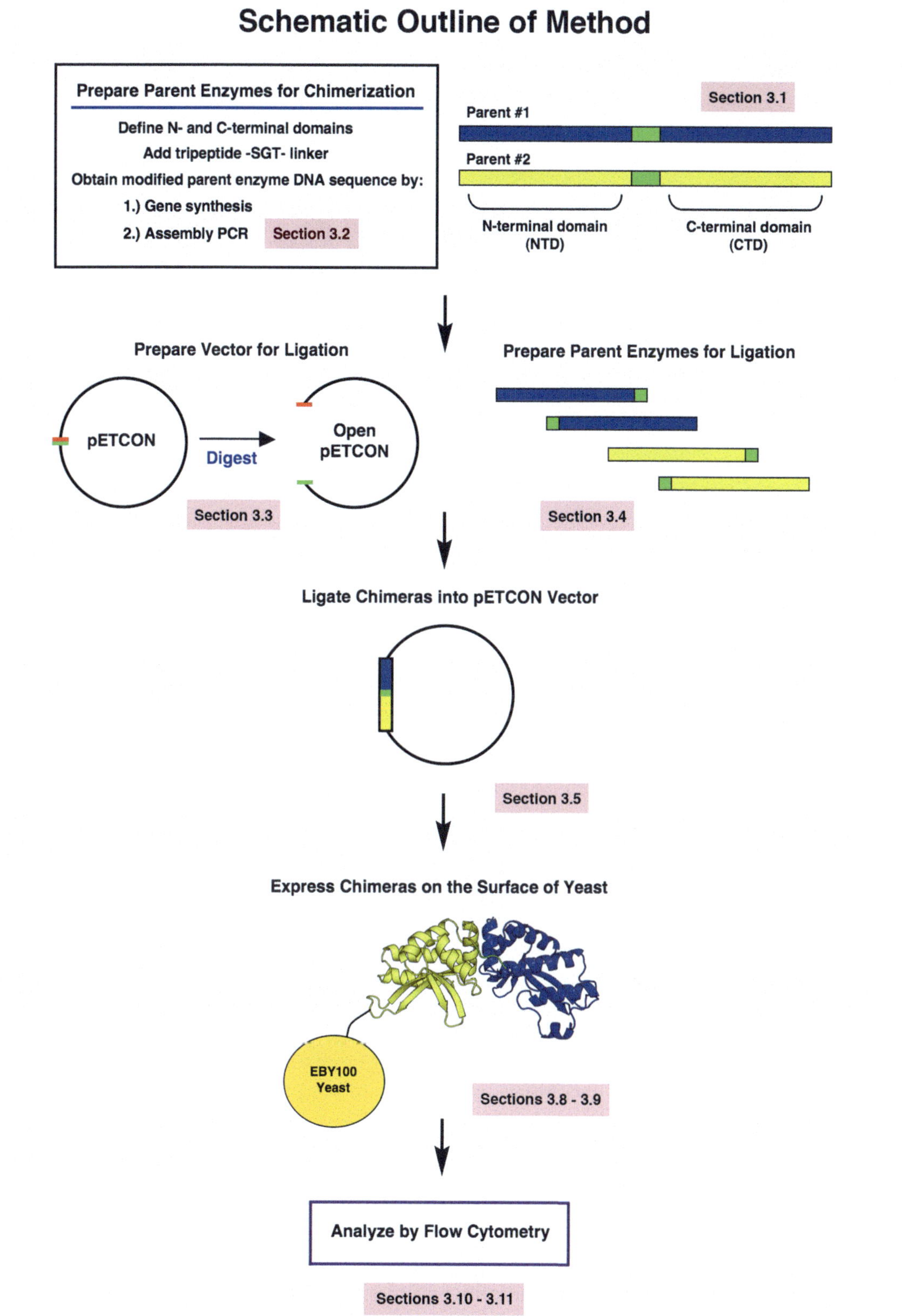

Fig. 1 Schematic outline of chimeric endonuclease generation. Sequence alignments are used to divide parent endonucleases into N- and C-terminal domains, and an -SGT- tripeptide linker is substituted at the center of the construct (Subheading 3.1). The resulting sequences are ordered as synthesized genes or generated by

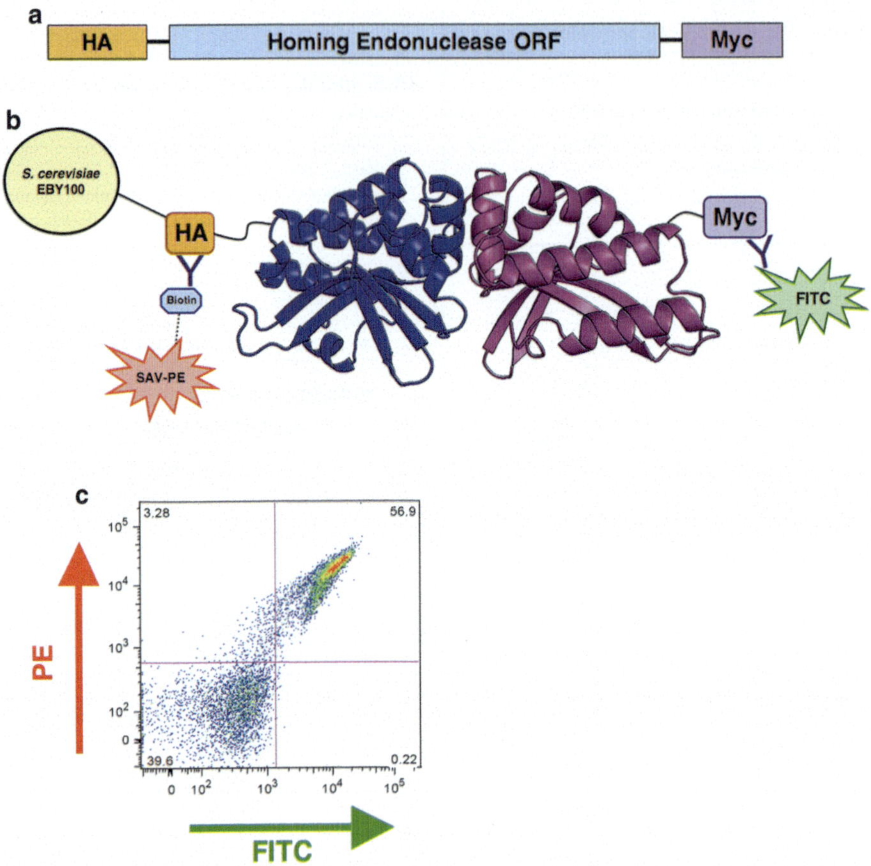

Fig. 2 Protein expression on the surface of yeast. (**a**) The pETCON vector expresses a cloned sequence on the surface of yeast with an N-terminal hemagglutinin (HA) tag and a C-terminal Myc tag. (**b**) Fluorescent antibody staining of the N- and C-terminal epitope tags allows for detection of stable, full-length protein. The N-terminal HA tag is stained with streptavidin–phycoerythrin (SAV-PE) via a biotin–streptavidin bridge, and the C-terminal Myc tag is stained with fluorescein isothiocyanate (FITC) conjugated anti-Myc antibody. (**c**) Sample FlowJo plot of surface expression staining. The *y*-axis measures signal from N-terminal biotin–streptavidin-linked PE fluorophore, while the *x*-axis measures signal from the C-terminal FITC fluorophore. Dual-stained cells in *the upper right-hand quadrant* represent stable surface expression of full-length protein

(Fig. 2), and introduction of fluorescently labeled DNA substrates allows for the rapid evaluation of both DNA binding and cleavage activities (Fig. 3). The activity observed by flow cytometry is validated by a complementary non-tethered in vitro cleavage assay (Fig. 4). Lastly, we provide a list of guidelines for further optimization of partially active chimeras.

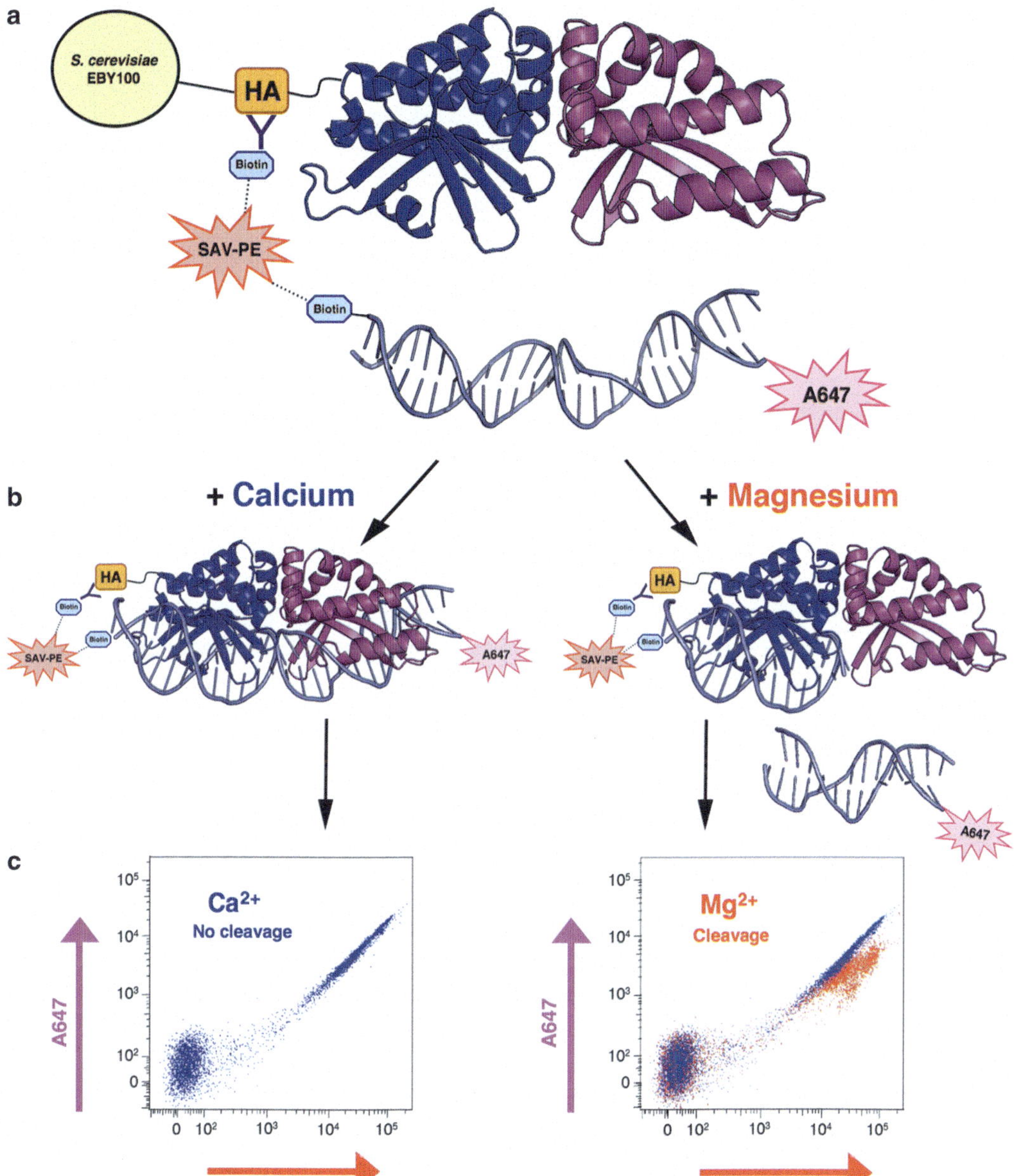

Fig. 3 Schematic representation of the flow-cytometric DNA cleavage assay. (**a**) An N-terminal hemagglutinin (HA) tag enables biotin–streptavidin tethering of a fluorescently labeled, double-stranded DNA substrate to the enzyme. (**b**) In the presence of calcium, the enzyme can bind the DNA substrate but not cleave it; with the addition of magnesium, the enzyme is able to bind and cleave the target substrate. When cleavage occurs, the N-terminal tethered PE fluorescence will be maintained while the A647 signal on the opposite end of the DNA target is lost. (**c**) The tethered, intact target substrate produces a characteristic co-linear PE versus A647 fluorescence profile on the flow cytometer (*left*). Cleavage of the tethered substrate leads to a loss of fluorescence in the A647 channel. Superposition of PE versus A647 plots from the Ca^{2+} and Mg^{2+} samples can be used to roughly quantify cleavage activity (*right*)

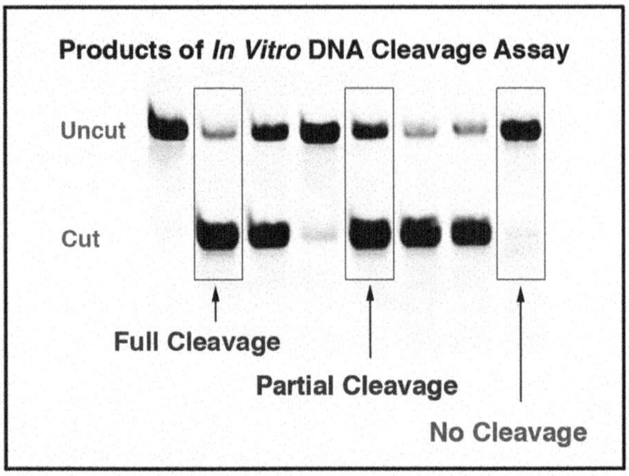

Fig. 4 Sample gel of in vitro cleavage assay products. Double-stranded DNA target substrates with an incorporated A647 fluorophore are incubated at 37 °C in the presence of free enzyme released from the yeast surface by DTT. The resulting cleavage products are run on an acrylamide gel and visualized by a fluorescence imager. The sample lanes shown here represent fully cleaved, partially cleaved, and non-cleaved target substrate. The non-tethered nature of this in vitro assay provides an important validation of cleavage activity observed in the flow-cytometric cleavage assay. In this assay, the enzyme must have sufficient binding affinity for its DNA substrate to produce observable cleavage products

2 Materials

Prepare all solutions using ultrapure RNAse- and DNAse-free water (0.22 μm filtered, deionized water) and analytical grade reagents. Cultures and reagents may be prepared at the bench, but care should be taken to use sterile components and aseptic technique.

2.1 Assembly PCR

1. Set of assembly oligonucleotide primers (*see* explanation of design in Subheading 3.2).
2. Forward and reverse primers for the full-length desired product (*see* Subheading 3.2).
3. Ultrapure water.
4. 1.5 mL microcentrifuge tubes.
5. Phusion® High-Fidelity DNA Polymerase with supplied 5× buffer (New England Biolabs).
6. dNTPs (10 mM).
7. Thermal cycler.
8. Agarose, molecular grade (Bioline).
9. Ethidium bromide solution (10 mg/mL) (Sigma-Aldrich).
10. Gel extraction purification kit (Zymo Research).

11. Restriction enzymes NdeI, KpnI-HF, and XhoI with supplied buffers and 100× BSA solution (New England Biolabs).
12. PCR purification kit (Qiagen).

2.2 Cloning of DNA Constructs

1. pETCON yeast surface expression vector (with NdeI and XhoI cloning sites) (Addgene reference).
2. DH5 alpha bacteria (either electro- or chemically competent, depending on method of bacterial transformation).
3. 0.1 cm electroporation cuvettes (if using electroporation) (Bio-Rad).
4. Electroporator (Bio-Rad).
5. Luria broth (LB) for bacterial culture.
6. Carbenicillin antibiotic (Bioline).
7. LB-carbenicillin bacterial agar plates (1:1,000 antibiotic).
8. Mini- or maxi-prep plasmid isolation kit (Qiagen).
9. Restriction enzymes NdeI, KpnI-HF, and XhoI with supplied buffers and 100× BSA solution (New England Biolabs).
10. Agarose, molecular grade (Bioline).
11. 1× Tris–Acetic acid–EDTA buffer (TAE) (Fisher Scientific).
12. Ethidium bromide solution (10 mg/mL) (Sigma-Aldrich).
13. Gel electrophoresis equipment.
14. Scalpel (for gel extraction).
15. Gel extraction DNA purification kit (Zymo Research).
16. UV gel illumination box with low intensity setting (for extraction of gel bands without damaging the DNA).
17. PCR purification kit (Qiagen).
18. Spectrophotometer.
19. T4 DNA Ligase with supplied buffer (New England Biolabs).
20. BigDye Terminator v3.1 sequencing mix (Applied Biosystems).
21. 5× sequencing buffer: 400 mM Tris, pH 9.0, 10 mM $MgCl_2$
22. pETCON sequencing primers:

Forward primer: 5′-GTTCCAGACTACGCTCTGCAGG-3′

Reverse primer: 5′-GATTTTGTTACATCTACACTGTTG-3′

23. Thermal cycler.

2.3 Dual-Labeled Double-Stranded DNA Substrates

1. Platinum® Taq High Fidelity DNA Polymerase, with 10× buffer and 50 mM $MgSO_4$ (Invitrogen).
2. Target site oligonucleotide template with flanking universal primer sites, standard desalting purification (Integrated DNA Technologies [IDT]) (Fig. 5).

Chimeric Target Site Template

Parent Enzyme Target #1	TTTCCACTTATTCAACCTTTTA
Parent Enzyme Target #2	AATGCTCCTATACGACGTTTAG
Chimeric Target	TTTCCACTTATACGACGTTTAG

Template for chimeric target with flanking universal primer sequences:

TGGACACGACTTGAGCTTTCCACTTATACGACGTTTAGCGTAGTGCTGTGCTGA

Universal forward primer with conjugated Biotin:

BIOTIN-TGGACACGACTTGAGC

Universal reverse primer with conjugated A647 fluorophore:

A647-TCAGCACAGCACTACG

Fig. 5 Reagents for PCR production of fluorescently labeled DNA target oligonucleotide substrate. Determination of the chimeric target recognition sequence requires the simple fusion of two parental half-sites (divided at the center of the native targets). If orientation of the parent enzyme on its DNA target site is NOT known, then one must test both the forward and reverse-complement sequences for that parent. To create the target site PCR template, the desired target sequence should be flanked by forward and reverse "universal primer" sequences (sample sequences are those currently used in our laboratories; alternate primer sequences may be substituted). Primers are ordered with A647 or biotin modifications for the PCR production of dual-labeled double-stranded DNA substrates

3. Biotin-labeled universal forward primer (IDT) (Fig. 5) (*see* **Note 1**).
4. A647-labeled universal reverse primer (IDT) (Fig. 5) (*see* **Note 1**).
5. 10 mM dNTPs.
6. 0.2 mL PCR strip tubes and/or 96-well PCR plates.
7. Thermal cycler.
8. Exonuclease I (New England Biolabs).
9. MultiScreen HTS-HV filter plate (Fisher Scientific).
10. illustra Sephadex G-100 (GE Healthcare): Prepare sephadex solution at 1 g/20 mL in water; allow at least 24 h for bead hydration; store at room temperature for a maximum of 4 months.
11. Tabletop centrifuge fit for spinning plates.

12. Odyssey infrared imaging system (Li-Cor Biosciences) or UV transilluminator.
13. NanoDrop or similar microvolume spectrophotometer (Thermo Scientific).

2.4 Polyacrylamide Gel

1. 30 % acrylamide–bis acrylamide (19:1) solution.
2. 10× Tris–borate–EDTA (TBE) buffer (*see* **Note 2**).
3. *N*,*N*,*N*′,*N*′-tetramethylethylenediamine (TEMED).
4. Ammonium persulfate (Acros Organics): 10 % w/v solution in water.
5. Plastic gel cassette, 1.0 mm.
6. Vertical gel electrophoresis apparatus.
7. 6× Ficoll loading buffer: 18 % (w/v) Ficoll-400 (Sigma-Aldrich) in 6× TBE, with NO added dyes (*see* **Note 3**).

2.5 Yeast Transformation

1. EBY100 yeast (Invitrogen).
2. 2× YPAD non-selective yeast media: 20 g Bacto yeast extract (BD), 40 g Bacto peptone (BD), 100 mg Adenine hemisulfate (Sigma-Aldrich), 50 g D-Glucose (Fisher Scientific), water to 1 l total volume, pH to 6.0. Filter-sterilize or autoclave and store at 4 °C (*see* **Note 4**).
3. Salmon sperm DNA (Sigma-Aldrich), 2 mg/mL solution in 1× TE buffer.
4. 1 M Lithium acetate solution in water.
5. 50 % w/v Polyethylene glycol in water, MW 3350 (PEG 3350) (Amaxa).
6. Plasmid DNA encoding a chimeric endonuclease in the pETCON yeast surface display vector.
7. 42 °C water bath.
8. Yeast selective growth media "SC–Ura–Trp": 6.7 g Yeast nitrogen base without amino acids (Sigma-Aldrich), 1.4 g yeast synthetic drop-out media supplement without Trp, Ura, His, Leu (Sigma-Aldrich), 76 mg Histidine, 380 mg Leucine, 4.34 g MES, and water to 900 mL. Adjust pH to 5.25 with HCl. Sterilize by autoclaving 20 min. Prior to use, add penicillin (100 i.u./mL), streptomycin (100 μg/mL), and kanamycin (25 μg/mL). Store at 4 °C.
9. 20 % w/v D-glucose (Fisher Scientific) solution, filter-sterilized. Store at 4 °C.
10. Selective growth media agar plates: add 20 g of bacteriological agar (BD) to 900 mL of SC–Ura–Trp selective growth media and autoclave for 20 min. Add 100 mL pre-warmed (55 °C) 20 % w/v D-glucose and penicillin (100 i.u./mL), streptomycin

(100 μg/mL), and kanamycin (25 μg/mL). Pour into petri dishes and let solidify at room temperature. Store plates at 4 °C.

11. Water-jacketed incubator.

2.6 Yeast Growth and Induction

1. EBY100 yeast transformed with surface-expression vector containing chimeric homing endonuclease.
2. SC–Ura–Trp selective growth media (for recipe, *see* Subheading 2.5).
3. 20 % w/v D-glucose (Fisher Scientific) solution, filter-sterilized. Store at 4 °C.
4. 20 % w/v D-(+)-raffinose pentahydrate (Sigma-Aldrich) + 0.1 % w/v glucose solution, filter-sterilized. Store at room temperature.
5. 20 % w/v D-(+)-galactose (Sigma-Aldrich) solution, filter-sterilized. Store at 4 °C.
6. Baffled Erlenmeyer flask(s).
7. Disposable 15-mL culture tubes.
8. Deep-well 96-well plate (flat bottom).
9. Shaking incubator.
10. Spectrophotometer.

2.7 Yeast Surface Display Flow-Cytometric DNA Binding and Cleavage Assays

1. Induced EBY100 yeast with surface expressed chimeric homing endonuclease (*see* Subheading 3.9).
2. 10× Yeast Staining Buffer (YSB): 1.8 M KCl, 0.1 M NaCl, 0.1 M HEPES, 2 % BSA, 1 % w/v D-(+)-Galactose, adjust pH to 7.5 with KOH (*see* **Note 5**). Filter-sterilize and store at 4 °C in a light-protected or foil-wrapped container.
3. 1× high-salt Yeast Staining Buffer (YSB + KCl): Dilute 10× YSB to 1× and add an additional 400 mM KCl for a final KCl concentration of 580 mM. Store at 4 °C in light-protected or foil-wrapped container.
4. 10× In-vitro Oligonucleotide Cleavage Buffer (IOCB): 1.5 M KCl, 0.1 M NaCl, 0.1 M HEPES, 0.05 M K-Glu (L-Glutamic Acid Potassium Salt Monohydrate), 0.5 % BSA, adjusted to pH 8.25 with KOH (*see* **Note 5**). Filter-sterilize solution and store at 4 °C in a light-protected or foil-wrapped container.
5. 1 M $CaCl_2$ solution. Filter-sterilize and store at room temperature.
6. 1 M $MgCl_2$ solution. Filter-sterilize and store at room temperature.
7. Biotin-labeled anti-HA antibody (Covance).
8. Streptavidin-PE (BD Biosciences).
9. FITC-conjugated chicken anti-cMyc antibody (Immunology Consultants Laboratory Inc.).

10. Costar 96-well V-bottom plate (Sigma-Aldrich).
11. BD FACScalibur or LSRII™ cytometer (BD Biosciences) or other cytometer with equivalent optics.
12. FloJo software (Tree Star Inc).

2.8 Non-tethered In Vitro Cleavage Assay and Gel

1. Induced EBY100 yeast with surface expressed chimeric homing endonuclease (*see* Subheading 3.9).
2. 1× IOCB (for recipe, *see* Subheading 2.7).
3. 1 M solution $CaCl_2$ (to be diluted to 5 mM).
4. 1 M solution $MgCl_2$ (to be diluted to 5 mM).
5. 0.2 M Dithiothreitol (DTT).
6. Polyacrylamide gel (for components, *see* Subheading 2.4).
7. Vertical gel electrophoresis apparatus.

3 Methods

Carry out all procedures on ice, unless otherwise specified.

3.1 Analysis and Preparation of Parent Homing Endonuclease DNA Coding Sequences

1. Codon optimize the DNA coding sequences of interest for expression in yeast (or dual optimize for expression in both yeast and bacteria if planning downstream assays in bacteria).
2. Search the codon-optimized DNA sequences for the following restriction sites: KpnI, NdeI, and XhoI. Replace any instances of these sites with alternate codons to remove the sites.
3. Align the amino acid sequences of interest with the I-OnuI sequence (or other reference sequence) to identify the positions of the two LAGLIDADG (Fig. 6).
4. Number *backwards* from the start of the second LAGLIDADG helix (in the I-OnuI reference structure, numbering begins with Asn167) (Fig. 6 inset).

   0 = Asn167

   1 = Pro166

   2 = Ile165

   3 = Asn164

   4 = Lys163

   5 = Asn162
5. Substitute a serine at position 5, a glycine at position 4, and a threonine at position 3. Use the sequence "GGTACC" for the Gly-Thr to introduce a KpnI restriction site. (Alternately, if structures are available, align the structure of interest to I-OnuI and replace the equivalent N–K–N residues with S–G–T.) (*see* **Note 6**).

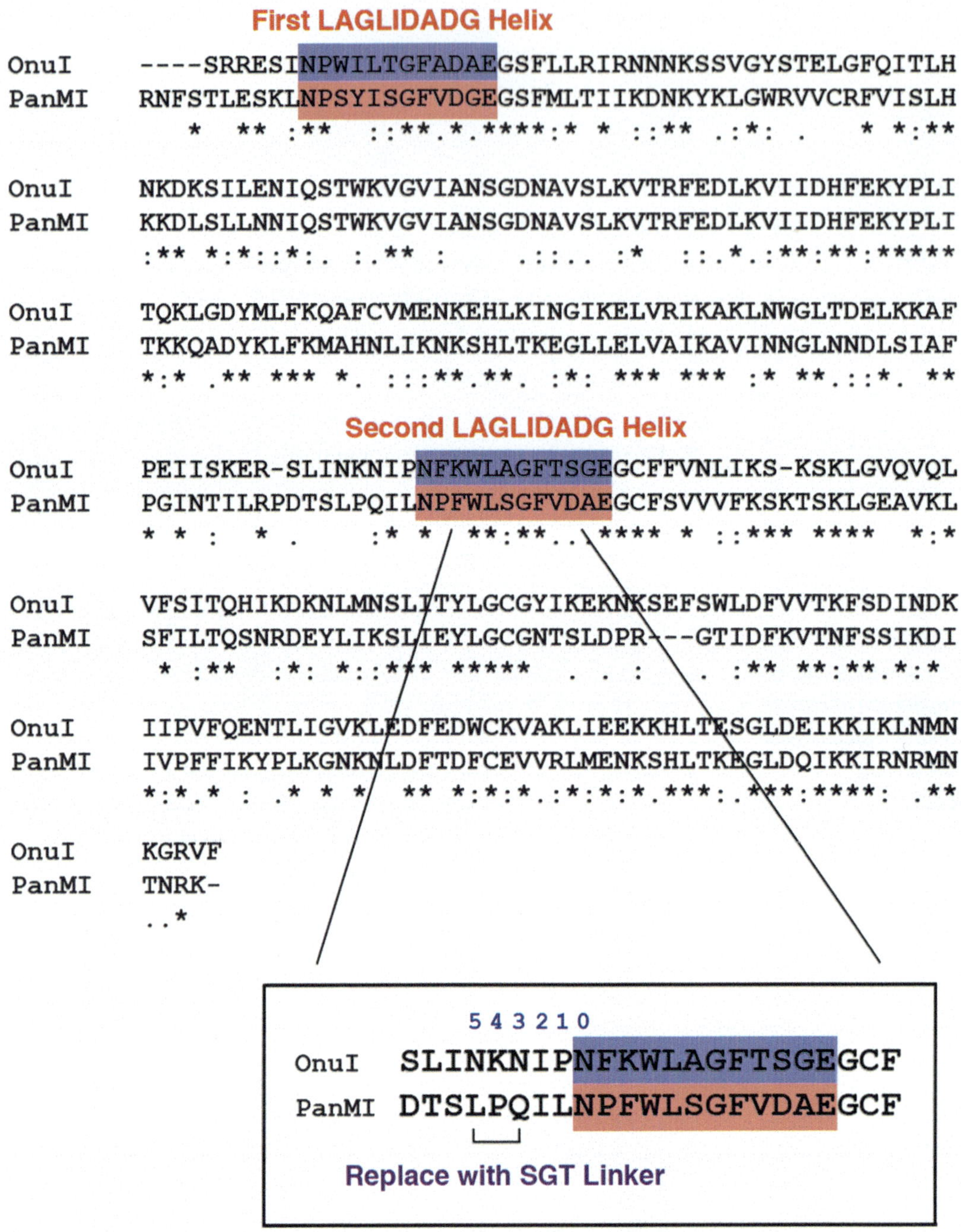

Fig. 6 Alignment of native endonuclease sequences. Alignment of a well-characterized reference sequence like I-OnuI is used to define the N- and C-terminal domains of the I-PanMI enzyme. The location of the second LAGLIDADG helix is used to guide the placement of the -SGT- tripeptide linker. Counting backwards from the residue at the top of the second LAGLIDADG helix (marked residue "0" in the *magnified inset*), the -SGT- linker is substituted at positions 3 through 5 (marked with a *bracket*). This location was chosen as a region where there are minimal interactions between the native linker and the adjacent protein domain

6. Add an NdeI restriction site (CATATG) to the 5′ end of the coding sequence and an XhoI restriction site (CTCGAG) to the 3′ end (for cloning into the pETCON vector).
7. Order the resulting sequence (we used synthesized genes from Genscript) or generate by assembly PCR (*see* Subheading 3.2). Clone into the pETCON yeast surface expression vector between the NdeI and XhoI restriction sites (*see* Subheading 3.3 for preparation of open vector).

3.2 Assembly PCR

1. Derive your desired homing endonuclease sequence, as described in Subheading 3.1 (codon-optimized for yeast expression and including the -SGT- tri-residue bridging linker between the N- and C-terminal halves). Position the sequence as it would appear inside the pETCON vector in order to see the sequence directly preceding and following the endonuclease reading frame.
2. Using Primer3 [7] (or similar primer design program), find forward and reverse primers for the N-terminal domain (*see* **Note 7**), including six or more base pairs preceding the NdeI site and six or more base pairs following the KpnI site (*see* **Note 8**). This is most easily accomplished by inputting a limited region including potential forward sites, and determining only the forward primer (uncheck "pick right primer or use right primer below"). Repeat for reverse primer.
3. Repeat **step 2** for the C-terminal domain (using the KpnI and XhoI restriction sites).
4. Input the coding sequences for each half of the chimera, including sequence adjacent to the restriction sites as described above, into the DNAWorks online assembly PCR design tool [8]. We used the following settings:
   (a) Annealing temperature: 61–63 °C.
   (b) Oligo length: 40–60 bp (*see* **Note 9**).
   Be sure to check the "Random" box, which allows the program to use oligonucleotides of varying lengths to find optimal overlaps. If left unchecked, the program will return a set of oligonucleotides of identical length (*see* **Note 10**).
   (c) The remaining options were left as default.
   (d) Paste your DNA sequence in the last field, and select "nucleotide" as the type of sequence.
   (e) Be sure to input your email address and a title for your job (do NOT include spaces in your title).
5. Once DNAWorks returns a completed set of solutions, scroll to the very bottom of the resulting logfile to find a table of statistics. The best set of oligonucleotides will have the LOWEST overall score. A low score will correspond to the

smallest number of misprimes. Note the number of the design run with the lowest score, and scroll back up the logfile to find that set of oligonucleotides.

6. Order the set of oligonucleotides. DNA synthesis on the 25 nM scale with standard desalting purification is adequate.
7. Resuspend the oligonucleotides at 100 μM concentration in ultrapure water.
8. Create a "pool" of all the oligonucleotides for each domain by combining 1 μL of each assembly primer. Add water to bring the final concentration of each primer in the pool to 1 μM. (Example: add 1 μL from each of 30 primers, each at a stock concentration of 100 μM. Then add 70 μL water to bring the total volume of the pool to 100 μL, and the final concentration of each primer to 1 μM) (*see* **Note 11**).
9. The assembly procedure consists of two separate PCR reactions. The first reaction allows the set of oligonucleotide primers to properly assemble. The second reaction amplifies the full-length assembly products. *See* Table 1 for PCR recipes and thermal

Table 1
Assembly PCR recipes and thermal cycler programs

Primary assembly reaction (50 μL)	*Secondary assembly reaction (50 μL)*
10 μL 5× Phusion buffer	10 μL 5× Phusion buffer
2 μL oligo pool	5 μL primary assembly reaction
2.5 μL dNTPs (10 mM)	2.5 μL dNTPs (10 mM)
1 μL Phusion enzyme	2.5 μL forward primer (10 μM)
34.5 μL water	2.5 μL reverse primer (10 μM) 1 μL Phusion enzyme 26.5 μL water
Thermal Cycler Program	
1. 98 °C for 1 min	1. 98 °C for 1 min
2. 98 °C for 10 s	2. 98 °C for 10 s
3. 58 °C for 15 s[a]	3. 63°Cfor 15 s[b]
4. 72 °C for 1 min	4. 72 °C for 1 min
5. Go to step 2, 29 more times	5. Go to step 2, 29 more times
6. 72 °C for 5 min	6. 72 °C for 5 min
7. 4 °C for ever	7. 4 °C for ever
8. End	8. End

[a]Adjust annealing temperature according to Tm of assembly oligonucleotides
[b]Adjust annealing temperature according to Tm of designed primers

cycler programs. Perform the primary assembly reaction for each domain. No purification of the reaction is necessary.

10. Perform the secondary assembly reaction for each domain (Table 1), using the forward and reverse primers you designed in **steps 2** and **3** above.
11. Run 1–5 μL of the secondary assembly reaction on a 1 % agarose gel with ethidium bromide. If there is a single, clean band at approximately 400 bp (representing one full-length domain), proceed directly to digestion. If there is significant contamination by products of other sizes, run the entire reaction on a 1 % agarose gel and purify by gel extraction of the correct band.
12. Digest the full-length assembly product with NdeI and KpnI-HF (for N-terminal domain), or KpnI-HF and XhoI (for C-terminal domain) to prepare for ligation into the pETCON vector (*see* **Note 12**).
13. Purify the digestion reaction with a PCR purification kit.
14. Clone into the pETCON yeast surface expression vector between the NdeI and XhoI restriction sites (*see* Subheading 3.3 for preparation of open vector).

3.3 Preparation of Open pETCON Vector DNA

1. Transform the pETCON vector DNA into DH5 alpha bacteria by standard electroporation (*see* **Note 13**) or chemical transformation protocols and plate on an LB + carbenicillin (carb) agar plate. Incubate overnight at 37 °C.
2. Culture a single colony from the transformation plate in LB + carb media.
3. Use a plasmid isolation kit (mini or maxi) to obtain a large quantity of the intact vector (a minimum of 10 μg of DNA should be enough for downstream reactions). Elute with sterile water.
4. Digest 5–10 μg of the intact vector with the restriction enzymes NdeI and XhoI. Digest at least 5–10 μg of the vector DNA (*see* **Note 14**).
5. Determine the efficiency of the double digest by running 1 μL of the digest products on a 0.5 % agarose gel (with ethidium bromide) in 1× TAE buffer for 30–60 min at 120–130 V. The correct size of the open vector is approximately 6,200 bp.
6. If more than one product size is visible (suggesting incomplete digestion), run the entire digestion reaction on a 0.5 % agarose gel (in multiple lanes or a single large lane). Excise the appropriate band for the linearized vector, and proceed with gel extraction purification (*see* **Note 15**). If only a single band of the correct size is visible, the reaction can be purified with a PCR purification kit. Elute with sterile water.
7. Quantify the purified open pETCON vector using a spectrophotometer.

3.4 Preparation of N- and C-Terminal Domain DNA for Ligation

This protocol is designed to prepare chimeras from several parent enzymes simultaneously.

1. Transform DH5 alpha bacterial cells (using standard electroporation or chemical transformation) with the parent homing endonuclease DNA coding sequence in the pETCON yeast surface expression vector, and plate on an LB + carb agar plate. Incubate overnight at 37 °C.
2. Culture a single colony from the transformation plate in LB + carb media.
3. Use a plasmid isolation kit (mini or maxi) to obtain a large quantity of the intact vector (a minimum of ten micrograms of DNA should be enough for downstream reactions). Elute with sterile water.
4. Set up double digest reactions with 5–10 μg of the intact vector DNA using the following restriction enzyme combinations: (a) NdeI and KpnI-HF (to cut out the N-terminal domain) (*see* **Notes 12** and **14**) and (b) KpnI-HF and XhoI (to obtain the C-terminal domain fragment).
5. Run the digested DNA in a large lane on a 0.8 % agarose gel (with 2 μL/100 mL ethidium bromide) in 1× TAE buffer at 120–130 V for 45–60 min.
6. Excise the correct-sized DNA fragments from the gel, using a low-intensity UV light box and minimizing the length of time that the product is exposed to UV light. For I-OnuI, the NTD fragment is 477 bp and the CTD fragment is 419 bp. Purify using a gel extraction purification kit, and elute with sterile water.

3.5 Ligation of N- and C-Terminal Domains into the pETCON Yeast Surface Expression Vector

1. Determine the volume of open pETCON vector you will use for each ligation reaction (use 100–200 ng vector).
2. Based on the size (in bp) and concentration (in ng/μL) of your purified N- and C-terminal half-inserts, calculate the nanograms of each half you will need to create a 3:1 molar ratio of insert to vector. The following online "Ligation Calculator" works well: http://www.insilico.uni-duesseldorf.de/Lig_Input.html.
3. Combine the N-terminal half-insert, C-terminal half-insert, and open pETCON vector with 1 μL 10× T4 ligase buffer and 0.5 μL T4 DNA ligase enzyme. Add water to a final volume of 10 μL.
4. Incubate the ligation reaction overnight at 16 °C, followed by heat inactivation at 65 °C for 10 min.

3.6 Transformation and Screening of Chimeric Homing Endonuclease Constructs

1. Transform the completed ligation reaction into bacterial cells (using either chemical transformation or electroporation). Use up to 4 μL of the ligation reaction for chemical transformation and up to 2 μL for electroporation.
2. Plate the transformed cells. In most cases, chemical transformation is sufficient to produce many colonies for screening.

Electroporation should be used if fewer than ten colonies are obtained.

3. Screen the resulting colonies by either: (a) colony sequencing (*see* **Note 16**) or (b) culture single colonies, isolate plasmids using a miniprep kit, and sequence the plasmid DNA (*see* **Note 17**).

3.7 Preparation of Dual-Labeled Double-Stranded DNA Substrates

1. In 0.2 mL PCR tubes, mix 0.08 μL Platinum High Fidelity Taq, 2 μL 10× Taq buffer, 1.3 μL 50 mM MgSO4, 0.4 μL 10 mM dNTPs, 2 nM (final concentration) target site template oligonucleotide, and 0.55 nM (final concentration) each of the A647 universal FP and biotin universal RP. Add H_2O to a final volume of 20 μL (*see* **Note 18**).
2. Thermal cycler program (for PCR amplification of target and incorporation of labels):

   90 °C × 1 min.

   40× (86 °C × 15 s, 48 °C × 15 s, 60 °C × 30 s).

   60 °C × 15 min.

   40× (70 °C × 30 s, decrease by 1 °C every cycle) (*see* **Note 19**).

   Hold at 4 °C.
3. Digest excess single-stranded DNA with Exonuclease I: Add 2 U of ExoI to each 20 μL PCR reaction in 2 μL total volume of water (*see* **Note 20**). Digest 4 h at 37 °C. This reaction can be stored at 4 °C overnight or at −20 °C for extended periods.
4. Load the hydrated sephadex G-100 suspension into the filter plate. For each 20 μL PCR reaction to be purified, add 500 μL total volume of suspension to a filter plate well. This is best accomplished by loading 320 μL sephadex suspension (using wide bore tips) into each necessary well of the filter plate, centrifuging briefly up to a speed of 500 × *g*, discarding water, and adding the remaining 180 μL sephadex (*see* **Note 21**). The plate should then be dehydrated by centrifugation at 2,000 × *g* for 7 min. Do not use any sephadex columns that contain cracks.
5. Load the 22 μL PCR + ExoI reaction directly to the center of each sephadex column. Secure a 96-well PCR plate below the filter plate to catch the purified flow-through, using tape if necessary. Centrifuge for 5 min at 2,000 × *g*. Approximately 12–14 μL of flow-through should be present in each recipient well.
6. Determine the concentration of purified target site substrates using a NanoDrop spectrophotometer. Target concentration should be approximately 10–25 ng/μL. A concentration greater than 25 ng/μL suggests inadequate ExoI digestion or sephadex purification. Sephadex purification can be repeated, but this will not help purify inadequately ExoI-digested samples.
7. Run a sample of the purified target oligonucleotides on a 15 % polyacrylamide gel. For a 7.5-mL gel, combine 0.75 mL 10×

TBE, 3.75 mL 30 % acrylamide–bis acrylamide (19:1), and 2.9 mL water. Add 100 μL 10 % APS and 10 μL TEMED, then immediately mix and pipette into the prepared gel cassette. Once the gel is set, load 0.5–1.0 μL purified target, diluted with 1.0 μL 6× Ficoll loading buffer and 4 μL water (*see* **Note 22**). Use 0.1 μL of the A647-labeled primer as a size standard. Run the gel for 90 min at 120 V.

8. Visualize the gel on a Licor Odyssey infrared imager, using the 700 nM laser. The gel should show a prominent single PCR product and minimal contamination by other bands or leftover primers. Alternatively, the gel can be stained with 1× SybrGold in 1× TBE for 20 min, washed in 1× TBE or water, and visualized on a UV transilluminator.

3.8 Yeast Transformation

(This transformation procedure is based on published protocols by Gietz and Scheizl) [9].

1. Thaw and spin down a frozen aliquot of EBY100 competent yeast cells (*see* **Note 23**).
2. Resuspend the pellet in the following transformation mixture: 50 μL denatured 2 mg/mL salmon sperm DNA (denature by heating at 95 °C for 5 min, then transfer immediately to ice), 36 μL 1 M LiAc, 260 μL 50 % PEG 3350, and 14 μL water plus plasmid DNA (up to 1 μg) (*see* **Notes 24** and **25**).
3. Incubate the yeast and transformation mixture at 42 °C for 40 min (*see* **Note 26**).
4. Fill the tube with SC–Ura–Trp + 2 % glucose media and spin down the cells. Remove supernatant.
5. Resuspend the yeast pellet in 1 mL SC–Ura–Trp + 2 % glucose media.
6. Plate 1–10 μL transformed yeast on selective growth media agar plates (SC–Ura–Trp + 2 % glucose) and incubate in a 30 °C water-jacketed incubator. Colonies of an appropriate size for picking should appear by 48–72 h.

3.9 Growth and Induction of Yeast

1. Transfer a single colony of transformed yeast into 1.5 mL SC–Ura–Trp + 2 % raffinose + 0.1 % glucose media (*see* **Note 27**).
2. Incubate overnight in a 15-mL culture tube at 30 °C with 250 RPM shaking (*see* **Note 28**) until the cells reach a density of 90–120 million/mL (*see* **Note 29**) and place on ice for up to 24 h.
3. Wash 30 million cells twice with water and transfer to 1.5 mL of SC–Ura–Trp + 2 % galactose media (*see* **Note 30**).
4. Incubate the galactose culture on the benchtop (room temperature with no shaking) for up to 16 h (*see* **Note 31**).

3.10 Yeast Surface Display Flow-Cytometric DNA Cleavage Assay

All components should be kept on ice throughout the assay, unless otherwise specified, including YSB and IOCB buffers. If possible, the centrifuge should also be kept at 4 °C.

1. Determine the density of induced yeast in the galactose culture. This can be accomplished using a hemocytometer and microscope or estimated by spectrophotometer (*see* **Note 29**). Aliquot 500,000 yeast per sample into a 96-well, V-bottom plate (*see* **Notes 32** and **33**).
2. Wash cells twice with 200 μL 1× YSB, centrifuging the V-bottom plate at 3,000 ×*g* for 3 min and discarding the supernatant. Yeast cells often not form a visible pellet until washed with YSB.
3. Gently resuspend cells at a concentration of 50 million/mL in 1× YSB with 1:250 dilution of anti-HA-biotin antibody (i.e., consider the anti-HA-biotin to be a 250× stock). Incubate at 4 °C for 30–60 min, mixing gently every 10–15 min (*see* **Note 34**).
4. During the anti-HA stain, prepare target substrates for conjugation in either 1.5-mL microcentrifuge tubes or plate format. For 500,000 cells (final cell density of 50 million/mL), use 25 μL total volume per well. SAV-PE should be diluted to 5 nM in the high salt 1× YSB + KCl buffer. Add the labeled ds-oligonucleotide target substrate to a final concentration of 40 nM (*see* **Notes 35** and **36**). Aliquot to plates (if necessary) and incubate in the dark and on ice for 30 min.
5. Following the anti-HA incubation, centrifuge cells for 3 min at 3,000 ×*g* and wash twice with 200 μL ice-cold high-salt 1× YSB + KCl.
6. Following the second wash, resuspend yeast cells with target site conjugates. Gently vortex or pipette to resuspend.
7. Incubate 30 min at 4 °C, in the dark or in a light-protected container, mixing briefly every 5–10 min.
8. During this incubation, make 5 mM $MgCl_2$ and $CaCl_2$ solutions in 1× IOCB, and pre-warm to 37 °C.
9. Following incubation with conjugated target substrate, wash yeast with 200 μL ice-cold high-salt 1× YSB + KCl, and centrifuge for 2 min at 3,000 ×*g*. Resuspend conjugated yeast in 200 μL ice-cold 1× IOCB (containing no divalent ions).
10. Create duplicate wells (one will contain calcium for no cleavage, and one will contain magnesium to allow cleavage) by transferring half of the resuspended volume into an adjacent set of wells. Add an additional 100 μL ice-cold 1× IOCB to each well (giving a final wash volume of 200 μL). Centrifuge for 3 min at 3,000 ×*g*. Discard the supernatant and blot/tap vigorously on paper towels to remove as much buffer as possible.

11. Add 30 μL 1× IOCB containing either Ca^{2+} or Mg^{2+} (pre-warmed to 37 °C) to each set of duplicate wells. Add Ca^{2+} IOCB first to minimize background cleavage events in the negative control, and work as quickly as possible.
12. Incubate 20–30 min at 37 °C.
13. Fill wells with ice cold 1× YSB to stop the reaction, centrifuge 3 min at 3,000 × *g*, and discard the supernatant.
14. Resuspend in 25 μL 1× YSB containing 1:100 dilution of anti-Myc FITC to give a final cell density of 100 million/mL.
15. Incubate 1–4 h at 4 °C, with foil wrap or in a refrigerator to protect from light, gently vortexing occasionally to keep the cells in suspension (*see* **Note 37**).
16. Wash cells twice with 1× YSB, and resuspend in a final volume of 60 μL. This will lead to an acquisition rate of 1,000–2,000 events/s depending on the acquisition settings for the cytometer.
17. Acquire data on a BD Biosciences LSRII with HTS. Record data for FSC-A, FSC-H, SSC-A, SSC-H, APC, PE, and FITC parameters.
18. Analyze the flow cytometry data using FloJo. Gate live cells (FSC-A by SSC-A), then singlets (FSC-H by FSC-A), then cells staining for both FITC and PE (representing full length expression of both the C′ and N′ termini) (Fig. 7). Visualize this final subset as APC (*y*-axis) versus PE (*x*-axis). Superimpose the Ca^{2+} and Mg^{2+} samples to observe any cleavage-induced shift in APC signal. Quantitative measurements of cleavage efficiency can be obtained by determining the median APC signal within a small gated subset of live, singlet, expressing cells. A greater Ca^{2+}-to-Mg^{2+} median APC ratio represents increased cleavage efficiency (Fig. 7d).

3.11 Yeast Surface Display Flow-Cytometric DNA Binding Assay

All components should be kept on ice throughout the assay, unless otherwise specified. If possible, the centrifuge should also be kept at 4 °C.

1. Determine the density of induced yeast. This can be done using a hemocytometer or by spectrophotometry (*see* **Note 29**).
2. Aliquot 100,000 yeast to appropriate wells in a 96-well plate (for 384-well format, *see* **Note 38**) and wash twice with 200 μL 1× IOCB containing 5 mM $CaCl_2$. Centrifuge at 3,000 × *g* for 3 min and discard the supernatant.
3. Resuspend cells in 30 μL 1× IOCB containing 5 mM $CaCl_2$, 1:100 anti-Myc FITC, and the desired concentration of target site oligonucleotide (*see* **Note 39**). Incubate at 4 °C for 2 h, vortexing gently every 30 min.

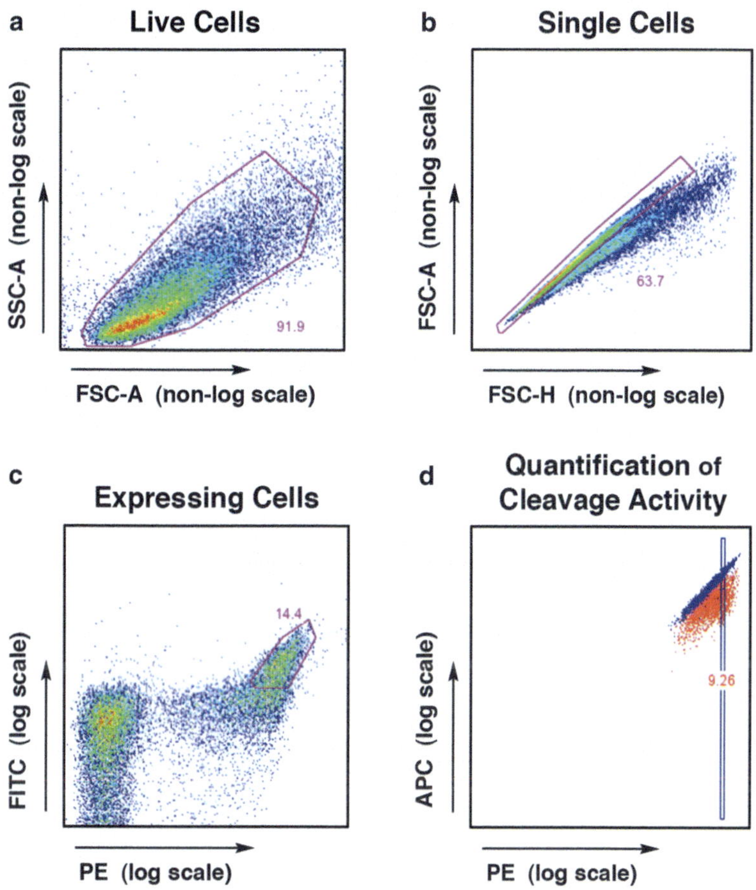

Fig. 7 FlowJo analysis of flow cytometry data from the tethered DNA cleavage assay. (**a**) Live cells are gated first, using FSC-A and SSC-A parameters, followed by (**b**) singlet cells using FSC-A and FSC-H. With proper induction conditions, approximately 50 % of cells will display the enzyme, and (**c**) the high-expressing population is gated using the PE signal (bound via a biotin–streptavidin bridge to the N-terminal HA tag) and FITC signal (bound directly by a fluorescence-conjugated antibody to the C-terminal Myc tag). Cleavage activity is visualized within this expressing population using a plot of (**d**) A647 fluorescence (observed in the APC channel, conjugated to the free end of the cleaved DNA) versus PE fluorescence (N-terminal HA tag). Cleavage activity can be quantified by gating on a PE-normalized subset of the expressing cells and calculating the ratio of median APC fluorescence for the uncleaved (Ca^{2+}) versus cleaved (Mg^{2+}) conditions

4. Wash twice with 1× IOCB + 5 mM $CaCl_2$.
5. Resuspend in 60 μL 1× IOCB + 5 mM Ca^{2+} for flow cytometry analysis. A final volume of 60 μL will lead to 500–1,000 events/s on the flow cytometer.
6. Acquire data on a BD Biosciences LSRII with HTS. Record data for FSC-A, FSC-H, SSC-A, SSC-H, APC, and FITC parameters.

If using HTS for collecting 96- or 384-well plate samples, set the machine to mix samples at minimum 3×.

7. Analyze the flow cytometry data using FloJo. Gate first the live cells, then singlets, then expressing cells, as described above. APC signal represents binding of the A647-labeled oligonucleotide to the surface-expressed homing endonuclease. Quantitative measurements of binding can be obtained by determining the ratio of median APC signal of the FITC-positive cells (which are expressing full-length enzyme) to the median APC signal of the FITC-negative cells (no enzyme expression).

3.12 Non-tethered in Vitro Cleavage Assay and Gel

All components should be kept on ice throughout the assay, unless otherwise specified.

1. Induce expression of the desired enzyme on the surface of yeast, as described above in Subheading 3.9 (*see* **Note 40**).
2. Determine the density of induced yeast (in galactose media) by spectrophotometer, as described above in 3.10 (*see* **Notes 29** and **41**).
3. Wash 5 million yeast cells with cleavage buffer (1× IOCB), spinning for 1 min at 10,000 × *g* or 3 min at 4,000 × *g* in a tabletop centrifuge. Resuspend in 200 μL IOCB (in plate) or 1 mL IOCB (in microcentrifuge tubes).
4. Divide the washed cells evenly into two separate tubes/wells. One sample will be run in the presence of calcium (control) and one in the presence of magnesium (cleavage).
5. Spin down the divided cells, and resuspend in the following mixtures: 50 μL total volume of 1× IOCB with 10 mM DTT, 5 mM $CaCl_2$ (control) or 5 mM $MgCl_2$ (cleavage), and 20–50 nM A647-labeled DNA substrate (*see* Subheading 3.7).
6. Mix thoroughly, and incubate at 37 °C for 5–60 min (*see* **Note 42**). Mix every 5–10 min throughout the incubation.
7. Transfer tubes/plates to ice to stop the reaction.
8. Spin the yeast at 4,000 × *g* for 3 min, or 10,000 × *g* for 1 min. Carefully pipette supernatant away from the cell pellets and transfer to a separate tube for storage (*see* **Note 43**).
9. Mix 2 μL of the 6× Ficoll loading buffer with 10 μL of the supernatant. Do not include any colored dyes. No DNA ladder is necessary. The calcium control lane(s) will provide an uncut control for size-comparison.
10. Carefully load the 12-uL samples into a 15 % acrylamide gel (*see* Subheading 3.7 for gel recipe) (*see* **Note 22**).
11. Run the gel for 60–90 min at 120 V.

12. Visualize the gel on a Licor Odyssey infrared imager, using the 700 nM laser. The Odyssey analysis software can be used for crude quantification of cleavage activity by manually gating cleaved and uncleaved bands for each sample, and determining the ratio of signal intensities.

3.13 Optimization of Partially Active Chimeras

In our experience, up to 50 % of I-OnuI family chimeras show significant cleavage activity with this direct domain fusion method [4]. However, certain enzyme combinations will be incompatible and require further engineering to achieve DNA cleavage activity. There are many different options for optimization of these enzymes and subsequent selection of the desired final outcome. Though we do not present step-by-step instructions for each option (as that is beyond the scope of this chapter), the following list can provide some initial guidance for beginning the optimization process. For chimeras with low cleavage activity, we have been able to improve stability and/or activity using the following strategies:

1. *Variation of residues at the DNA-distal end of the LAGLIDADG helices.* The LAGLIDADG helices form a major portion of the chimeric interface, and they show remarkable structural conservation between family members. However, this conservation breaks down at the DNA-distal end of the helices (Fig. 8a). We have found that manipulation of these distal helical residues can greatly increase the overall stability and resulting activity of a low-performing chimera [4]. Assembly PCR (*see* Subheading 3.2) is a great method to use for introduction of variation at one or more positions in a chimeric enzyme. Degenerate codons can be used to allow more than one residue at a given position (Fig. 9 inset), and the resulting library can be screened for the desired cleavage activity.
2. *Error-prone PCR.* Introduction of random variation across the entire coding sequence can often produce an enzyme with improved stability and/or activity.
3. *Parent interface substitution.* Another way to potentially improve an incompatible chimeric interface is to engineer the entire interface to match one of the two parent enzymes (Fig. 8b) [4]. This strategy can help overcome problems associated with steric clashes between interacting residues, but only if the surrounding core residues can support the new set of residues.
4. *Library of interface residues.* To find a functional solution for an important, but poorly functioning chimera, variation can be introduced along the entire chimeric interface. Again, assembly PCR with degenerate codons can be used to construct this type of interface library, which can then be screened for active enzymes. Care must be taken to limit the size of the overall library to that which can be reasonably screened by available yeast or bacterial cleavage selection methods.

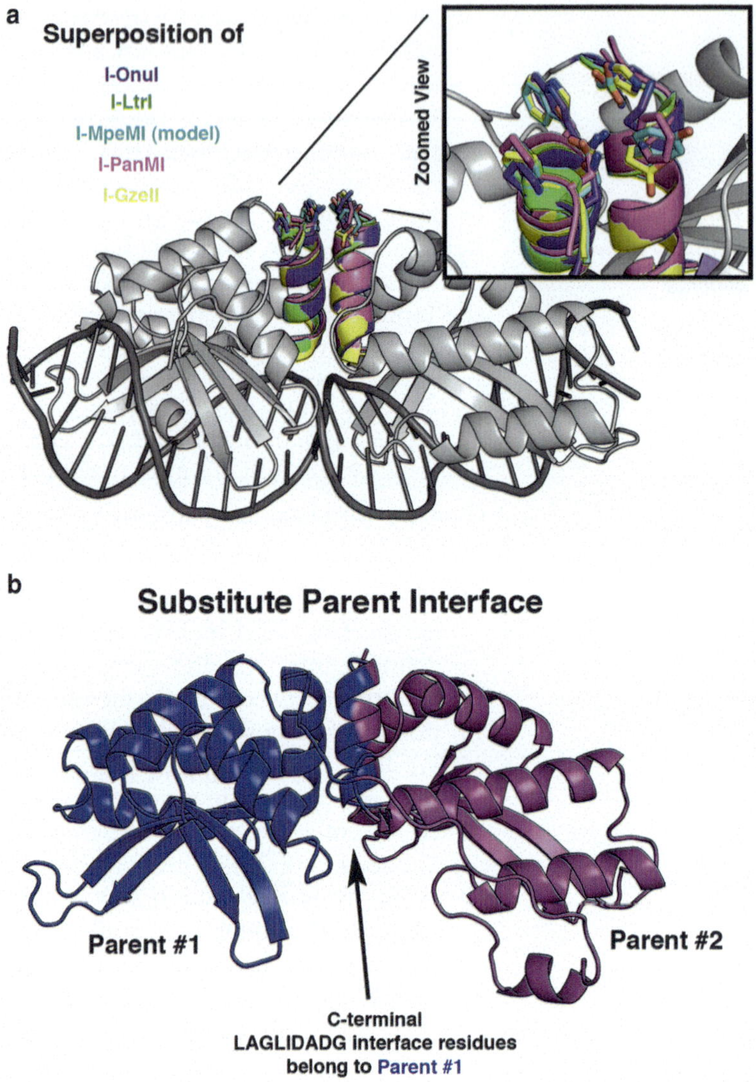

Fig. 8 Structural representation of areas for optimization of partially active chimeras. (**a**) The I-OnuI structure (*gray*) is shown with superimposed LAGLIDADG helices from five different enzymes: I-OnuI helices are colored *purple*, I-LtrI are *green*, I-MpeMI (homology model) are *cyan*, I-PanMI are *magenta*, and I-GzeII are *yellow*. Side chains are depicted as "sticks" at the DNA-distal end of the LAGLIDADG helices. A magnified view highlights the lack of conservation at these positions. (**b**) A significant portion of the chimeric interface is formed between the two adjacent LAGLIDADG helices. Residues from a single parent enzyme (*blue*) can be substituted onto BOTH sides of the LAGLIDADG interface to help overcome problems with an otherwise incompatible chimeric interface

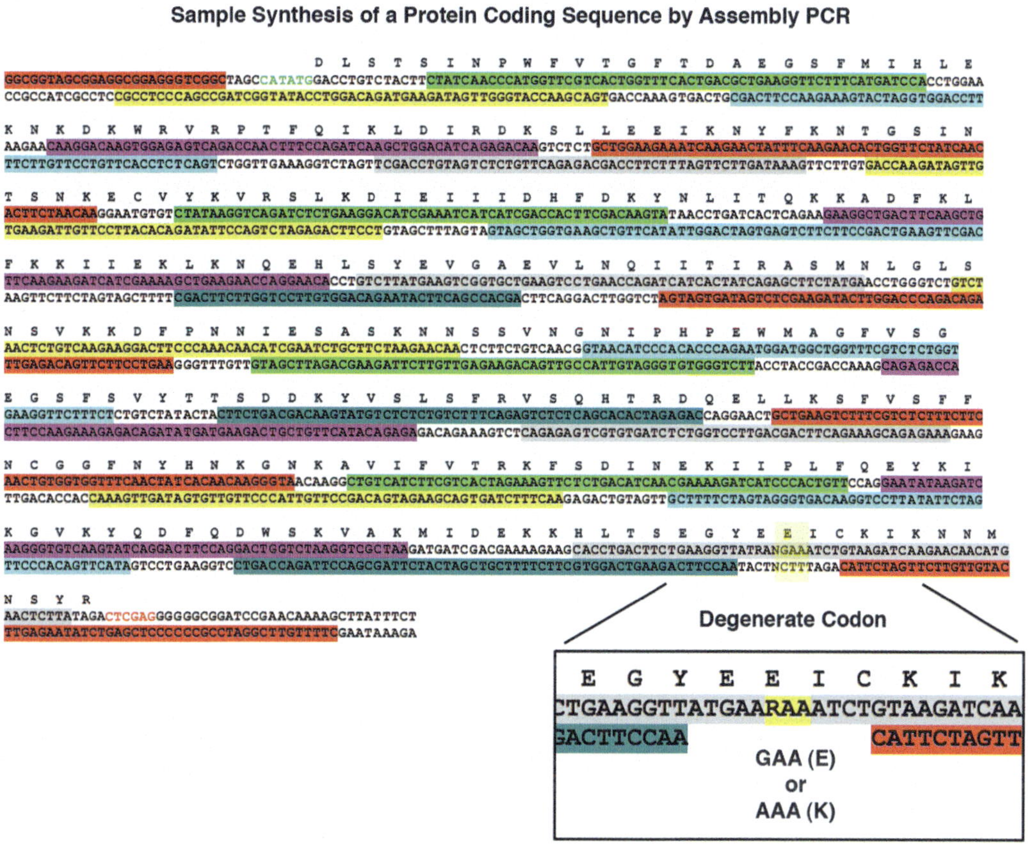

Fig. 9 Example of assembly PCR primers and introduction of variation via degenerate codons. Each *colored selection* represents a single assembly primer; the sum of all primers is designed to produce the entire coding sequence shown. An NdeI restriction site (CATATG) has been added to the N-terminal end of the sequence, and an XhoI restriction site (CTCGAG) has been added to the C-terminal end. The *magnified inset* towards the C-terminal end of the sequence gives an example of introducing variation using the degenerate codon "RAA." The "R" base designates the introduction of either a guanine (G) or adenine (A) base at that position, resulting in a translated protein sequence with either glutamic acid (E) or lysine (K)

4 Notes

1. The biotin and A647 labels can be affixed to either the forward or the reverse primer. Maximum cleavage signal is achieved empirically by testing each endonuclease with both labeling schemes. For reasons that remain speculative, a more pronounced cleavage shift may occur with the A647 label on one end of the oligonucleotide versus the other.
2. 10× TAE buffer can be used in place of 10× TBE buffer. If this substitution is made, be sure to use 1× TAE as the gel running buffer in place of 1× TBE.

3. Double-stranded oligonucleotide targets are visualized by the incorporated fluorescent A647 tag, and single-stranded, unincorporated A647 primer is used as a control to determine approximate purity of the double-stranded target. Dyes used to track DNA location on the gel also fluoresce, and should therefore be avoided. Loading of the colorless DNA solution into the gel lanes is difficult, so we suggest using a marker to trace each lane on the outside of the gel cassette and number the wells for reference. Watch the gel loading tip carefully to make sure it is correctly inserted into the desired well, and load the DNA slowly.
4. If autoclaving, add the glucose *after* autoclaving. To increase the shelf life of this media, make a 50× stock of the adenine hemisulfate and add it at 1× concentration just prior to use. Store the 50× stock at −20 °C (upon thawing, there will be a small amount of precipitation which will not go back into solution).
5. Solution should be adjusted to pH 7.5 using KOH. This limits introduction of additional sodium ions.
6. The position of this flexible S–G–T tripeptide was chosen to replace a short portion of the linker sequence making minimal to no interactions with the adjacent structure (to minimize detrimental effects on overall protein stability or enzyme activity).
7. These primers will be used to amplify full-length gene assembly PCR product for the N-terminal domain, and should have reasonably well-matched melting temperatures (Tm).
8. The sequence flanking the restriction sites can, theoretically, be set to any sequence. However, matching the sequence to the actual vector allows for downstream use of recombination-based cloning techniques, such as InFusion (Clontech), if standard sticky-end ligation methods are not yielding adequate product. For this purpose, we prefer to include 15 or more vector-matched base pairs on either side of the NdeI and XhoI restriction sites.
9. Using oligonucleotides over 70 bp in length greatly increases the error rate associated with synthesis of longer oligonucleotides. We kept our lengths at 60 bp or less to allow ordering of 25 nM-scale standard oligonucleotides from IDT. If there are large sections of the gene in which no subsequent alterations or variation is desired, ultramers may be considered in place of oligonucleotides.
10. If too many variables are included in a design run, the program will time-out and return NO results. This is especially true when you have the "Random" box selected. If you are NOT getting successful design runs, decrease the number of

variations possible. One way to do this is to significantly decrease the range of oligo lengths.

11. Save an aliquot of this pool in the event that any subsequent steps fail or need to be repeated, to avoid having to re-mix the full set of assembly primers.
12. The high fidelity (HF) version of the KpnI restriction enzyme is used for its compatibility with NdeI or XhoI enzymes in NEB Buffer #4.
13. Transformation of intact, supercoiled plasmid DNA is a high efficiency reaction. If using electroporation for this step, a single aliquot of electrocompetent cells can be used for up to ten transformations. To do this, we dilute an aliquot of thawed competent cells (commonly 50 μL) to the appropriate volume for ten individual transformation reactions (500 μL total volume), using sterile, ice-cold 10 % glycerol in water. Also, only a small amount of plasmid DNA is needed. Dilute the plasmid DNA 1:10 with water, and use 1 μL diluted DNA for the transformation.
14. The restriction enzyme NdeI (from New England Biolabs) is not as efficient as the other enzymes at cleaving DNA purified by plasmid isolation kits (as described in the technical details of the NEB product catalog). Compensate for this by using a larger volume of this enzyme in the reaction or by letting the digest run for a longer duration.
15. Undigested vector will significantly contaminate subsequent transformations, and this step should be performed carefully. If a considerable number of clones are identified as undigested vector in subsequent sequencing reactions, a new vector digestion and purification should be performed.
16. Colony sequencing: lightly touch a single colony with a small pipette tip, and transfer the tip to a PCR tube containing 6 μL of sterile, PCR-grade water. Heat to 96 °C for 6 min to lyse the cells. If the water was cloudy after adding the bacterial colony, it should now be clear. If the mixture is still cloudy after heating/lysis, this suggests that too many bacteria have been transferred, and the subsequent sequencing steps may yield poor data and possibly clog the sequencing machine. To the lysed cells, add 0.25 μL BigDye mix, 0.25 μL 10 mM primer, 2 μL 5× sequencing buffer, and water to a final volume of 10 μL. Proceed with PCR sequencing reaction. If using an external sequencing service, permission should be requested before submitting colony sequencing reactions.
17. After sequencing your clones, if contamination by undigested vector is a problem, use the following trick: Start with pETCON vector containing a stuffer sequence which contains a unique restriction site NOT found in the pETCON vector

sequence. It is best to use an enzyme which is compatible with NEB Buffer #1, because this buffer composition most closely matches that of the DNA ligation buffer. In our case, we started with pETCON vector containing a homing endonuclease ORF with a PacI restriction site. After the ligation step and PRIOR to transformation into bacteria, we added 10 μL of 1× NEB buffer #1 and 2× BSA to the ligation reaction (giving a new reaction volume of 20 μL), and 0.3 μL of the PacI restriction enzyme. Digesting for 1 h (followed by heat inactivation of the restriction enzyme) eliminates any undigested original vector from the ligation reaction, and the DESIRED ligation product remains intact. For the subsequent transformation step, we used DNA from the 20 μL PacI-treated ligation reaction (no purification necessary).

18. Target oligonucleotide substrate can be made in large batches and stored at −80 °C in a light-protected container. One 20 μL PCR reaction should yield approximately 12–14 μL of purified 300–500 nM substrate. When scaling up production of these substrates, maintain 20 μL PCR reaction volumes, and increase the number of reactions simultaneously run.
19. This gradual decrease in final temperature allows for high efficiency annealing into double-stranded DNA target oligonucleotide (leaving minimal single-stranded or mis-annealed target).
20. ExoI is diluted with water to a total volume of 2 μL per sample for ease and accuracy of transfer to the PCR reaction. Do not add any of the supplied ExoI buffer.
21. Special care should be taken to avoid any bubbles in the sephadex suspension when mixing or aliquoting to the filter plate. Bubbles will lead to cracks within the final, centrifuged sephadex columns, and cracked columns should be discarded. We find that careful pipetting, using wide bore tips or standard p200 tips cut at approximately the 50 μL gradation, reduces frequency of cracking. We also find that allowing 30 min between pipetting and centrifugation can significantly reduce column cracking.
22. While it is more difficult to load samples without dye in the loading buffer, this allows for clear visualization of the target oligonucleotide and A647-labeled primer. Colored dyes will fluoresce under the Licor excitation wavelength and confound the image.
23. Frozen competent cells are prepared according to the published protocol by Gietz and Schiestl [10]. Add 2.5×10^9 EBY100 cells from an overnight 2× YPAD culture to 500 mL fresh 2× YPAD media. Grow at 30 °C to a density of at least 20 million/mL. Pellet the cells and wash with sterile water. Resuspend the washed cell pellet in 5 mL of 5 % v/v glycerol + 10 % v/v DMSO

in water. Aliquot 50-μL volumes to microcentrifuge tubes. Pack the tubes into a styrofoam rack with lid (or similar form of insulation) and place at −80 °C. (The insulation allows for gradual freezing of the cells.)

24. When transforming a library of variant homing endonucleases, increase the number of yeast and volume of the transformation mixture according to Gietz and Schiestl [11].
25. When transforming high quality plasmid DNA, a single frozen aliquot of competent yeast cells in the described volume of transformation mixture can be used for multiple reactions. In this case, divide the resuspended cells (*prior* to addition of DNA) into up to 15 equal volumes and add up to 1 μL total volume of plasmid DNA to each aliquot. Proceed to the incubation step.
26. We have found that an incubation time of 40–42 min at 42 °C provides the highest transformation efficiency with lowest cell death. Longer incubation times can lead to significant cell death. If using a high-quality plasmid, a shorter incubation time of 20 min will suffice for the generation of transformed clones.
27. Raffinose cultures can be successfully started using a single colony from a selective media + glucose plate. Alternatively, we have found that an initial overnight incubation in YPAD media (at 30 °C with 250 RPM shaking) can substantially increase induction efficiency, and the absence of selective media at this stage does not result in significant plasmid loss.
28. When using vertical tube racks inside a shaking incubator, position the 15-mL culture tubes at a slant to allow for maximum aeration, and do not use more than 1.5 mL of media.
29. On our spectrophotometer, the density of a yeast culture can be estimated by mixing a 1:10 dilution of yeast in water and measuring the resulting OD_{600}. A simple calculation of $OD_{600} \times 300$ provides an estimated value for density of the culture in millions of cells per mL. The validity of this estimate should be checked when using a different instrument.
30. Care should be taken to wash yeast from the raffinose culture at least twice before transferring to the galactose media. This limits carry-over of raffinose and/or glucose.
31. Induced galactose cultures should be kept on ice or at 4 °C. Well-folded homing endonucleases will be stably expressed on the yeast surface for several days, although the total expression levels and catalytic activity may decrease slightly over time, depending on the endonuclease.
32. If running a large number of samples, the assay can be performed in a 384-well conical-bottom plate, with 50,000 cells/well. Volumes for staining and wash steps for a 384 well plate

are: 8 μL anti-HA stain, 8 μL conjugated DNA-SAV-PE stain, and 10 μL anti-Myc-FITC stain. All washes should use a minimum 100 μL buffer.

33. If running a small number of samples, the assay can be performed using 1.5 mL microcentrifuge tubes. In this format, cells can be spun down in a tabletop centrifuge at speeds up to 10,000 × *g* for 1 min. Perform 4 °C incubation steps on slow rotator, if possible.
34. A rotator at 4 °C can be used to assure continued suspension for thorough staining.
35. Include the volume of DNA target substrate in calculations for total conjugation reaction, as the dilution volume should be around 1:8 to 1:14, and is therefore substantial. The resulting slight decrease in IOCB salt concentration is not problematic at this point, and can be disregarded.
36. If using 1.5 mL microcentrifuge tubes, pipette the DNA target substrate onto the side of the tube not contacting SAV-PE, then gently vortex to mix the SAV-PE and DNA quickly. Likewise, in plate format the DNA target substrate should be pipetted onto the side of plate wells not contacting the SAV-PE mixture, if possible, and quickly mixed by vortex or multichannel pipette.
37. Yeast can be left in anti-Myc FITC stain overnight, if necessary. Due to relatively low affinity of this antibody, FITC-stained cells should not be washed or diluted greater than 2× if cells are to sit for more than 4 h prior to acquisition.
38. 25,000 yeast can be assayed in a 384-well plate format. Use a total volume of 20 μL. Low cell number is important to ensure that the effective concentration of DNA target substrate is not altered in affinity titration experiments.
39. For the I-OnuI family of homing endonucleases, specific binding can be detected from 100 pM to 50 nM. Higher concentrations of DNA target substrate may be non-specifically bound, and lower concentrations can be difficult to detect. This protocol can also be used to determine binding along a titration of various DNA concentrations.
40. Yeast can be used for up to 3 days following induction with preservation of most cleavage activity (when stored at 4 °C). However, because enzyme activity degrades over time, one should collect data from fresh samples for optimal activity.
41. The total concentration of enzyme used in this assay cannot be reliably standardized between populations, as efficiency of surface expression depends on the enzyme and varies considerably. Standard induction of a stable LHE should yield up to 10^5 individual enzymes per yeast cell [12]. Therefore this assay

is used to approximate relative activities. In order to determine more exact values for K_m or K_{cat}, or compare enzymes with a high degree of sensitivity, recombinant enzyme should be produced in and purified from bacteria.

42. Highly active, stable enzymes should show considerable cleavage after only 5 min. Cleavage of the DNA substrate should be completed by 1 h, and further incubation rarely yields any detectable increase in cleavage product. Shorter incubation times should be used when observing subtle differences between enzymes with comparable activities.
43. Supernatant can be stored at −20 °C in a light-protected container.

Acknowledgement

This work was supported by NIH grants RO1CA133832, RL1GM133833, 5RL1GM84433-04, and U19AI096111.

References

1. Takeuchi R, Lambert AR, Mak AN-S et al (2011) Tapping natural reservoirs of homing endonucleases for targeted gene modification. Proc Natl Acad Sci U S A 108:13077–13082
2. Chevalier BS, Kortemme T, Chadsey MS et al (2002) Design, activity, and structure of a highly specific artificial endonuclease. Mol Cell 10:895–905
3. Epinat JC, Arnould S, Chames P et al (2003) A novel engineered meganuclease induces homologous recombination in yeast and mammalian cells. Nucleic Acids Res 31:2952–2962
4. Baxter S, Lambert AR, Kuhar R et al (2012) Engineering domain fusion chimeras from I-OnuI family LAGLIDADG homing endonucleases. Nucleic Acids Res 40:7985–8000
5. Volná P, Jarjour J, Baxter S et al (2007) Flow cytometric analysis of DNA binding and cleavage by cell surface-displayed homing endonucleases. Nucleic Acids Res 35:2748–2758
6. Jarjour J, West-Foyle H, Certo MT et al (2009) High-resolution profiling of homing endonuclease binding and catalytic specificity using yeast surface display. Nucleic Acids Res 37:6871–6880
7. Rozen S, Skaletsky HJ (2000) Primer3 on the WWW for general users and for biologist programmers. In: Krawetz S, Misener S (eds) Bioinformatics methods and protocols: methods in molecular biology. Humana, Totowa, NJ, pp 365–386
8. Hoover DM, Lubkowski J (2002) DNAWorks: an automated method for designing oligonucleotides for PCR-based gene synthesis. Nucleic Acids Res 30:e43
9. Gietz RD, Schiestl RH (2007) High-efficiency yeast transformation using the LiAc/SS carrier DNA/PEG method. Nat Protoc 2: 31–34
10. Gietz RD, Schiestl RH (2007) Frozen competent yeast cells that can be transformed with high efficiency using the LiAc/SS carrier DNA/PEG method. Nat Protoc 2:1–4
11. Gietz RD, Schiestl RH (2007) Large-scale high-efficiency yeast transformation using the LiAc/SS carrier DNA/PEG method. Nat Protoc 2:38–41
12. Boder ET, Wittrup KD (1997) Yeast surface display for screening combinatorial polypeptide libraries. Nat Biotechnol 15:553–557

Chapter 15

Bioinformatics Identification of Coevolving Residues

Russell J. Dickson and Gregory B. Gloor

Abstract

Positions in a protein are thought to coevolve to maintain important structural and functional interactions over evolutionary time. The detection of putative coevolving positions can provide important new insights into a protein family in the same way that knowledge is gained by recognizing evolutionarily conserved characters and characteristics. Putatively coevolving positions can be detected with statistical methods that identify covarying positions. However, positions in protein alignments can covary for many other reasons than coevolution; thus, it is crucial to create high-quality multiple sequence alignments for coevolution inference. Furthermore, it is important to understand common signs and sources of error. When confounding factors are accounted for, coevolution is a rich resource for protein engineering information.

Key words Coevolution, Covariation, Protein alignment, Structure prediction, Protein design, Protein engineering

1 Introduction

Families of aligned proteins are a rich resource of structural, functional, and, evolutionary information. A standard intuition about homologous protein families is that evolutionarily conserved features—structural or sequential positions and motifs—can be seen as functionally important; if not, they would have changed through the random meanderings of evolutionary change [1, 2]. The critical point is that mutations may occur in a probabilistically random fashion, but the process by which they are maintained is non-random. Mutations that are detrimental to fitness are less likely to be retained. Thus, conserved features are hypothesized to correspond with functional importance [3].

Intramolecular coevolution—coevolution within a protein sequence—is potentially as robust a heuristic as conservation to determine functionally important positions. While conservation indicates catalytically critical residues (like Serine in a Serine Protease) or critical residue properties (like the acidic residues

David R. Edgell (ed.), *Homing Endonucleases: Methods and Protocols*, Methods in Molecular Biology, vol. 1123, DOI 10.1007/978-1-62703-968-0_15, © Springer Science+Business Media New York 2014

occupying the catalytic site of a LAGLIDADG homing endonuclease), it cannot easily indicate context-dependent conservation or conserved *interactions.*

Coevolution analysis is a statistical method that identifies important structural and functional interactions from sequence. Coevolution is detected by identifying covariation between homologous sequence positions within a protein family. Coevolving positions have been shown to be much more likely to be in contact than chance; the coevolution-contact relationship is supported by the hypothesized mechanisms of protein evolution.

While coevolution tells us much about a protein, it depends entirely on the protein alignment from which it extracts information. Alignment errors resulting from both sequence collection and sequence shift result in false positive results in not only coevolution inference [4, 5, 6], but also phylogenetic inference [7, 8]; many coevolution statistics cannot differentiate between covariation due to error or real evolutionary signal.

The sensitivity of coevolution statistics to error creates a necessity for protein family alignments of the highest quality. Errors in analysis can easily propagate and lead to false conclusions. There is a major tradeoff in sequence selection because the inclusion of homologous proteins with non-identical function (e.g., paralogues) can lead to false-positive identification of coevolving pairs [6]; conversely, it is crucial to maximize the number and diversity of the sequence alignment in order to produce accurate predictions [9]. Furthermore, the alignment of the putatively orthologous members of the protein family is equally crucial as alignment errors are known to produce false-positive results [4, 5, 6].

Manual curation of an alignment—inspecting and revising an alignment by shifting and deleting sequences after the initial alignment procedure—is a crucial step to obtaining the highest quality coevolutionary signal. Several recent studies have shown that human curators outperform automated algorithmic solutions on protein structure and alignment problems [10, 11]. The drawback of manual curation of a protein alignment is the additional time needed and the potential for human error—in a large alignment it is easy to overlook alignment errors. The compromise solution between automation and accuracy is covariation-guided curation, where software identifies the potentially erroneous region, but a curator is required to make an ultimate decision on the final alignment. LoCo [6] uses a covariation-based heuristic, local covariation, to identify systematic misalignments within a modified Jalview [12, 13] interface for real-time alignment editing.

Herein we describe how to collect sequences using PSI-BLAST, build a structure-guided master–slave sequence alignment using Cn3D, refine that alignment using LoCo, calculate coevolutionary statistics using the MIpToolset, and finally visualize the network of coevolving pairs in contact map and network format.

This analysis pipeline is thorough and complete and has lead to the identification of putative coevolving pairs in many protein families. However, the pipeline can also be viewed as somewhat modular, in that it is possible for an expert user to replace a given step of the analysis with a different methodology. For example, Hidden Markov Model-based homologue collection [14, 15] may provide a different set of homologous protein sequences than PSI-BLAST [16], and precomputed alignments like those in CDD [17] and PFAM [18] can be used as time-saving resources; likewise, there are many different sequence alignment methods including MAFFT [19], or PRANK [20] which could be used in place of MUSCLE [21] to align sequences when no structural information is available. Nonetheless, we must advise caution when making such substitutions in the analysis pipeline. Coevolution analysis is not robust to the many sources of error that routinely arise in homology detection and sequence alignment [4, 6]. A thorough analysis in early steps will yield more and more reliable coevolution data.

2 Materials

Since this is a bioinformatics approach, the only physical requirement is a computer with a working Internet connection. All software and databases are freely available online. Some required software expects a Unix-like interface, so Mac OS X and Linux users should find the setup straightforward; Windows users will have to emulate this by installing additional software for certain steps.

2.1 Computer and General-Purpose Software

1. Computer with an operating system that has a Unix-like interface. Common examples of Unix-like operating systems include Mac OS X and Linux. This solution has been tested on Mac OS X 10.5 through 10.8 and Ubuntu Linux. Windows users, *see* **Note 1**.
2. GCC, the GNU compiler collection, is used to build some software outlined later (http://gcc.gnu.org/). The gcc compiler and make utility are necessary to build some platform-agnostic tools from source code (*see* **Note 2**).
3. Perl: a programming language commonly used in bioinformatics applications (http://www.perl.org/get.html) and preinstalled on most Unix-like operating systems.
4. Java: a common portable programming language (http://www.java.com/en/).

2.2 Sequence Collection Tools and Databases

1. Target sequence: the FASTA sequence of a protein you are studying or of a representative member of the family that you are studying (*see* **Note 3**).

2. Target structure (optional): the PDB ID of the 3D structure of the target sequence or an orthologue of the target sequence. If a structure is available, we will build an alignment using section 3.3. If a structure is not available, we will use sequence-only tools for building the alignment in 3.4.
3. PSI-BLAST (Position-Specific Iterated Basic Local Alignment Search Tool) is a search tool designed to infer sequences that are homologous to a query sequence [16] (ftp://ftp.ncbi.nih.gov/blast/executables/LATEST/). It is available as part of the BLAST+package. Executables are available for many different operating systems.
4. There are many databases available for sequence collection. Suggested databases for this pipeline include nr and optionally nr_env (ftp://ftp.ncbi.nih.gov/blast/db/) (*see* **Note 4**).
5. readseq is a Java-based bioinformatics file format conversion tool [22]. Download *readseq.jar* from (http://iubio.bio.indiana.edu/soft/molbio/readseq/java/).

2.3 Sequence Alignment Tools

1. Cn3D is a structure-guided sequence alignment tool (http://www.ncbi.nlm.nih.gov/Structure/CN3D/cn3dinstall.shtml) used to build and curate the CDD database [23].
2. MUSCLE is an iterative sequence alignment tool (http://www.drive5.com/muscle/) [21] which seems to have parameters well suited for creating protein alignments for coevolution analysis at the time of publication of this manuscript.

2.4 Alignment Curation Tools

1. LoCo [6] (http://sourceforge.net/projects/locoprotein/) is a modified version of the popular Jalview [12, 13] alignment tool. LoCo is used to curate (i.e., validate and improve) protein alignments (*see* **Note 5**).

2.5 Coevolution Analysis

1. The MIpToolset [4, 5] (http://sourceforge.net/projects/miptoolset/) is a collection of C and Perl programs that calculates several popular Mutual Information-based coevolution metrics, *MI* [9, 26], *MIp/Zp* [27], *Zpx* [4], and ΔZp [4].
2. Graphviz is visualization software used to display networks of coevolving residues [29] (http://www.graphviz.org).
3. R is a statistical programming language which can be used to plot contact map and contact map figures [28] (http://www.r-project.org).

3 Methods

3.1 Building and Installing Bioinformatics Tools

1. Download LoCo from (http://sourceforge.net/projects/locoprotein/) and the MIpToolset from (http://sourceforge.net/projects/miptoolset/). Unzip these programs into the desired install location (e.g., your *bin* or *Applications* folder).

2. Build LoCo by following the directions in *README_LoCo.txt*, which is in the main LoCo directory. In brief, open a Terminal window on your computer. Change directory (*cd*) to the location where LoCo was unzipped. Then change directory to the *dist/MIpToolset_jalview/MIP_C_CODE/*directory and then type *make* to build the program.
3. The MIpToolset is built similarly; instructions are found in the *README* file and additional information is found in the *DOCUMENTATION* folder. Inside the main MIpToolset directory is a directory called *MIP_C_CODE*. Change directory to the *MIP_C_CODE* subdirectory and run *make*, by typing *make* into the terminal. This will read the *makefile* and build two programs which calculate coevolution statistics and inter-residue distances, *MIp* and *dist_pdb*. An MIpToolset wrapper Perl script will access *MIp*, *dist_pdb*, and other Perl scripts when calculating covariation scores.
4. Additionally, you should make sure to add the installation directory to your *$PATH* environment variable by following the directions according to your operating system.

3.2 Collecting Protein Sequences and Structures

1. Search for protein sequences that are homologous to your target protein by running the PSI-BLAST program installed in **step 3** of Subheading 2.2 (*see* **Note 6**).
2. Store your target sequence in FASTA format in a file called *"target.fa"* (*see* **Note 7**).
3. Download the *nr* and *nr_env* databases and store them in an accessible directory like*/data/* (*see* **Note 4**).
4. Select an initial *Expectation value (E value)* cutoff by using the following heuristic: *1E-X* where *X* is *(the length of the target sequence/10)*. Round *X* to the nearest whole number. For example, if the protein is 200 residues long, choose *1E-20* (*see* **Note 7**).
5. Run PSI-BLAST by executing the following from the terminal with several substitutions:

   blastpgp -i target.fa -j 12 -d DATABASE_PATH -m 4 -e E_VALUE -I t -o psiblast.out.

   Substitute *DATABASE_PATH* with the location of the nr database and the name of the database from **step 3** of Subheading 3.2; for example, if the database is stored in/*data/*then replace *DATABASE_PATH* with*/data/nr*. Replace *E_VALUE* with the value selected in **step 4** of Subheading 3.2; for example, *1E-10* if the protein length is 100. Therefore, an example with substitutions is:

   blastpgp -i target.fa -j 12 -d/data/nr -m 4 -e 1E-10 -I t -o psiblast.out.

6. Ensure that the PSI-BLAST converged by inspecting the *psiblast.out* file created by the *blastpgp* command.
7. Convert your blast output into FASTA format using *readseq.jar*. This program is a java archive (*.jar*) file and is run from the command line terminal using Java. In the following, it is assumed that the readseq.jar file is in the current directory; if it is not, simply replace *readseq.jar* with the relative or absolute path to the file. Type the following into the command line terminal:

   java -jar ~/Downloads/readseq.jar -degap=- -f 8 psiblast.out.

   which should create the file *psiblast.out.fa*.
8. Inspect your alignment to determine whether you have enough sequences to continue. Ideally, you want at least 150 orthologous sequences with less than 90 % sequence identity, so you should have approximately 300 sequences in the dataset as a minimum to ensure sufficient sequences following the downstream filtering steps. Consider completing Subheading 3.2 again using *nr_env* in place of *nr* to collect even more sequences.

3.3 Building a Structure-Guided Sequence Alignment (Alternately, Use Subheading 3.4 for a Sequence-Only Alignment if No Structure Is Available)

1. Network Load your target structure in Cn3D by selecting the *Network Load* option from the *File* menu and entering either the PDB or MMDB identifier. This will render the protein structure in the open *Structure Viewer* window and open the corresponding sequence in the *Alignment Viewer* window (*see* **Note 8**).
2. Adjust the appearance of your target structure by selecting *Style* → *Rendering shortcuts* → *Tubes,* which aids in comparing structural alignment, and *Style* → *Coloring shortcuts* → *Sequence Conservation* → *Fit,* which colors the sequence and structure according to the position specific scoring matrix (PSSM) of the sequence alignment (*see* **Note 9**).
3. Open the *Import Viewer* window by selecting *Imports* → *Show Imports.* Arrange your windows so that the Structure, Alignment, and Import windows are all visible. Import new structures into the *Import Viewer* window by selecting *Edit* → *Import Structure* and then Via *Network* to enter the PDB or MMDB identifier, or *From a File* followed by a .pdb file (Fig. 1) (*see* **Note 10**).
4. Attempt to *Merge* all the new structures from the *Import Viewer* into the *Alignment Viewer* by selecting *Alignments* → *Merge All.* There may be conflicts when your new structures are imported (residues highlighted in pink in the *Import Viewer*); conflicts will prevent structures from being added to the *Alignment Window* via the *Merge All* command.
5. Cn3D separates an alignment into *block* and *gap* sections, defined by the row labeled *blocks* located above the alignments

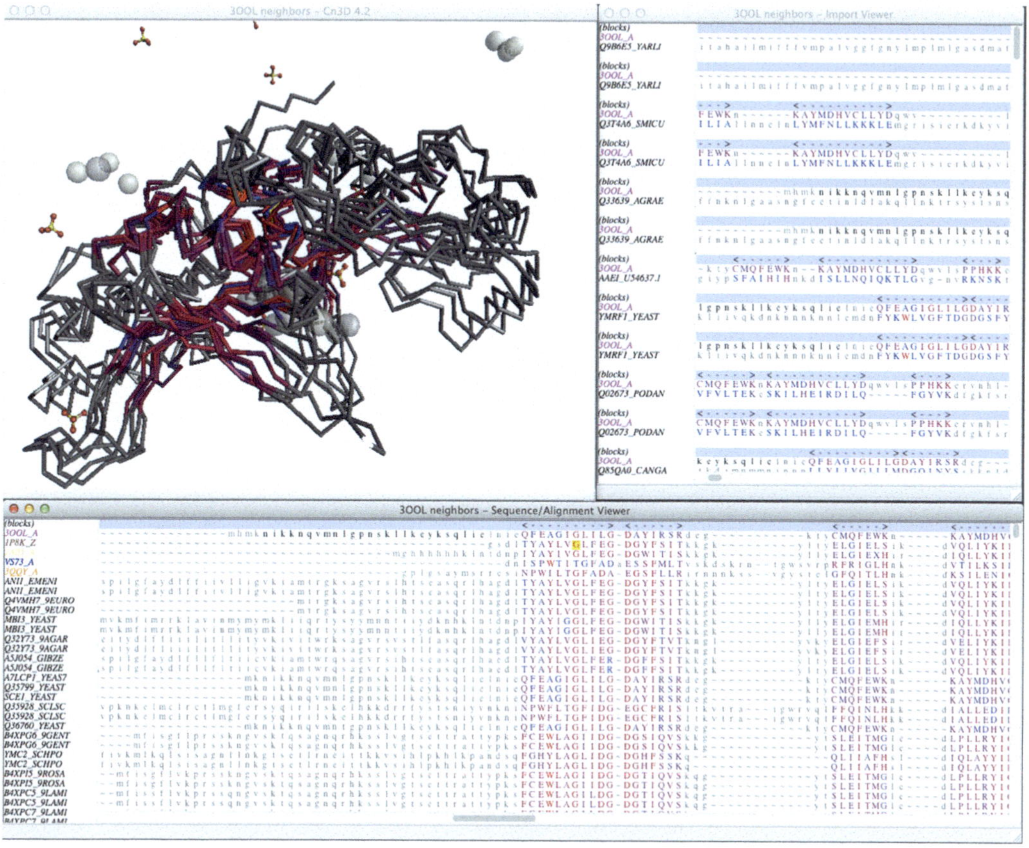

Fig. 1 Screenshot of the three windows comprising the Cn3D workspace showing an analysis of the LAGLIDADG Homing Endonuclease family. The *Structure Viewer* window (*top left*) shows a protein structure alignment of the protein family. The *Sequence Viewer* window (*bottom*) shows the sequence alignment which corresponds to the structure alignment with additional sequences from the protein family. The *Import Viewer* window (*top right*) shows sequences which have been imported into Cn3D, but have not been added to the sequence alignment

in the *Alignment Viewer*. View the *blocks* in the *Alignment Viewer* by selecting *Edit → Enable Editor*. Each block defines a structurally conserved segment of the alignment which cannot accept insertions or deletions. Adjust the each block in the *Alignment Viewer* so that the new structures in the *Import Viewer* no longer conflict; *Split, Merge, Create, Delete* (under the *Edit* menu), and *Horizontal Drag* (under the *Mouse Mode* menu) the blocks to resolve all pink conflicts in the *Import Viewer* window (*see* **Note 11**).

6. Transfer the structures from the *Import Viewer* into the *Alignment Viewer* by selecting *Alignments → Merge All* in the *Import Viewer* window. Save your work to a file by selecting *Save* from the *File* menu in the *Structure Viewer* window. After saving, reopen the file to view the imported structures.

Fig. 2 An example of a shift error added to a segment of a LAGLIDADG alignment. A vertical column in the sequence alignment was selected which highlights the corresponding residues in *yellow* in the *Structure Viewer*. One of the structure's helices is out of alignment by one residue turn. This should be corrected by shifting the corresponding sequence in the sequence alignment

Cn3D will not render the newly imported structures until the structures are *merged* into the *Alignment Viewer* and the file is saved and reopened.

7. Select *File* → *Realign Structures* to superimpose the newly merged structures in the *Structure Viewer* as defined by the sequence relationships in the *Alignment Viewer*. Repeat this step whenever a change is made to the blocks or any structure in the *Alignment Window*, to update the structure alignment (*see* **Note 12**).
8. Now that all structures are visible, revise the blocks to reflect the structural diversity of the protein family. Each block defines a conserved and critical structural feature that cannot accept insertions or deletions. Adjust the blocks in the *Alignment Viewer* so that they correspond to the structurally conserved core of the protein. Create gaps between the blocks in regions of structural divergence where insertions or deletions would be tolerated. As in **step 5**, use the options in the *Edit* and *Mouse Mode* menus to make changes (*see* **Note 13**).
9. Structure alignment algorithms are vulnerable to "shift errors," where major structural elements are superimposed, but individual residues are not (Fig. 2); for example, two beta sheets may be superimposed, but in a configuration where aligned residues' side chains point in opposite directions.

Search the alignment for shift errors by highlighting aligned residues (*Mouse Mode* → *Select Column* in the *Alignment Viewer*) and examining the corresponding structural alignment in the *Structure Viewer*; highlighted residues are highlighted in yellow in both windows. Select alternate residues in the *Structure Viewer* by double clicking. Use both sequence homology and structural evidence to revise the structure alignment. Be thorough, as errors at this step will be propagated through the rest of the alignment creation process.

10. Import the sequences collected by *PSI-BLAST* by selecting *Edit* → *Import Sequences* from the *Import Viewer* window; then select *From File* and finally the name of the *PSI-BLAST* fasta file. These imported sequences will behave as the structures did in this window.
11. Align the new sequences to the target structure by selecting *Algorithms* → *Block Align N*; this procedure aligns sequences in the *Import Viewer* to the PSSM of the sequence alignment in the *Alignment Viewer*. After the block align procedure is complete, merge the sequences that fit with the existing block model (*see* **Note 14**).
12. Sort the sequences by selecting *Edit* → *Sort Rows* → *By Score* and then *Edit* → *Sort Rows* → *Float PDBs* in the *Alignment Viewer* window. Inspect the alignment in the *Alignment Viewer*. The *Fit* coloring scheme will color residues red that fit will with the PSSM for that position and blue for residues that fit poorly. Each sequence in the *Alignment Viewer* contributes to the PSSM, so poorly aligned sequences should be moved back to the *Import Viewer* by selecting *Imports* → *Realign Rows from List*.
13. In the *Import Viewer* perform a *Block Align N* to realign the sequences to the expanded PSSM defined by the *Alignment Viewer*. Once again *Merge All* the sequences into the Alignment Viewer.
14. Repeat **steps 12** and **13** iteratively until the alignment in the *Alignment Viewer* contains as many homologous sequences that will fit with both the block model and the sequence fit (*see* **Note 15**).

3.4 Creating a Sequence-Only Alignment (Alternately, Use Subheading 3.3 for a Structure-Guided Alignment)

1. Run *muscle* on the collected sequences by opening a command line terminal and running: *muscle -in psi_blast.fasta -out alignment.fasta* where *psi_blast.fasta* is the file containing the FASTA-formatted sequences you collected in Subheading 3.2 (*see* **Note 16**).

3.5 Curating and Validating an Alignment

1. Follow LoCo instructions to start the software; enter the LoCo directory and run the shell script by typing:

   ./run_loco.sh

   Open your alignment from the *File* menu by selecting *Input Alignment - from file* (*see* **Note 17**).
2. Set the color scheme from the Color menu; it is good to start with a general-purpose color scheme like Zappo. Zappo colors the sequence by amino acid properties so patterns can emerge visually (*see* **Note 18**).
3. Sequences are manipulated in LoCo (and Jalview) as follows: Selected sequences are moved up and down within the alignment using the up and down arrow keys. Sequences are deleted from the alignment by selecting a sequence name and pressing *delete/backspace* on the keyboard. Specific regions of the sequence are altered by selecting the residue(s) in a red box by left clicking and dragging; the selected area is where the sequence alterations will occur. To delete the selected residues or gap characters, press *delete/backspace.* To realign a selected region, hold *control* on the keyboard while left clicking and dragging; you must select some gap characters when realigning sequence in order to have empty space to move the residues into. These operations will be used during the curation process (*see* **Note 19**).
4. Identify and remove incomplete sequences by examining the alignment and looking for sequences that are not long enough to span the length of the alignment. Truncated sequences in the N and C termini should be removed (*see* **Note 20**).
5. Inspect each column for gap characters in the overview window and delete and sequences that are missing crucial parts of the protein. The inclusion of a gap character at a position is the implicit acknowledgement of structural uncertainty at that position; in order for a position to be analyzed, sequences that contain gaps at that position must be deleted (*see* **Note 21**).
6. Under the alignment, there is a heuristic for judging the quality of the alignment at that position, Local Covariation. Local Covariation identifies regions of a protein that are likely to be misaligned. Use Local Covariation as a guide to inspect positions in the protein (Fig. 3).
7. In each position that has *high* local covariation (a yellow-colored bar) complete this procedure. Select the region of local covariation (which includes the contiguous positions with a yellow bar and five positions to the right) and select *Calculate—Calculate Tree—Neighbour Joining Using % Identity* from the menu bar. Then select *Calculate–Sort—by Tree Order*. Inspect your newly clustered region for positions

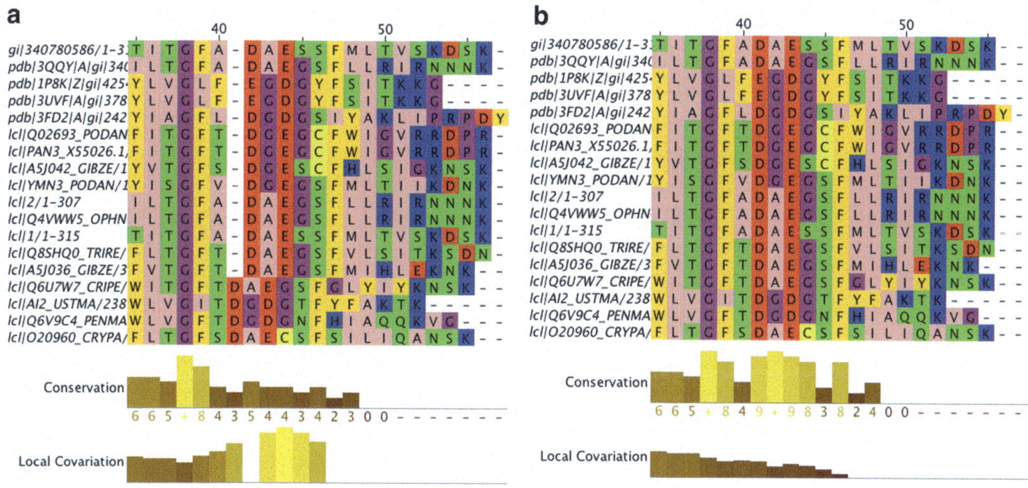

Fig. 3 (**a**) A screenshot of the LoCo workspace area which includes a misalignment. The misalignment can be identified by the high local covariation seen in the *Local Covariation* histogram below the sequence alignment itself. (**b**) A screenshot of the corrected alignment shows improved conservation, reduced local covariation, and improved alignment according to the Zappo color scheme

that appear to be shifted out of alignment and realign them as outlined in **step 3**. If the realignment is correct, the local covariation score should go down (*see* **Note 22**).

8. Another potential source of local covariation signal is the inclusion of paralogous sequences—sequences that are similar by because of shared ancestry, but have diverged due to a gene duplication event and now have a new function. Removing paralogous sequences can also decrease the local covariation score and increase downstream accuracy.

9. Finally, identify and remove any sequences that do not conform to known properties of the protein. For example, serine proteases should contain a serine in the catalytic position, or LAGLIDADG homing endonucleases should contain acidic residues in the catalytic positions.

10. Restore the target sequence to the beginning of the alignment by selecting it and using the up arrow key to move it back to the top. Save a copy of this alignment.

11. Trim your alignment by selecting the column which contains the first residue in the target structure and click *Edit → Remove Left*. Trim the C-terminus by selecting the column containing the final residue in the target sequence and clicking *Edit → Remove Right*.

12. Remove redundant sequences by selecting the entire alignment (*Select → Select All*) and then clicking *Edit → Remove Redundancy*. Then adjust the redundancy threshold to *90* and

then click *Remove*. Save this alignment; it will be used as input for the coevolution analysis. Be careful not to remove your target sequence when removing redundancy as it is required for ensuring that numbering is correct in Subheading 3.5. It is crucial that at least 125 sequences are in the final alignment.

3.6 Coevolution Analysis

1. Ensure that your alignment has enough sequences and ungapped positions to perform a coevolution analysis. At this point the alignment should contain at least 125 sequences and at least 50 ungapped positions.
2. Ensure MIp.pl, MIp, and dist_pdb have been compiled and placed in the $PATH shell variable. You can verify this by typing each of the program names into your shell interface which will print out a help message for MIp.pl and a brief message from MIp and dist_pdb.
3. Run the following if you have a pdb file:

   MIp.pl -i alignment.fasta -o coevolution.txt -d T -p pdb_file.pdb -e 0.001 -a 1.

   Run the following if you do not have a pdb file

   MIp.pl -i alignment.fasta -o coevolution.txt -d F -e 0.001 -a 1 (*see* **Note 23**).
4. Verify that MIp.pl has executed successfully by examining the main output file, *count* file and three *.dot.* files.
5. Open the MIpToolset output file and verify that the *aa_c* and *aa_d* columns correspond to the *PDB_i* and *PDB_j* columns respectively. These columns represent the residue identity at the assigned number in the sequence and structure. If these values do not align, this means that the sequence and structure begin numbering in two different places; not all proteins begin their numbering at 1. Use the *-a* option to set the offset between the sequence and PDB numbering (*see* **Note 24**).

3.7 Visualizing and Interpreting Results

1. The raw data is produced as a tab-delimited table so it is viewable in a standard spreadsheet program or statistical programming language. Excel users can append *.xls* to the end of their coevolution output (e.g., *coevolution.txt.xls*) to make Excel interpret the file as an Excel Spreadsheet. The data can be explored by sorting in descending order in the *Zp*, *Zpx*, and *ndz* (which is ΔZp) columns (*see* **Note 25**).
2. MIpToolset will create files with the extension *.dot* appended to the end of the given output file name. Open these files in *graphviz* to see the network of potentially coevolving residues as defined by the coevolution statistic. Labeled circles represent positions in the protein alignment numbered according to the target sequence/structure. Lines connecting circles represent covariation; line thickness indicates the strength of the coupling (Fig. 4) (*see* **Note 26**).

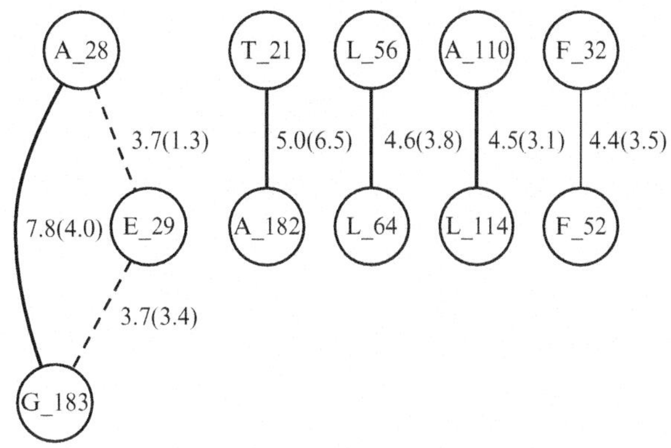

Fig. 4 A network representation of coevolving residues. This network represents a small portion of a full network of coevolving residues from the LAGLIDADG Homing Endonuclease protein family. Each node in the network corresponds with a residue position in the sequence alignment, labeled by both residue number and identity. Each edge in the network represents a potential interaction via coevolution. Edge thickness corresponds to the coevolution score (in this case the coevolution statistic *Zpx*). Edges are labeled by coevolution score, followed by inter-residue distance measured in angströms in *parentheses*

3. Visualize the coverage and accuracy of your results by plotting the data in *contact map* format. Create an *XY Scatter*-style plot where the x and y axes are defined as positions in the protein and open circles represent positions less than 6 Å apart in structure. Then plot smaller filled circles as the top-scoring pairs according to the selected coevolution statistic. The coevolution prediction coverage and correspondence with residue contacts provides an indication of the quality of the predictions by the coevolution statistic (Fig. 5) (*see* **Note 27**).

4 Notes

1. Windows users have a number of options for creating a Unix-like interface. One option is to install Linux on the Windows machine, e.g., Ubuntu (http://www.ubuntu.com/). Another option is to install Cygwin, a program that emulates a Unix-like interface without installing a new operating system (http://www.cygwin.com/). Both of these options work with this method; however, a native unix-like environment is preferred, and the Cygwin workaround is not supported. If you are going to use the Cygwin workaround, make sure that you install the sequence analysis tools in a directory where Cygwin has the ability to read and write, like your home directory inside the Cygwin installation; also, make sure that Cygwin and the other bioinformatics tools are not blocked by your antivirus software.

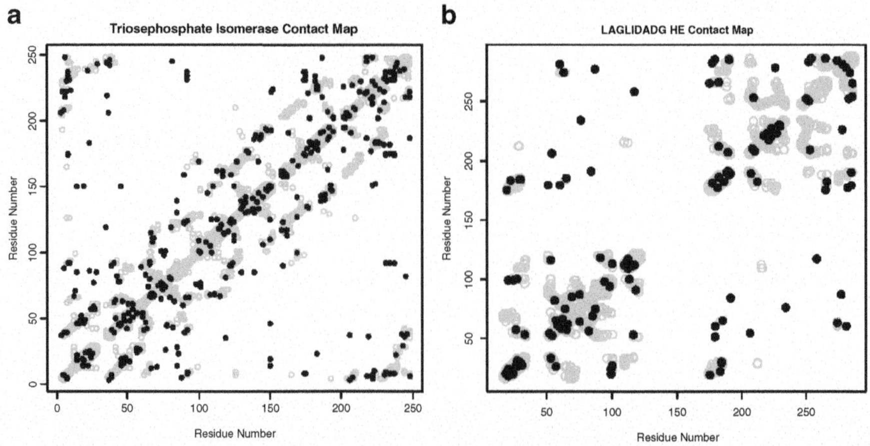

Fig. 5 Predicted contact map visualization of a coevolution network. Contacting residues are labeled *grey*. Predicted potentially coevolving residues (*Zpx*>3) are printed on top of contacting residues and colored *black*. The correspondence between contact and predicted coevolving pair is apparent in this visualization. (**a**) Represents near-optimal results obtained from analyzing a Triosephosphate Isomerase alignment with optimal characteristics. (**b**) Represents typical-use results obtained from a LAGLIDADG HE alignment with fewer sequences, and a less-confident alignment

2. gcc is a program that builds software tools from source code. There are many ways to install gcc on your respective platform. For Mac OS X users, gcc is included as part of an add-on to the Xcode tool available on the Mac App Store, presently. Download Xcode from the App Store (https://itunes.apple.com/ca/app/xcode/id497799835); once Xcode is installed, the *Command Line Tools* are available from the *Downloads* section of *Preferences*. For Ubuntu users, type *apt-get install gcc* into a terminal window to download gcc. For those using the unsupported Windows Cygwin workaround, gcc is installed using the same *setup.exe* that installs Cygwin itself. Install gcc as is instructed for your respective operating system.
3. Your target sequence is ideally the wild-type sequence for your protein family of interest from the species of interest. In most cases this will be the human sequence, but in others it may be from the model organism you are studying or corresponds with a crystal structure you are interested in.
4. BLAST databases are split into multiple files, but addressed through the BLAST command line interface through the database name only. For example, the nr protein database is split into numbered files labeled *nr.00.tar.gz*, *nr.01.tar.gz*, *nr.02.tar.gz*. These files can be downloaded through your Web browser, favorite FTP client, or through the update_blast.pl Perl script included in the aforementioned BLAST+ suite. Details are available from NCBI: ftp://ftp.ncbi.nlm.nih.gov/blast/documents/blastdb.html.

5. Discriminating homology (sequences similar by common ancestry) and non-homologous analogy (sequences similar by convergence or coincidence, i.e., homoplasy) is a major challenge of bioinformatics. It is critical to understand that sequence similarity and not sequence annotation must be used to collect homologues. Annotation errors or incomplete annotations will lead to an erroneous and incomplete final protein family alignment. The ubiquitous BLAST tool [31] can be used to infer homology by sequence similarity, but struggles to identify highly diverged homologues because it uses generic substitution matrices like *BLOSUM* [32]. PSI-BLAST uses a similar search strategy, but builds position-specific scoring matrices tailored to the protein family over multiple iterations of searching [16]. The PSI-BLAST strategy improves detection of divergent homologues.
6. PSI-BLAST is similar to BLAST in that it is a software tool which searches for homologous protein sequences; however, PSI-BLAST uses an iterative search strategy which improves detection of sequentially dissimilar homologues. Briefly, PSI-BLAST employs a *Position-Specific Scoring Matrix* (PSSM) which defines how favorably an amino acid will be scored on a position-by-position basis as defined by the growing alignment of matches. PSI-BLAST performs multiple rounds of searching; an updated PSSM calculated and used each round. The fact that custom scoring matrices are used for each position in the protein rather than a generic scoring matrix generated by averaging across many positions in many proteins gives PSI-BLAST an advantage in detecting homology [33].
7. While the precise cutoff value for your PSI-BLAST may need to be revised, the starting point provided in **step 3** of Subheading 3.2 is a good heuristic. If the search returns too many dissimilar sequences (i.e., paralogues), then make the search more strict. Conversely, if the search returns too few sequences, make the search less strict. To increase the speed of the search for the ideal cutoff, adjust the E value by several orders of magnitude and observe the effect; much time is wasted changing an E value by a factor of 2 or 3. Determining whether sequences are orthologous or paralogous is difficult and requires knowledge about the protein family. A lower score compared to the rest of the protein family may indicate homology with functional divergence. Other evidence includes (1) bidirectional "top matches" between two organisms, (2) including only one sequence per organism, and (3) the absence of critical functional residues may indicate alternate function and thus paralogous sequences.
8. If your protein family is well studied, there may be an existing *Conserved Domain* [17] or *PFAM* [28] available as a starting point.

Search the *Conserved Domain Database* (*CDD*) (http://www.ncbi.nlm.nih.gov/Structure/cdd/cdd.shtml) for your target protein family. If an entry exists, open the *CDD* structure alignment by expanding the *Structure* section, selecting the maximum value for *Aligned Rows* and clicking the *Structure View* button. You will still need to perform the curation steps as even these database alignments can contain errors.

9. *Cn3D* uses a *Position Specific Scoring Matrix* (like *PSI-BLAST*) to indicate alignment quality and align sequences that do not have structural information. Each position in the alignment has a different scoring matrix which defines which residues are deemed favorable or unfavorable. The PSSM is calculated based on all the sequences included in the *Alignment Viewer*.
10. When choosing structures to import, the goal should be diversity. An attempt should be made to include structures which encompass the diversity of the protein family.
11. The strength of Cn3D as tool for creating protein family alignments is its ability to explicitly define regions of structural similarity and divergence. The underlying hypotheses of coevolution analysis are structural in nature, as inter-residue contact is used as a proxy for coevolution when benchmarking covariation statistics. Thus, a structure-based view of homology is critical when creating alignments. The goal of creating a Cn3D alignment is to maximize the number of structurally defined positions that are conserved among the entire protein family—this is the function of the *Block* regions. Defining a position within a *block*, the aligner asserts that this position is structurally defined and required for membership in the protein family. Understandably, parameters of the protein family will dictate the number and length of each *block* region: a recently derived, well-conserved, orthologous protein family will contain fewer, longer *blocks*, with fewer shorter gaps, while a more diverse and diverged family will contain fewer, shorter *blocks* and numerous longer gaps. As a general rule, the core of the protein should be easily defined in blocks as there is less structural divergence; conversely, segments of unstructured surface loops typically are not included in *blocks*, as they accept insertions and deletions, and are less likely to be structurally superimposable, and thus, defined. Structurally speaking, alignment within diverged gap regions is not meaningful because homology cannot be inferred. The Cn3D documentation advises that only regions within *blocks* are considered "aligned"; this is why positions that contain gaps are not included in a coevolution analysis by the MIpToolset.
12. The structure alignment superimposition is defined entirely by the sequence alignment window. The structures are rigid and will be rotated and translated relative to one another in order

to minimize the distance between alpha carbons of residues which occupy the same column in the alignment window. The menu command *File → Realign structures* must be selected every time you want the *Structure Viewer* window to be updated to reflect the current *Sequence Viewer* alignment. Normally, the structural superimposition reflects all positions in block regions in the sequence alignment; however, it is also possible to further restrict the structure alignment to highlighted/selected positions. This option is very useful in determining whether a segment of the alignment is structurally acceptable: an incorrectly aligned segment may appear to be aligned correctly because other correctly aligned positions will "outvote" the poorly aligned segment and create the illusion of correct structural superimposition between the rigid structures. To align based on a single segment, highlight it and answer *Yes* to the prompt; to align based on all *block* positions, answer *No*.

13. The placement of *block* regions is critical for generating a high quality alignment. While the *blocks* that are already defined in the Conserved Domain Database (CDD) [17] are generally very good, they are subject to human error and potential misalignments have been identified [4, 6] that could be attributed to human error in this important step: setting the *blocks* to reflect structural conservation. Blocks should extend across the conserved core of the protein, through any region that would not accept an insertion or deletion without drastically affecting the protein fold. Secondary structural elements should individually be included as a single block; although rarely alpha helices can accept a single residue insertion which will require a *block* to be split. Regions of structural divergence should not be included in *blocks*. These include surface loops, long unstructured linker regions, and sometimes the N- and C-termini. It is not uncommon to see a secondary structural element at a terminus need to be excluded from a *block* because a wild-type orthologue does not contain this structure.

14. It is possible for the majority of sequences to not merge because the structures available in the structure alignment imply a block model that is too restrictive. If a conflict exists in a region of structural conservation where a gap is structurally possible, though not observed, consider splitting the block to allow the small insertion or deletion in that region. Likewise, watch out for structured, though non-critical N- and C-terminal features which can be excluded from the block model; though, do not delete important features at either terminus because partial sequences will not merge. As the curator, it is your job to make these important interpretations for your protein family.

15. With each iteration, the PSSM encompasses more diversity within the protein family and (when done properly) includes fewer errors. Multiple iterations are required because dissimilar members of the protein family are likely misaligned in the initial iterations, but will be correctly placed once the PSSM represents the entire protein family accurately.
16. While it is still possible to obtain a high quality set of coevolution predictions using an alignment built from sequence alone, it is more difficult than using a structure based alignment. A greater emphasis will need to be spent on Subheading 3.4, Curation, if **step 3.4** is used instead of **3.3**. As well, it is difficult to evaluate the accuracy of the predictions; the standard benchmark is to use the fraction of pairs of covarying positions above a threshold that are in contact as a proxy for true coevolution. Some even infer indirect coevolution between non-contacting pairs if other comparably scoring pairs are in contact [34]. Without structural information, this evaluation is impossible.
17. For some users, an example file will automatically open many more windows than is necessary for LoCo to function on an unrelated example protein family. This will happen every time upon start up unless it is disabled. Uncheck the box located in *Tools* → *Preferences* → *Visual* → *Open File* to disable this example file upon start up of the tool. Close and reopen the program.
18. It is possible to get a holistic view of the alignment by using the *Overview Window*. This window provides rapid navigation by clicking on the region you wish to travel to. As well, it provides a way of looking for alignment abnormalities without scrolling through the entire alignment in detail. Some problems will be more apparent more quickly in the *Overview Window*. Look for sequences with long gaps or abnormal gap placement. As well, look for sequences that do not fit with the colored "motifs" of the other sequences. A sequence that looks like it "doesn't belong" likely contains a shift error or not homologous. The ideal setup for curating an alignment is across two monitors: one monitor for editing the alignment and one for examining the entire alignment in the *Overview Window*.
19. If the *Local Covariation* bar is empty for the entire length of your alignment, this may be an indication that the *MIp* program did not build properly or does not have write permissions in its current directory. This is especially problematic for (unsupported) Windows users who are emulating a Unix-like interface using *Cygwin*. Consult the known_issues.txt and README files at (http://sourceforge.net/projects/locoprotein/files/) for more assistance. As well, make sure that your target (i.e., first) sequence does not contain any non-canonical amino acids, as this can affect numbering. The easy solution to this problem is to use the up and down arrow keys to shift

a different sequence into the top (target) position temporarily. (Remember that you can use the arrow keys to shift sequences up and down but never left or right; this will cause the *entire sequence* to shift visually and cause alignment problems.)

20. Do not grow too attached to the sequences in your alignment. Do not hesitate to delete the ones that are incomplete or seem otherwise erroneous. Incorrect sequences are sources of error in downstream analysis.
21. Often, precomputed alignments in major databases contain too many gaps to be used for coevolution analysis. For example the *Full* PFAM alignments for the single-chain two-domain LAGLIDADG homing endonuclease (PF00961) contains sequences that only cross one of the two domains. Columns that contain gaps are not included in the coevolution analysis.
22. Remember that coevolution scores are not calculated on gapped positions. When realigning a segment of an alignment to reduce the Local Coevolution score, be careful you do not simply erase the signal by introducing new gaps. Look for alternate alignments that are supported by sequence, structure, and Local Covariation-based evidence.
23. Some PDB files contain additional formatting options which are not parsed properly by the MIpToolset. If PDB and alignment residue identities do not match, search the PDB file for missing lines or additional *HETATM* lines interspersed throughout the *ATOM* definition lines.
24. Positions that contain gaps are excluded from analysis by MIp; be sure to inspect the FASTA alignment file. There are many reasons why the distance information does not match the alignment file (as indicated in by a mismatch between the residues in the coevolution output file). PDB files may omit or change some residues which are present in the wild-type sequence. They also may include additional *HETATM* lines mid-chain, or label *ATOMS* as *HETATMs*. You may get different results depending on where you obtain the PDB file as well; experience has shown that PDB files obtained from the RCSB Protein Data Bank may be slightly different than PDB files obtained from NCBI. As well, be sure that the first sequence in the FASTA alignment is the same as the structure input as a PDB.
25. Do not misunderstand the use of *Z*-scores in the statistic Zp. MIp is approximately normal if generated on random sequence which is why MIp is transformed into *Z*-scores in the statistic Zp. If a pair of positions has a score that falls outside the normal distribution, it implies that the pair is coevolving. *Zpx* scores are comparable to Zp scores, though slightly more accurate. ΔZ is a heuristic and does not have a rigorous statistical framework.

26. A network representation of covarying positions is an excellent way of gaining a holistic view of the potentially coevolving positions. Every node in the network represents a position in the protein numbered according to the target sequence. Every *edge* (line) in the graph corresponds with a comparatively high covariation score; thicker lines represent stronger coupling. The actual covariation score and distance label each edge—i.e., *4.5(5.0)* represents a pair with a covariation score of 4.5 and a distance of 5.0 Å between the two closest non-hydrogen atoms of the respective residues in the target structure. If no structure data is specified, the distances will appear as *0.0*. Interestingly, some covarying pairs may be in contact between protein chains; intra-chain distances will indicate that the positions are not in contact, but will appear in contact when viewing the quaternary structure of the protein.
27. Contact Maps are an excellent way to view the relationship between covarying positions and the protein's secondary and tertiary structure. Here we define contact as any non-hydrogen atom less than 6 Å apart. In a contact map, the *X* and *Y* axes represent the position in the target sequence; thus, the points that run along the diagonal of the plot represents *"sequence-local interactions"* and off-diagonal points represent *"sequence-distant interactions."* Contact always occurs along the diagonal because a protein is a single chain. But other common interactions are visible in the contact map as well, like interactions between termini, which manifest as contacts in the furthest corners of the contact map. Furthermore, parallel interactions, like parallel beta-sheet, will appear as off-diagonal contacts that run parallel to the diagonal; antiparallel interactions, like antiparallel helices, will appear as contacts that run antiparallel to the diagonal.

References

1. Kimura M (1968) Evolutionary rate at the molecular level. Nature 217:624–626
2. Kimura M, Ota T (1974) On some principles governing molecular evolution. Proc Natl Acad Sci U S A 71:2848–2852
3. Kleinstiver BP, Fernandes AD, Gloor GB, Edgell DR (2010) A unified genetic, computational and experimental framework identifies functionally relevant residues of the homing endonuclease I-BmoI. Nucleic Acids Res. doi:10.1093/nar/gkp1223
4. Dickson R, Wahl L, Fernandes A, Gloor G (2010) Identifying and seeing beyond multiple sequence alignment errors using intramolecular protein covariation. PLoS ONE 5: e11082
5. Dickson RJ, Gloor GB (2013) The MIp toolset: an efficient algorithm for calculating Mutual Information in protein alignments. arXiv, Ithaca, NY
6. Dickson RJ, Gloor GB (2012) Protein sequence alignment analysis by local covariation: coevolution statistics detect benchmark alignment errors. PLoS ONE 7:e37645
7. Talavera G, Castresana J (2007) Improvement of phylogenies after removing divergent and ambiguously aligned blocks from protein sequence alignments. Syst Biol 56:564
8. Privman E, Penn O, Pupko T (2012) Improving the performance of positive selection inference by filtering unreliable alignment regions. Mol Biol Evol 29:1–5

9. Martin LC, Gloor GB, Dunn SD, Wahl LM (2005) Using information theory to search for co-evolving residues in proteins. Bioinformatics 21:4116–4124
10. Kawrykow A et al (2012) Phylo: a citizen science approach for improving multiple sequence alignment. PLoS ONE 7:e31362
11. Khatib F, DiMaio F, Cooper S (2011) Crystal structure of a monomeric retroviral protease solved by protein folding game players. Nat Struct Mol Biol 18:1175–1177. doi:10.1038/nsmb.2119
12. Clamp M, Cuff J, Searle SM, Barton GJ (2004) The Jalview Java alignment editor. Bioinformatics 20:426–427
13. Waterhouse AM, Procter JB, Martin DMA, Clamp M, Barton GJ (2009) Jalview Version 2–a multiple sequence alignment editor and analysis workbench. Bioinformatics 25:1189–1191
14. Söding J (2005) Protein homology detection by HMM-HMM comparison. Bioinformatics 21:951–960
15. Eddy SR (2011) Accelerated profile HMM searches. PLoS Comput Biol 7:e1002195
16. Altschul SF et al (1997) Gapped BLAST and PSI-BLAST: a new generation of protein database search programs. Nucleic Acids Res 25: 3389–3402
17. Marchler-Bauer A et al (2009) CDD: specific functional annotation with the Conserved Domain Database. Nucleic Acids Res 37: D205–D210
18. Punta M et al (2012) The Pfam protein families database. Nucleic Acids Res 40:D290–D301
19. Katoh K, Toh H (2008) Recent developments in the MAFFT multiple sequence alignment program. Brief Bioinform 9:286–298
20. Loytynoja A, Goldman N (2008) Phylogeny-aware gap placement prevents errors in sequence alignment and evolutionary analysis. Science 320:1632–1635
21. Edgar RC (2004) MUSCLE: a multiple sequence alignment method with reduced time and space complexity. BMC Bioinform 5:113
22. Gilbert D (2002) Sequence file format conversion with command-line readseq.. doi:10.1002/0471250953.bia01es00
23. Hogue CW (1997) Cn3D: a new generation of three-dimensional molecular structure viewer. Trends Biochem Sci 22:314–316
24. Wang Y, Geer LY, Chappey C, Kans JA, Bryant SH (2000) Cn3D: sequence and structure views for Entrez. Trends Biochem Sci 25: 300–302
25. Ash RB (1965) Information theory. Courier Dover, New York
26. Cover TM, Thomas JA (1991) Elements of information theory. Wiley, New York
27. Dunn SD, Wahl LM, Gloor GB (2008) Mutual information without the influence of phylogeny or entropy dramatically improves residue contact prediction. Bioinformatics 24: 333–340
28. R Development Core Team (2008) R: A language and environment for statistical computing. R Foundation for Statistical Computing, Vienna, Austria. ISBN 3-900051-07-0, http://www.R-project.org.29.
29. Ellson J, Gansner E, Koutsofios L, North S, Woodhull G (2002) Graphviz—open source graph drawing tools. Springer, Heidelberg, pp 594–597
30. Bromham L (2009) Reading the story in DNA. Oxford University Press, USA
31. Altschul SF, Gish W, Miller W, Myers EW, Lipman DJ (1990) Basic local alignment search tool. J Mol Biol 215:403–410
32. Henikoff S, Henikoff JG (1992) Amino acid substitution matrices from protein blocks. Proc Natl Acad Sci U S A 89:10915–10919
33. Altschul SF (1998) Generalized affine gap costs for protein sequence alignment. Proteins 32: 88–96
34. Burger L, van Nimwegen E (2010) Disentangling direct from indirect co-evolution of residues in protein alignments. PLoS Comput Biol 6:e1000633

Chapter 16

Identification and Analysis of Genomic Homing Endonuclease Target Sites

Stefan Pellenz and Raymond J. Monnat Jr.

Abstract

Homing endonucleases (HEs) are highly site-specific enzymes that enable genome engineering by introducing DNA double-strand breaks (DSB) in genomic target sites. DSB repair from an HE-induced DSB can promote target site gene deletion, mutation, or gene addition, depending on the experimental protocol. In this chapter we outline how to identify potential genomic target sites for HEs with known target site specificities and the different experimental strategies that can be used to assess site cleavage in living cells. As an example of this approach, we identify potential human genomic target sites for the LAGLIDADG HE I-CreI that, by nine different selection criteria, may be new "safe harbor" sites for gene insertion.

Key words Homing endonuclease (HE), DNA double-strand break (DSB), Position weight matrix (PWM), Position-specific scoring matrix (PSSM), Genomic target site, Safe harbor site (SHS)

1 Introduction

Homing endonucleases (HEs) are valuable reagents for genome engineering in many organisms because of their exceptionally long DNA target sites and high specificity of site cleavage [1–3]. HEs can be used to cleave specific genomic target sites to promote gene disruption, modification, or addition at the cleavage site. This engineering capability may be broadly generalizable to many genes and genomic regions, as HEs with different target specificities continue to be identified, and it has become easier to engineer new target site specificities for existing HEs [4]. Existing HEs can also be used to facilitate the most common gene therapy goal, which is therapeutic gene insertion. This chapter provides protocols for the identification and analysis of potential target sites for well-characterized HEs in sequenced genomes to facilitate basic science and enable therapeutic gene insertion.

The starting point for the identification of potential genomic HE target sites is a detailed knowledge of HE cleavage specificity. Homing endonucleases exhibit some flexibility in their DNA recognition

David R. Edgell (ed.), *Homing Endonucleases: Methods and Protocols*, Methods in Molecular Biology, vol. 1123, DOI 10.1007/978-1-62703-968-0_16, © Springer Science+Business Media New York 2014

sequence, i.e., they are able to tolerate some base pair changes within their target site without losing site-specific activity. This target site degeneracy can be quantified in the form of binding [5] or cleavage [6, 7] profiling of a target site: HE pair or the interrogation of complex target site libraries (ref. Chapter 11). The resulting binding or cleavage profiles can be integrated to generate an HE-specific target site position weight matrix (PWM) or position-specific scoring matrix (PSSM) to enable subsequent target site searches (ref. Chapter 11). Within a PWM, a numerical value reflecting binding or cleavage efficiency is assigned to each nucleotide at every target site position. PWM values are typically referenced to the native base at that position which is assigned to an activity of 100 %. PWM can also be generated that reflects the informational content of target site base pair positions (ref. Chapter 11).

Two useful, web-accessible tools can be used to convert HE-specific PWMs into lists of genomic target sites. The LAGLIDADG homing endonuclease database and engineering server (LAHEDES) [8] includes a growing list of LAGLIDADG HE target site PWM data and can be used directly to search for the best potential target sites in short DNA sequences, e.g., an individual gene. The NCBI's BLAST server [9] can be used with LAHEDES output to identify the best target sites for a given HE in genomic sequences. The use of these two search options in sequence is fast, as illustrated below, and typically identifies dozens or a few hundred potential genomic target sites in the human genome depending on how stringent the initial LAHEDES PWM search is and the quality of the genomic sequence being searched.

The quality of the starting genomic sequence, both in terms of accuracy and completeness, can strongly determine search output. The presence and nature of genomic variation, ranging from single nucleotide polymorphism (SNP) variants through short insertion–deletion variants (indels) to large-scale structural and copy number variants (CNVs) [10], can strongly influence the likelihood of experimental success. For example, global error rate estimates for the current, extensively analyzed and well-documented hg19 human genome build range from 1 × 10e-6 to 1 × 10e-4. This translates into thousands to hundreds of thousands of potential sequence differences between the genome of a person or human cell line and the corresponding genome sequence. Thus, it is essential that potential genomic HE target sites identified as described below be experimentally verified before embarking on any HE-enabled genome engineering protocol.

A simple way to verify potential HE target sites is to amplify and sequence the target site(s) from genomic DNA and use the same amplified fragment(s) as a substrate for HE digestion in vitro. This approach verifies the site is present, documents sequence differences between the genome sequence and genomic target, and provides a direct measure of the functional consequences of sequence differences between the native HE and genomic target sites. The cleavage

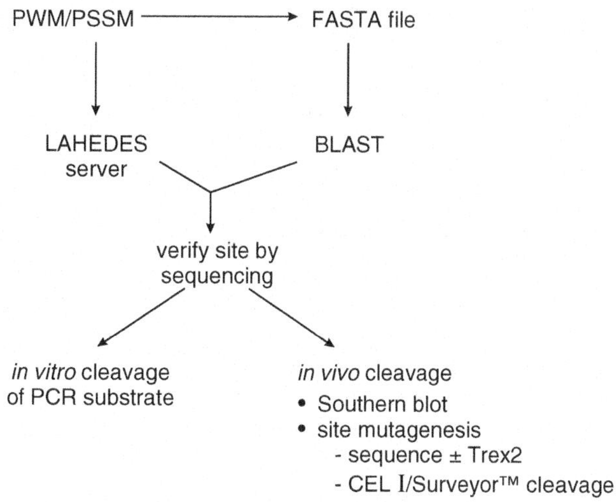

Fig. 1 Identification and analysis of genomic homing endonuclease target sites. Target site searches are driven by the use of HE-specific PWMs (position weight matrices, *upper left*) to drive target site searches in short sequences or in genomic DNA. Site searching and the generation of site libraries for FASTA conversion to facilitate BLAST searching (*top right*) of sequenced genomes or large blocks of sequence are facilitated by the LAHEDES (LAGLIDADG Homing Endonuclease Design and Engineering Server). Sites once identified by searches need to be verified, then can be further analyzed by in vitro (*lower left*) or in vivo (*lower right*) cleavage or sequence-based assays

sensitivity of a genomic HE target site in vivo can be assessed by direct measures of cleavage activity such as Southern blot analysis. Indirect measures of site cleavage in vivo are also very useful and can be more easily multiplexed than Southern blotting. For example, the efficiency of site cleavage in vivo can be estimated by determining how frequently a target site is mutated after HE expression. This approach takes advantage of the error-prone nature of both canonical and alternative nonhomologous DNA end-joining (NHEJ) pathways [11, 12] and can be rapidly implemented across many potential target sites to provide a minimum estimate of site cleavage efficiency. Additional data on the molecular nature of misrepair products can be obtained by DNA sequencing of small numbers of target sites. The sensitivity of all of these assays can be further enhanced by co-expressing an HE with the exonuclease TREX2 to promote error-prone rejoining up to ~25-fold [13] or by the use of highly accurate duplex sequencing protocols that can reliably identify target site mutations at frequencies as low as the background mutation frequency ($\leq 1 \times 10e\text{-}6$; [14]). The interrelationship of site searches and experimental site validation and analysis are shown schematically in Fig. 1. Protocols for these site analysis approaches are outlined below.

2 Materials

1. Oligonucleotide primers: these are designed to allow specific amplification of genomic target sites for target site confirmation by sequencing and for mutation analyses. Genomic primers of approximately 20 bases with melting temperatures around 55 °C work well for the amplification of human genomic target sites and can be designed using Primer3, Primer3Plus, Primer-BLAST, or other widely available PCR primer design tools [15, 16].
2. Genomic DNA purification kit.
3. PCR cleanup kit.
4. Spectrophotometer to measure DNA concentration.
5. HE cleavage buffer: optimized for the specific HE(s) to be used.
6. HE digestion stop solution: again, optimized for a specific HE(s) to be used.
7. Image analysis software: e.g., ImageJ (http://rsb.info.nih.gov/ij/) or equivalent.
8. Nylon hybridization membrane for Southern blot analysis.
9. Chemiluminescent kit.
10. Surveyor™ Mutation Detection Kit (Transgenomic, Omaha).
11. TA cloning vector with a high fidelity PCR polymerase that leaves 3′ A-tails.
12. Luria broth bacterial agar plates with ampicillin (50 μg/ml): protocols for this and other standard molecular biology and microbiology protocols can be found in several widely available protocols manuals.
13. IPTG, 1.2 g in 50 ml of H_2O, filter sterilized and stored at 4 °C.
14. X-gal, 100 mg in 2 ml *N,N′*-dimethylformamide, stored away from light at −20 °C.

3 Methods

Two types of target site searches are useful, either alone or in sequence, depending on whether you are looking for potential HE target sites in a specific gene or small number of genes or for sites located in a sequenced genome. The first, more limited search strategy can be efficiently implemented by making use of the HE-specific search matrices and search function contained in the LAHEDES homing endonuclease web server (http://homingendonuclease.net/). The second search protocol is to identify potential HE-specific target sites in sequenced genomes. This search protocol makes use of the NCBI's BLAST search engine (BLAST: http://blast.ncbi.nlm.nih.gov/) together with a list of high-quality HE-specific target site sequences generated from the LAHEDES HE web server. The protocols for each search type are given below.

3.1 LAHEDES Server HE Target Site Searches

The LAHEDES web server facilitates HE target site searches in single genes or small number of genes. These searches make use of the previously defined HE-specific PWMs contained in the LAHEDES server that can be found by following "Browse>PWM Browser." Custom PWMs can also be defined by following "Entry'>'Custom PWM Entry." Weights or values for cleavage or binding activity at each target site position across all nucleotide combinations should sum to 1.0 to ensure proper handling of the new matrix in searches. An example of a LAHEDES search using a predefined search matrix is given below.

1. Open the LAHEDES web server in a browser window (*see* **Note 1**).
2. Go to "Search'>'PWM search."
3. Enter the query sequence in FASTA format into the input box. The current version of the server can accommodate searches in typical human genes (~100 kb of contiguous sequence), though lacks the capacity to do genome-scale searches.
4. Select the HE you wish to search against your target sequence and the corresponding HE-specific PWM you would like to use.
5. Select the number of search results you want returned.
6. Run the search.

Figure 2 shows this sequence of steps in outline and provides an example of the output from a search of a 5,020 bp long query sequence using two different PWMs for mCreI, the monomerized version of the canonical LAGLIDADG homing endonuclease I-CreI [6]. These PWMs are based on I-CreI/mCreI single base pair profiling and degeneracy data or represent the output of a straight identity search. Search output in each case is in the form of a tabular list of target sites in the query sequence, their location and orientation, and the location of base pair differences between the input DNA target site sequence and the mCreI target site. Quantitative assessment of the target site matches is given by a target quality score and number of mismatches compared to the wild-type sequence.

3.2 BLAST Server Genomic HE Target Site Searches

The NCBI's BLAST server [9] can be used to search query sequences against large target sequences, e.g., entire genomes. BLAST searches can be set up using HE-specific PWM data once it has been converted into a FASTA file format. The example

a

open LAHEDES web server → Search> PWM search → enter query sequence in FASTA format → select HE and preferred PWM → select number of results → run search

b

LAHEDES - PWM (PSSM) search

gene	position	-11-10-9 -8 -7 -6 -5 -4 -3 -2 -1 1 2 3 4 5 6 7 8 9 10 11	score	mismatches
WT sequence	N/A	G A A A A C G T C G T G A G A C A G T T T C	7.97976e-8	0
>231	f2500/5020	t A A t c t G T C a T a t G A C A G a T T a	5.82473e-7	8
>231	r2500/5020	t A A A c t G T C a T a t G A C A G T T T a	1.43114e-07	8
>231	f3112/5010	t A A t A t t T t G T G t a A t A a T T T t	6.32896e-11	10
>231	f1243/5020	t A A g A t t c C a c a g a A C A c T T T g	3.33558e-11	12
>231	r263/5020	G A A A A C a T C G T G A G A C g G c c T t	2.77944e-11	9

c

LAHEDES - identity search

gene	position	-11-10-9 -8 -7 -6 -5 -4 -3 -2 -1 1 2 3 4 5 6 7 8 9 10 11	score	mismatches
WT sequence	N/A	G A A A A C G T C G T G A G A C A G T T T C	0.00311348	0
>231	r2500/5020	t A A t c t G T C a T a t G A C A G T T T a	3.11348e-11	8
>231	f2500/5020	t A A A c t G T C a T a t G A C A G a T T a	3.11348e-11	8
>231	r60/5010	G g A A A a G T C c T G c a A C A a T a T t	3.11348e-11	8
>231	f4055/5020	G A A t A C G g C c T G g c A C A t T g c C	3.11348e-11	8
>231	f3047/5020	G A t A A g c T g G c a t a A t A G T T T C	3.11348e-12	9

Fig. 2 LAHEDES search steps, options, and outputs. (**a**) Outline of steps for a target site search in a short sequence using one of the predefined HE PWMs contained in the LAHEDES server. (**b**) Search output of potential I-CreI/mCreI target sites in a 5,020 bp query sequence from human genomic region chr4 58974113 to 58979132 in assembly GRCh37.p10 using a PWM/PSSM of the homing endonuclease mCreI that incorporates single base pair cleavage degeneracy information [7]. The search results display the HE native site sequence at top ("WT sequence") followed by a list of target sites in the query sequence (here designated ">231"). The position and strand on which the potential sites are located together with target site coordinates are listed next, followed by site sequences referenced to the 22 bp I-CreI/mCreI target site. Lower case, *red letters* indicate base pair positions where there is a difference between the target and the mCreI target site sequence. Nucleotide positions that match central four nucleotides of the native site are in *upper case* and *blue*. "Score" and "mismatch" columns give a quantitative assessment of site quality and the number of base differences, respectively. (**c**) An equivalent search using an "identity" PWM that identifies the closest matches between the native I-CreI/mCreI target site and the target or query sequence. This emphasizes the value of using matrices that incorporate site degeneracy data for searches

below illustrates how to search for I-CreI/mCreI sites in the human genome.

1. Develop a list of all HE-specific target sites you wish to BLAST search from PWM data. Cleavage degeneracy matrices are often the most useful for this step, as they are typically the best populated with data and reflect the most common goal of genomic target site identification which is to cleave and/or

modify these genomic target sites in vivo. The more stringent your site selection is at this step (e.g., only for sites that have a high likelihood of being cleaved with high efficiency), the fewer the sites your BLAST search is likely to return (*see* **Note 2**).

2. Generate a text file in which each line consists of a candidate target site sequence. This list should include all possible combinations of nucleotide positions and base pairs that exceed a defined functional threshold as outlined in **step 1**.
3. Convert this list of potential target sites sequences to FASTA format by preceding each sequence with a ">" and a unique site identifier.
4. Open the BLAST web server in a browser window.
5. From the list of BLAST Assembled RefSeq Genomes, choose "Human."
6. Upload your file of FASTA-compatible candidate target sites as the "Query Sequence."
7. Run BLAST with the following parameters (*see* **Note 3**):

   Database: Genome (reference only).

   Optimize for: Somewhat similar sequences (blastn).

   Max target seqs: 50.

   Short queries: Adjust for short sequences.

   Expect threshold: 1.

   Word size: 7.

   Match/mismatch: 4, –5.

   Gap cost: Existence: 12/Extension: 8.
8. In the results page that opens, chose each of the query sequences from the "Results for" drop-down menu.
9. Check the "Alignments" for query sequences that align perfectly to the target sequence (Fig. 3).
10. Follow the Sequence ID hyperlink to the NCBI reference sequence.
11. In the window that opens, expand the box "Change region shown"; chose "Selected region"; enter the coordinates for the hit in the BLAST results window and verify that the displayed sequence and the sequence for the BLAST hit are identical (Fig. 3).
12. Expand the region shown by changing the "selected region" to 2,500 bp upstream and downstream of the candidate target site.
13. Save this sequence by selecting "Send'>'Complete Record'>'File'>'Format:FASTA'>'Create File" to capture your putative target sites.

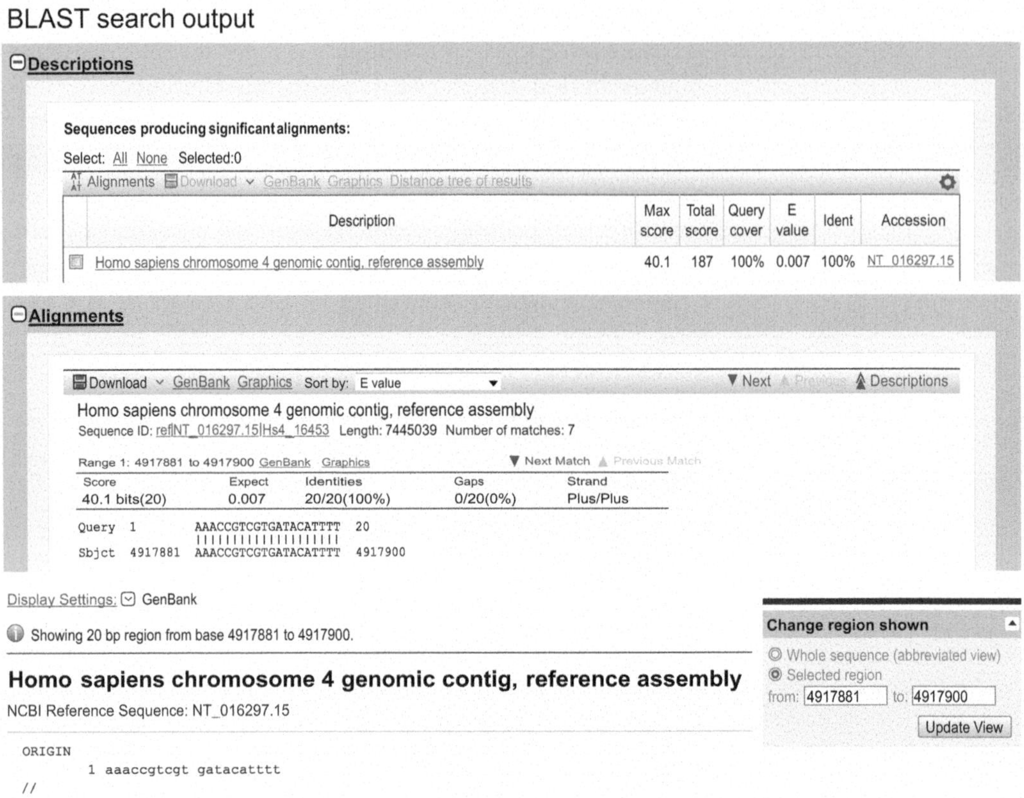

Fig. 3 BLAST search results for a potential I-CreI/mCreI human genomic target site. The *upper* and *center panel* shows a BLAST search hit. The alignment reveals a perfect match for one of the candidate target sequences on chromosome 4. The "Accession" hyperlink opens the identified target site match in a new window (*lower panel*). By changing the region shown in the *gray box* at right, it is possible to recover the flanking sequences of the candidate target site. This example was obtained using Build 36.3 of the "reference only" database

3.3 Sequence Verification of Genomic HE Target Sites

Search results are *potential* target sites: their existence and sequence need to be confirmed in your cells or host organism of interest before proceeding. This is most easily done by using the flanking genomic sequence captured in your BLAST search above to design oligonucleotide primers that can be used to amplify the putative site region from genomic DNA as a PCR product of >500 bp. Ideally, the target site is in the middle of the PCR product. This way, successful cleavage of the target site results in replacement of one substrate band by a second, smaller product band doublet.

1. Design PCR primers flanking your putative genomic HE target site using search output from Subheading 3.2 and the primer design tools listed in Subheading 2. Design a third sequencing primer that is located ~100 bp upstream or downstream of the putative HE target site.

2. Prepare genomic DNA using a suitable molecular biology kit.
3. Use the PCR primer pair from **step 1** above to amplify the region of interest from your genomic DNA sample, and run an aliquot on an agarose check gel with flanking size standards to determine whether the predicted size product has been generated and how many other potentially contaminating PCR products are present.
4. Gel-purify your target site band of interest, and use this as a template together with your site-specific sequencing primer to determine the DNA sequence of the putative target site and flanking genomic DNA.
5. Compare the sequence of the target site region of your genomic PCR product with both the reference sequence and your predicted target site sequence to confirm that the site exists and has the expected sequence. If unexpected sequence differences are present between your sequenced genomic site and the reference genome you are starting, PWN can be used to assess their potential functional consequences. You may be able to assess whether a sequence difference is a known human genomic sequence variant by using the dBSNP database (http://www.ncbi.nlm.nih.gov/projects/SNP/).

3.4 In Vitro Cleavage Analysis of Genomic HE Target Sites

Sequence confirmation of genomic HE target sites is reassuring, but often does not directly predict the cleavage sensitivity of the site, especially if there are multiple base pair changes between the genomic and native HE target site sequences. The PCR product from Subheading 3.3 above can be used as substrate in a cleavage reaction to confirm target site cleavage sensitivity. Digesting the PCR product with different concentrations of HE and including a native target site as a control in addition will reveal how cleavage sensitive a genomic target site sequence is relative to the native site. This protocol is shown in outline in Fig. 4.

1. Clean up the PCR product using a suitable purification protocol or kit.
2. Determine the concentration of the PCR product using a spectrophotometer. Calculate the molar concentration of the PCR product.
3. Prepare control and experimental sample reactions for each substrate using ~100 ng of PCR substrate in a final reaction volume of 15 μl. Each sample reaction should contain the same amount of substrate to simplify interpretation. A good starting point for the molar ratio of enzyme to substrate is equimolar (1:1), followed by a second reaction with ten times more enzyme than substrate.

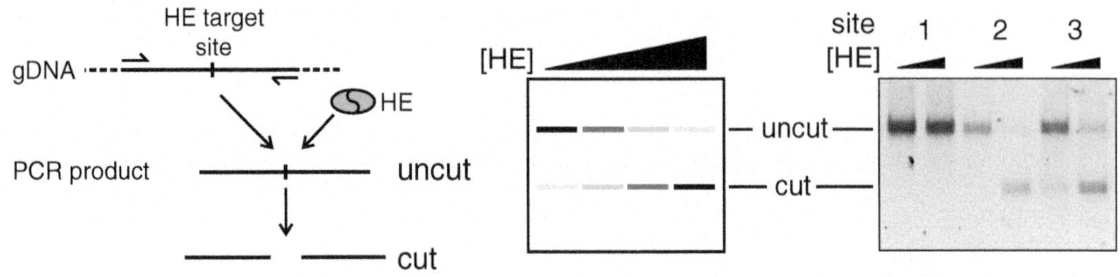

Fig. 4 In vitro cleavage verification of a potential HE genomic target site. (**a**) Schematic outline of Protocol in Subheading 3.4, in which a putative genomic target site is PCR-amplified from genomic DNA, then digested with a cognate HE. (**b**) Schematic and gel photo of panel (**a**) in which target site PCR substrate DNA is digested with an increasing amount of HE to verify site cleavage sensitivity. The gel examples show a cleavage-resistant target site (*left*) and partially (*center*) and largely (*right*) cleavage-sensitive sites. For each reaction 15 fmol of the substrate PCR product was digested with equal amounts of enzyme (*left lane* for each target) or ten times more enzyme (*right lane*)

Sample reaction:

Substrate: 15 fmol.

Reaction buffer 10×: 1.5 μl.

Homing endonuclease: 15 fmol.

H_2O: add to 15 μl.

4. Incubate the reaction mix for 1 h at 37 °C.
5. Depending on the homing endonuclease, add 1/10 volume stop buffer to the reaction (*see* **Note 3**).
6. Separate the digestion products on an agarose gel.
7. Determine the intensity of the bands corresponding to the digested and undigested PCR product bands with reference to the native site control using ImageJ or other image analysis software.

3.5 Southern Blot Analysis of In Vivo Target Site Cleavage

Southern blot analysis of target site cleavage in vivo is still a "gold standard" assay for HE activity in living cells. Cleavage time course profiles can provide a good sense of steady-state cleavage levels, and integration of these data over time can be used to estimate cleavage efficiency and investigate other aspects of HE-induced DSB repair such as repair kinetics and the genetic or functional requirements for DSB repair (see, e.g., [17]). Southern blot analysis can detect low frequencies of target site-specific cleavage (~0.5 % of potential target sites) that are difficult or impossible to detect by other strategies. The following is a general protocol that provides an overview of major steps in Southern blot analysis. For additional technical detail and more explicit protocols, see [18, 19] or one of the widely available molecular biology methods manuals.

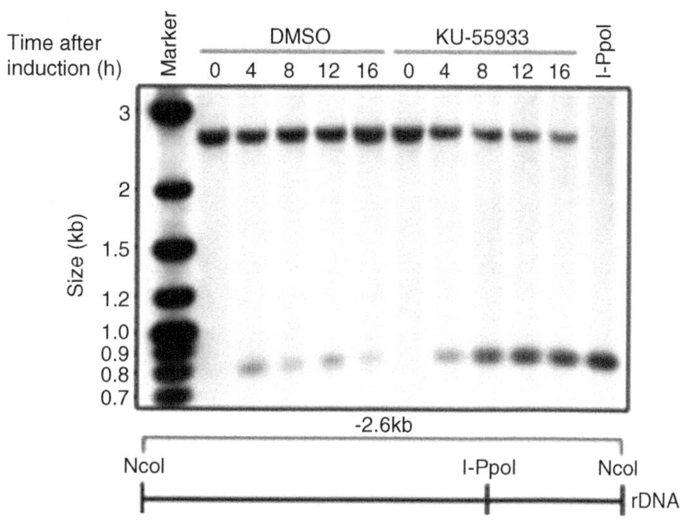

Fig. 5 Southern blot analysis of in vivo HE target site cleavage. Southern blot analysis of genomic DNA from cells expressing the I-PpoI HE in the presence or absence of the ATM inhibitor KU-55933 for the indicated times. The I-PpoI target is shown in a representation of the NcoI digestion product. The probe used for the Southern blot anneals to the smaller, 0.8 kb long cleavage fragment. This blot panel was previously published as fig. 16.4e in [18]

An example of a Southern-based analysis of cleavage of a genomic target site for the I-PpoI HE is shown in Fig. 5.

1. Express your HE in host cells by transfecting or infecting cells with an expression vector. Prepare a mock-infected control sample and any desired time point samples. Short time courses (24–36 h post-transfection) work well with most HEs unless your aim is to drive target site mutagenesis, when longer time points up to 72 or more hours may be advantageous (*see* **Note 4**).
2. Prepare genomic DNA from transfected cells using a genomic DNA purification kit.
3. Cut the genomic DNA with a restriction enzyme(s) to generate a target site fragment for blot analysis. The best starting fragments are ~5 kb long and have the HE cleavage site located asymmetrically in the resulting restriction fragment (*see* **Note 5**).
4. Determine the concentration of the digestion products using a spectrophotometer.
5. For each time point load 5 μg of digested genomic DNA/lane of a 1 % agarose gel. Separate the digestion products by electrophoresis in 1× TBE buffer for 16 h at 20 V on a 10-cm long agarose gel. Include as controls a genomic DNA sample that you have cleaved in vitro with your HE or a restriction endonuclease that is close to or in the HE target site. Include size standards.

6. Blot the separated DNA fragments from your agarose gel onto a nylon hybridization membrane.
7. Prepare labeled probe(s) to detect and quantify site-specific cleavage events. The PCR product used for target site sequence verification above can be used as a probe if it is free of repeats and known to have little or no cross-hybridization issues. This probe fragment can be captured by cloning if desired for future use. Probes can be designed to anneal to one or both of the HE/restriction endonuclease digestion products and can be labeled with radioactivity or a nonradioactive detection system (e.g., biotinylation) depending on your experience and radiation licensing status.
8. Detect probe hybridized to your membrane-bound digestion product. Radioactive probes can be detected directly by imaging on film or phosphorimager screens. Biotinylated probes can be detected using the chemiluminescent kit.

3.6 Analysis of In Vivo Target Site Cleavage by Site Amplification and Cleavage

A simpler, though less sensitive, approach to assess in vivo site cleavage is to amplify the target site from cells after HE expression to determine their cleavage sensitivity. This approach takes advantage of the fact that HE target sites cleaved in vivo may undergo error-prone repair [11, 12]. The mutagenic "footprints" of error-prone DSB repair can be detected by HE cleavage, restriction endonuclease cleavage, or mismatch nuclease cleavage of target site DNA fragments PCR-amplified from HE-expressing cells. All of these methods can be further enhanced by co-expressing an HE with the TREX2 3′ repair exonuclease in vivo: TREX2 degrades free DNA ends, antagonizes the error-free religation of cleaved target sites, and thus promotes the generation of mutant target site repair products. TREX2 co-expression is discussed first below.

3.6.1 Co-expression of HEs and TREX2 in Human Cells

Several different types of expression systems can be used to co-express an HE and the exonuclease TREX2 [13]. The open reading frames can be cloned together in one expression plasmid, separated either by an internal ribosome entry site (IRES) or a 2A ribosome skipping sequence [20, 21] to ensure co-expression of the two gene products. It is advantageous to integrate a fluorescent protein (e.g., mCherry) into the same expression plasmid downstream of the two ORFs to allow for easy screening (and, if desired, sorting) of cells that co-express an HE and TREX2. Alternatively, the HE and TREX2 proteins can be expressed from two different plasmids that are co-transfected at the same time.

The presence and frequency of HE target site mutations can be assayed by digesting genomic DNA target site sequences with the cognate HE or with a restriction enzyme that cleaves within the HE target site as outlined above in Subheading 3.4. Target site PCR products from HE-expressing and control cells can also be annealed to generate mismatches between mutant and control target sites that can be detected with the mismatch-cleaving nuclease CEL I (available commercially as the Surveyor™ nuclease cleavage assay). The CEL I/Surveyor™ endonuclease [22] is a member of the plant-derived CEL nuclease family [23] that cuts DNA at nucleotide mismatches.

3.6.2 CEL I/Surveyor Cleavage of Target Site PCR Products

1. Prepare two cell cultures: one will be transfected to express HE (and TREX2, if desired), and the other will be mock-transfected to serve as a control.
2. Prepare genomic DNA from both cultures using the genomic DNA purification kit.
3. PCR amplify the putative homing endonuclease target site region from both samples.
4. Clean up the PCR products using a suitable purification protocol or kit.
5. Verify the quality of the PCR products on a 1 % agarose gel.
6. Determine the concentration of the PCR products using a Nanodrop spectrophotometer.
7. Mix, denature, and anneal equimolar amounts of the two template populations (the HE-expressing experimental and mock-transfected control) then digest with CEL I nuclease following the protocol included in the Surveyor™ Mutation Detection Kit manual.
8. Separate the digestion products on a 1 % agarose gel.
9. Determine the intensity of the bands corresponding to the digested heteroduplex and undigested homoduplex DNA molecules using ImageJ or other image analysis software.
10. Calculate the percentage of heteroduplex, mutant-containing DNA molecules by dividing the intensity of the digested band by the total of the digested and undigested bands.

3.7 Analysis of In Vivo Target Site Cleavage by Site Sequencing

Target site sequencing from cells expressing an HE (and, if desired, TREX2) can provide additional information beyond the above protocols on the frequency and molecular nature of target site mis-repair and mutagenesis events. Sequencing is potentially the most revealing of the target site analysis methods beyond Southern blot analysis and can be performed on small numbers of cloned target sites or by high throughput DNA sequencing (HTS) with bar coding if desired. The protocol below is designed for the analysis of small numbers (tens to dozens) of mutant sites. The use of HTS to analyze target sites is covered in Chapter 12 (*see* **Note 6**).

1. Prepare genomic DNA from experimental (±HE/±TREX2) and control cell cultures using a genomic DNA preparation kit.
2. Design PCR primers to amplify the HE target site that anneal ~250 bp upstream and downstream of the target site. Design a

second set of sequencing primers that anneal within the predicted PCR product and are located ~100 bp from the target site region.

3. Use the flank primer pair to PCR amplify target sites from cellular DNA samples with a high fidelity PCR polymerase that leaves 3′ A-tails.
4. Clean up the PCR products using a suitable purification protocol or kit.
5. Clone the PCR products into protocol vector suitable for TA cloning.
6. Transform the ligation products into an *E. coli* strain that allows for blue/white selection, e.g., DH5α, and plate on LB plates with ampicillin/IPTG/X-Gal.
7. Sequence 96 white colonies using the DNA sequencing primer(s) designed in **step 2** above (*see* **Note 7**).
8. Compile and compare the sequencing results from experimental and control samples with the genomic target site sequence defined in Subheading 3.3 above.

3.8 Worked Example: Identification of Potential Human Genomic "Safe Harbor" Sites Cleaved by the LAGLIDADG HE I-CreI/mCreI

HEs are being used in a growing number of organisms to target the disruption (or "knockouts") or modification of specific genes. Another less common though practically important genome engineering goal is to use HE cleavage of a genomic "safe harbor" site to facilitate transgene insertion without disrupting adjacent gene structure or expression. The inserted transgene may have therapeutic value or may provide a convenient and consistent way to "tag" the same site in different cells with a molecular bar code or other easily selected or scored marker gene such as a fluorescent protein coding cassette.

This section provides an example of how the protocols described above can be used to identify potential genomic cleavage sites for a HE based on sequential PSSM/PWM and BLAST searching, and then determine whether these sites could serve as new genomic "safe harbor" sites (SHS) for a range of genome engineering applications [24, 25]. The outline of this series of experiments is shown in Fig. 6.

1. Identify potential genomic SHS by LADHEDES PWM analysis: We used the protocol outlined in Subheading 3.1 to identify 128 I-CreI/mCreI target sites predicted by PWM data to be highly cleavage sensitive. The design criteria used with PWM data to generate this site list required that individual base pair differences, when combined in all possible target site combinations, did not reduce the predicted cleavage sensitivity of any site below 90 %.
2. BLAST search high-quality target site variants against the human genome: The list of 128 potential target sites from **step 1** was converted into FASTA format as described in Subheading 3.2

a

best potential sites from degeneracy data		BLAST search output		best potential SHS's
128 sites	⟶	29 sites / 37 locations	⟶	3 sites

b

position in hg19	site	criteria match
chr4:58,976,613 - 58,976,632	AAACTGTCATA**t**GACAGATT	8/9
chr2:48,830,185 - 48,830,204	AAACTG**a**CATAAGACAGATT	5/9

Fig. 6 Search for potential I-CreI "safe harbor" sites (SHSs) in the human genome. (**a**) The human genome was searched for high-quality I-CreI target site variants by the sequential use of LADHEDES I-CreI PWM data and BLAST. This search yielded 128 possible sites, of which 29 were identified in the human genome at 37 different locations. Only three of these sites, predicted to be highly cleavage sensitive by the LADHEDES I-CreI degeneracy PWM, met ≥8 of the 9 SHS criteria detailed in Table 1. (**b**) Two examples of human genomic I-CreI target sites that have high potential (*upper row*, 8 of 9 SHS criteria met) or low potential (*lower row*, 5 of 9 SHS criteria met) to serve as new human genomic SHS that could be specifically targeted with I-CreI or mCreI

then used to BLAST search the human genome sequence. A total of 29 of the 128 sites on our starting list were found at a total of 37 locations in the human genome.

3. Verify predicted target sites by amplification, sequencing, and cleavage analysis: The protocols in Subheadings 3.3 and 3.4 were next used to verify the sequence and predicted cleavage sensitivity of 6 of the 29 different target site sequences identified in **step 2** (results not shown).

4. Determine suitability to serve as a safe harbor site (SHS): There are no generally accepted criteria for SHS identification, so we assembled a list of nine different, stringent SHS scoring criteria in order to rank order the 29 different sites identified in **step 2** above. These criteria included uniqueness, accessibility, and likely safety as assessed by site proximity and activity measures. Table 1 summarizes these criteria and the most useful data sources including UCSC Genome Browser tracks to facilitate additional SHS assessments. Three of the 29 potential I-CreI/mCreI SHS from **step 2** met 8 of these 9 criteria and were judged to be of high value as potential new human genomic safe harbor sites (Fig. 6).

5. Next experimental steps: The next step to verify the utility of all 29 and the three highest scoring SHS candidates is to assess their cleavage sensitivity in vivo using the protocols outlined in Subheadings 3.6 and 3.7 in cells expressing mCreI ± TREX2 protein. Target sites that appear the most cleavage sensitive in vivo from these data will be used to design donor cassettes that include flank homology arms to facilitate homology-dependent, site-specific recombination, together with two initial transgene constructs that express either a drug-resistance marker or a fluorescent protein marker (*see* **Note 8**).

Table 1
Criteria for human genomic "safe harbor" sites (SHS)

	SHS criterion	Useful UCSC browser track	Refs.
Unique/ consistent accessible	Uniqueness (one copy in human genome)	None (BLAST search result)	–
	Not located in copy number variation (CNV)/segmental duplication region	*Variations and repeats/segmental dups*	[35, 36]
	Located in open chromatin	*Regulation/ENC DNase/FAIRE*	[37, 38]
Safety	Proximity to genes (>50 kb from the 5′ end of any gene)	*Genes and gene prediction tracks/ RefSeq genes*	[39]
	Proximity to miRNA/other functional small RNAs (>300 kb away from any miRNA)	*Genes and gene prediction tracks/ sno/miRNA*	[40–44]
	Proximity to cancer-related genes or mutations (>300 kb from any cancer-related gene)	*Phenotype and disease associations/ COSMIC*	[45, 46]
Functional silence	Low transcriptional activity	*mRNA and EST tracks/human mRNAs*	[47, 48]
	Located outside known replication origins (no origin within >50 kb)	*Regulation/UW Repli-seq/peaks*	[49, 50]
	Location outside ultraconserved elements (>50 kb from UCEs)	*Regulation/Vista Enhancers*	[51]

4 Notes

1. The LAHEDES server works well with most common Internet browsers, i.e., Internet Explorer, Mozilla Firefox, or Safari.
2. The stringency of the initial genomic sites search and correspondingly the number of potential target sites returned can be adjusted depending on the search aim. Our starting search in Subheading 3.8 focused on only those base substitutions that have near-native levels of activity (e.g., 90–95 % of the activity observed on the native target site base pair) in order to identify a small number of genomic target sites that had a high likelihood of being cleavage sensitive as DNA target sites and perhaps in chromatin as well.
3. The parameters for BLAST searches should again, at least initially, be kept restrictive to identify the most potentially useful genomic targets. Once these sites are defined, the expected threshold and/or the seed (word) length can be increased, and

penalties for mismatches and gaps reduced, to provide a more exhaustive site search. This type of secondary "relaxed" search can give a useful sense of potential genomic site numbers and their distribution (or "landscape") for a given HE and thus the potential for off-target or "collateral damage" by an HE with a defined specificity.

4. Product release may be a rate-limiting step for some HEs (e.g., I-CreI and derivatives). This requires the use of a stop buffer containing a denaturant such as SDS to unambiguously identify cleavage products.
5. A time course should be performed with sampling at least every 12 h over a 48 h interval to identify the time point with the highest fraction of cleaved molecules. Alternatively, addition of the ATM inhibitor KU-55933 [26] to 10 μM in the growth medium during HE expression interferes with DSB repair and increases the steady-state level of cleavage products and, by extension, mutant target sites.
6. Restriction fragments of ~5 kb run and transfer well in Southern blot analyses. Digest excess genomic DNA (10–20 μg) when possible to guarantee that enough sample is available to do equal lane loadings to detect cleavage products. The location of an asymmetric HE cleavage site allows both products to be visualized if the hybridization probe that is used covers both of the flanking DNA segments. An alternative, mentioned in Subheading 3.4, is to use a substrate band with the HE site placed in or near the center to double the intensity of the produce band. This placement is more difficult to achieve with restriction-generated as opposed to PCR-amplified substrates.
7. PCR suppression is an alternative technique to distinguish intact versus cleaved and native versus mutated target sites following homing endonuclease expression in vivo. While this approach can work, it is less sensitive than the methods described, may reveal only a minority of misrepair events, and is more prone to false positive and negative results [27–29].
8. There are many variations on this general protocol that can include, e.g., an enrichment step for mutant target sites if the goal of sequencing is to define a mutant repair spectrum [3, 30, 31].
9. Successful in vivo cleavage of the targeted SHS can be monitored by insertion of a selectable marker or a fluorescent protein into the generated DSB exploiting the cellular homology-directed repair (HDR) [25]. The reporter gene ORF can be preferentially inserted in the SHS by adding flanking homology arms of 400–800 bp adjacent to the intended homing endonuclease target site to facilitate homology-directed repair. Shorter homology arms of 50–100 bp length have been shown to work, but are less effective [32].

For standard laboratory and mammalian cell culture techniques, refer to Molecular Cloning—A Laboratory Manual [33], Current Protocols in Molecular Biology [34], and Protocols Online (http://www.protocol-online.org/).

References

1. Rouet P, Smih F, Jasin M (1994) Introduction of double-strand breaks into the genome of mouse cells by expression of a rare-cutting endonuclease. Mol Cell Biol 14:8096–106
2. Choulika A, Perrin A, Dujon B, Nicolas JF (1995) Induction of homologous recombination in mammalian chromosomes by using the I-SceI system of *Saccharomyces cerevisiae*. Mol Cell Biol 15:1968–73
3. Monnat RJ Jr, Hackmann AF, Cantrell MA (1999) Generation of highly site-specific DNA double-strand breaks in human cells by the homing endonucleases I-PpoI and I-CreI. Biochem Biophys Res Commun 255:88–93
4. Arnould S, Perez C, Cabaniols JP, Smith J, Gouble A, Grizot S, Epinat JC, Duclert A, Duchateau P, Paques F (2007) Engineered I-CreI derivatives cleaving sequences from the human XPC gene can induce highly efficient gene correction in mammalian cells. J Mol Biol 371:49–65
5. Zhao L, Pellenz S, Stoddard BL (2009) Activity and specificity of the bacterial PD-(D/E)XK homing endonuclease I-Ssp6803I. J Mol Biol 385:1498–1510
6. Li H, Pellenz S, Ulge U, Stoddard BL, Monnat RJ Jr (2009) Generation of single-chain LAGLIDADG homing endonucleases from native homodimeric precursor proteins. Nucleic Acids Res 37:1650–1662
7. Li H, Ulge UY, Hovde BT, Doyle LA, Monnat RJ (2012) Comprehensive homing endonuclease target site specificity profiling reveals evolutionary constraints and enables genome engineering applications. Nucl Acids Res 40:2587–2598
8. Taylor GK, Petrucci LH, Lambert AR, Baxter SK, Jarjour J, Stoddard BL (2012) LAHEDES: the LAGLIDADG homing endonuclease database and engineering server. Nucleic Acids Res 40:W110–W116
9. Altschul SF, Madden TL, Schaffer AA, Zhang J, Zhang Z, Miller W, Lipman DJ (1997) Gapped BLAST and PSI-BLAST: a new generation of protein database search programs. Nucleic Acids Res 25:3389–3402
10. 1000 Genomes Project Consortium (2010) A map of human genome variation from population-scale sequencing. Nature 467: 1061–1073
11. Shrivastav M, De Haro LP, Nickoloff JA (2008) Regulation of DNA double-strand break repair pathway choice. Cell Res 18: 134–147
12. Wang M, Wu W, Wu W, Rosidi B, Zhang L, Wang H, Iliakis G (2006) PARP-1 and Ku compete for repair of DNA double strand breaks by distinct NHEJ pathways. Nucleic Acids Res 34:6170–6182
13. Certo MT, Gwiazda KS, Kuhar R, Sather B, Curinga G, Mandt T, Brault M, Lambert AR, Baxter SK, Jacoby K et al (2012) Coupling endonucleases with DNA end-processing enzymes to drive gene disruption. Nat Methods 9:973–975
14. Schmitt MW, Kennedy SR, Salk JJ, Fox EJ, Hiatt JB, Loeb LA (2012) Detection of ultra-rare mutations by next-generation sequencing. Proc Natl Acad Sci U S A 109:14508–14513
15. Rozen S, Skaletsky H (2000) Primer3 on the WWW for general users and for biologist programmers. Methods Mol Biol 132:365–386
16. Untergasser A, Nijveen H, Rao X, Bisseling T, Geurts R, Leunissen JA (2007) Primer3Plus, an enhanced web interface to Primer3. Nucleic Acids Res 35:W71–W74
17. Berkovich E, Monnat RJ Jr, Kastan MB (2007) Roles of ATM and NBS1 in chromatin structure modulation and DNA double-strand break repair. Nat Cell Biol 9:683–690
18. Berkovich E, Monnat RJ, Kastan MB (2008) Assessment of protein dynamics and DNA repair following generation of DNA double-strand breaks at defined genomic sites. Nat Protocols 3:915–922
19. Southern E (2006) Southern blotting. Nat Protoc 1:518–525
20. Donnelly ML, Gani D, Flint M, Monaghan S, Ryan MD (1997) The cleavage activities of aphthovirus and cardiovirus 2A proteins. J Gen Virol 78(Pt 1):13–21
21. Luke GA, de Felipe P, Lukashev A, Kallioinen SE, Bruno EA, Ryan MD (2008) Occurrence, function and evolutionary origins of "2A-like"

sequences in virus genomes. J Gen Virol 89: 1036–1042

22. Qiu P, Shandilya H, D'Alessio JM, O'Connor K, Durocher J, Gerard GF (2004) Mutation detection using Surveyor nuclease. Biotechniques 36:702–707
23. Oleykowski CA, Bronson Mullins CR, Godwin AK, Yeung AT (1998) Mutation detection using a novel plant endonuclease. Nucleic Acids Res 26:4597–4602
24. Papapetrou EP, Lee G, Malani N, Setty M, Riviere I, Tirunagari LMS, Kadota K, Roth SL, Giardina P, Viale A et al (2011) Genomic safe harbors permit high b-globin transgene expression in thalassemia induced pluripotent stem cells. Nat Biotech 29:73–78
25. Silva G, Poirot L, Galetto R, Smith J, Montoya G, Duchateau P, Paques F (2011) Meganucleases and other tools for targeted genome engineering: perspectives and challenges for gene therapy. Curr Gene Ther 11:11–27
26. Hickson I, Zhao Y, Richardson CJ, Green SJ, Martin NM, Orr AI, Reaper PM, Jackson SP, Curtin NJ, Smith GC (2004) Identification and characterization of a novel and specific inhibitor of the ataxia-telangiectasia mutated kinase ATM. Cancer Res 64:9152–9159
27. Seyama T, Ito T, Hayashi T, Mizuno T, Nakamura N, Akiyama M (1992) A novel blocker-PCR method for detection of rare mutant alleles in the presence of an excess amount of normal DNA. Nucleic Acids Res 20:2493–2496
28. Orum H, Nielsen PE, Egholm M, Berg RH, Buchardt O, Stanley C (1993) Single base pair mutation analysis by PNA directed PCR clamping. Nucleic Acids Res 21:5332–5336
29. Rand KN, Ho T, Qu W, Mitchell SM, White R, Clark SJ, Molloy PL (2005) Headloop suppression PCR and its application to selective amplification of methylated DNA sequences. Nucl Acids Res 33:e127
30. Argast GM, Stephens KM, Emond MJ, Monnat RJ Jr (1998) I-PpoI and I-CreI homing site sequence degeneracy determined by random mutagenesis and sequential in vitro enrichment. J Mol Biol 280:345–353
31. Scalley-Kim M, McConnell-Smith A, Stoddard BL (2007) Coevolution of a homing endonuclease and its host target sequence. J Mol Biol 372:1305–1319
32. Orlando SJ, Santiago Y, DeKelver RC, Freyvert Y, Boydston EA, Moehle EA, Choi VM, Gopalan SM, Lou JF, Li J et al (2010) Zinc-finger nuclease-driven targeted integration into mammalian genomes using donors with limited chromosomal homology. Nucleic Acids Res 38:e152
33. Sambrook J, Russell DW (2001) Molecular cloning – a laboratory manual. Cold Spring Harbor Laboratory Press, Cold Spring Harbor, NY
34. Ausubel FM, Brent R, Kingston RE, Moore DD, Seidman JG, Smith JA, Struhl K (2013) Current protocols in molecular biology, John Wiley & Sons, Hoboken, NJ
35. Bailey JA, Gu Z, Clark RA, Reinert K, Samonte RV, Schwartz S, Adams MD, Myers EW, Li PW, Eichler EE (2002) Recent segmental duplications in the human genome. Science 297:1003–1007
36. Bailey JA, Yavor AM, Massa HF, Trask BJ, Eichler EE (2001) Segmental duplications: organization and impact within the current human genome project assembly. Genome Res 11:1005–1017
37. Ho L, Crabtree GR (2010) Chromatin remodelling during development. Nature 463:474–484
38. Geiman TM, Robertson KD (2002) Chromatin remodeling, histone modifications, and DNA methylation—how does it all fit together? J Cell Biochem 87:117–125
39. Pruitt KD, Tatusova T, Maglott DR (2005) NCBI reference sequence (RefSeq): a curated non-redundant sequence database of genomes, transcripts and proteins. Nucleic Acids Res 33:D501–D504
40. Griffiths-Jones S (2004) The microRNA registry. Nucleic Acids Res 32:D109–D111
41. Griffiths-Jones S, Grocock RJ, van Dongen S, Bateman A, Enright AJ (2006) miRBase: microRNA sequences, targets and gene nomenclature. Nucleic Acids Res 34:D140–D144
42. Griffiths-Jones S, Saini HK, van Dongen S, Enright AJ (2008) miRBase: tools for microRNA genomics. Nucleic Acids Res 36:D154–D158
43. Lestrade L, Weber MJ (2006) snoRNA-LBME-db, a comprehensive database of human H/ACA and C/D box snoRNAs. Nucleic Acids Res 34:D158–D162
44. Weber MJ (2005) New human and mouse microRNA genes found by homology search. FEBS J 272:59–73
45. Forbes SA, Bhamra G, Bamford S, Dawson E, Kok C, Clements J, Menzies A, Teague JW, Futreal PA, Stratton MR (2008) The catalogue of somatic mutations in cancer (COSMIC). Curr Protoc Hum Genet Chapter 10, Unit
46. Forbes SA, Bindal N, Bamford S, Cole C, Kok CY, Beare D, Jia M, Shepherd R, Leung K, Menzies A et al (2011) COSMIC: mining

complete cancer genomes in the catalogue of somatic mutations in cancer. Nucleic Acids Res 39:D945–D950

47. Benson DA, Karsch-Mizrachi I, Clark K, Lipman DJ, Ostell J, Sayers EW (2012) GenBank. Nucleic Acids Res 40:D48–D53

48. Kent WJ (2002) BLAT–the BLAST-like alignment tool. Genome Res 12:656–664

49. Hansen RS, Thomas S, Sandstrom R, Canfield TK, Thurman RE, Weaver M, Dorschner MO, Gartler SM, Stamatoyannopoulos JA (2010) Sequencing newly replicated DNA reveals widespread plasticity in human replication timing. Proc Natl Acad Sci U S A 107: 139–144

50. Thurman RE, Day N, Noble WS, Stamatoyannopoulos JA (2007) Identification of higher-order functional domains in the human ENCODE regions. Genome Res 17:917–927

51. Pennacchio LA, Ahituv N, Moses AM, Prabhakar S, Nobrega MA, Shoukry M, Minovitsky S, Dubchak I, Holt A, Lewis KD et al (2006) In vivo enhancer analysis of human conserved non-coding sequences. Nature 444:499–502

Chapter 17

Redesigning the Specificity of Protein–DNA Interactions with Rosetta

Summer Thyme and David Baker

Abstract

Building protein tools that can selectively bind or cleave specific DNA sequences requires efficient technologies for modifying protein–DNA interactions. Computational design is one method for accomplishing this goal. In this chapter, we present the current state of protein–DNA interface design with the Rosetta macromolecular modeling program. The LAGLIDADG endonuclease family of DNA-cleaving enzymes, under study as potential gene therapy reagents, has been the main testing ground for these in silico protocols. At this time, the computational methods are most useful for designing endonuclease variants that can accommodate small numbers of target site substitutions. Attempts to engineer for more extensive interface changes will likely benefit from an approach that uses the computational design results in conjunction with a high-throughput directed evolution or screening procedure. The family of enzymes presents an engineering challenge because their interfaces are highly integrated and there is significant coordination between the binding and catalysis events. Future developments in the computational algorithms depend on experimental feedback to improve understanding and modeling of these complex enzymatic features. This chapter presents both the basic method of design that has been successfully used to modulate specificity and more advanced procedures that incorporate DNA flexibility and other properties that are likely necessary for reliable modeling of more extensive target site changes.

Key words Protein–DNA interactions, Computational design, Rosetta, Specificity, In silico prediction, Gene targeting, Direct readout

1 Introduction

Direct interactions between amino acids and DNA nucleotides are an important determinant of the substrate preference of a DNA-binding protein. A position in a binding site where the protein displays a preference for one nucleotide over the others is considered to have high specificity [1–3]. These positions are often characterized by strong direct interactions that are disrupted when the favored base is replaced (Fig. 1, *see* **Notes 1** and **2**). Being able to redesign interface residues to alter this specificity would enable the targeting of a DNA-binding protein to a site of interest (*see* **Note 3**). This technology is particularly useful for targeting genome-specific

David R. Edgell (ed.), *Homing Endonucleases: Methods and Protocols*, Methods in Molecular Biology, vol. 1123, DOI 10.1007/978-1-62703-968-0_17, © Springer Science+Business Media New York 2014

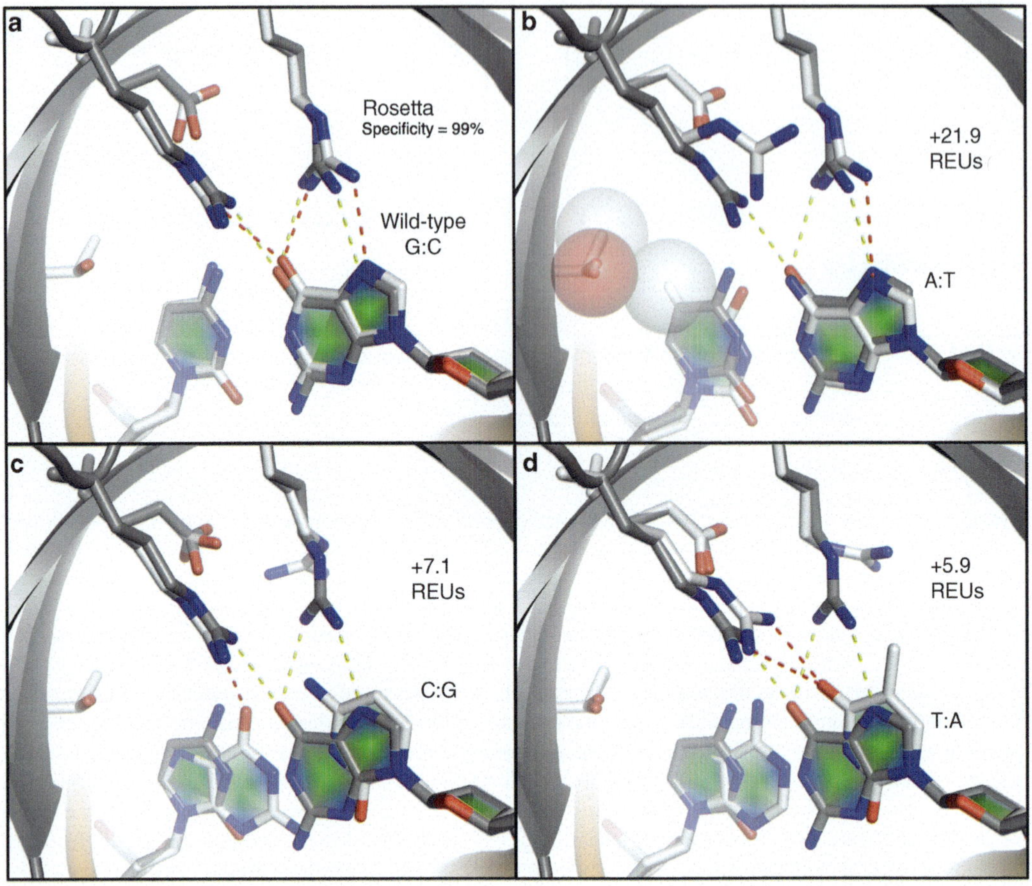

Fig. 1 The predicted role of direct interactions in protein–DNA specificity. Each panel represents the structure with the mean energy from a set of 56 repacks done with Rosetta. The wild-type base pair is the G:C at position −6 or 409 (crystal structure numbering) in the 2Q0J pdb. The native 2Q0J structure is shown in *gray*, with native hydrogen bonds shown in *yellow*, and the Rosetta structures are shown in *white*, with predicted hydrogen bonds shown in *red*. The wild-type base pair has a very high predicted specificity when compared to the three alternative base pairs. (**a**) The repacked structure for the wild-type base pair maintains the energetically favorable direct hydrogen bonds that are in the crystal structure. (**b**) The A:T base pair loses hydrogen bonding, and the methyl group of the thymine nucleotide has significant repulsion (highlighted by *spheres*) with a neighboring threonine side chain. (**c**) The C:G base pair loses hydrogen bonding. (**d**) The T:A base pair loses hydrogen bonding

DNA cleavage enzymes, such as LAGLIDADG endonucleases [4, 5], to sites that are relevant to genome engineering and gene therapy applications [6–9].

Computational methods for engineering specificity offer an efficient alternative to more labor-intensive experimental procedures, such as directed evolution [10–13]. Additionally, experimental and in silico approaches are not mutually exclusive, as the predicted results can also be used to guide the design of libraries for screening and enhance the likelihood of successfully identifying an active variant [14–16]. The Rosetta program for macromolecular modeling and design [17] has been successfully used to alter the specificity of several LAGLIDADG homing endonucleases [18–21]. Describing the protocols used to computationally redesign endonuclease specificity with Rosetta is the focus of this chapter.

The main design algorithm in Rosetta searches protein sequence and rotameric [22] space to find a set of amino acids that is compatible with the DNA sequence being targeted (*see* **Note 4**). Each attempted amino acid combination is evaluated with a physically based energy function in order to identify the lowest-energy sequence [17, 23]. The majority of the previous endonuclease redesign successes [18–21] utilized a standard fixed-backbone algorithm, in which both the protein and DNA backbones in the starting crystal structure are not flexible. This chapter details this basic method and additionally introduces some alternative DNA interface design protocols. These advanced methods include flexibility on one or both sides of the protein–DNA interface [21, 24], explicit design for specificity using a genetic algorithm [19, 21], and the use of libraries of native-like interactions (called motifs) to guide rotamer sampling [23, 25]. These approaches provide ways to diversify design results over the fixed-backbone approximation available in release versions of Rosetta.

Computational design is an efficient approach for altering specificity, as long as the particular problem of interest is feasible with the currently available technology. The majority of the previously published successes were limited to single base-pair switches in the twenty base-pair target sequence characteristic of LAGLIDADG endonucleases [18–20]. The exception that stands out was design for a triple base pair. However, the crystal structure of this variant revealed extensive DNA movement and interface coordination that were not predicted by the standard Rosetta modeling [21]. Directed evolution methods have produced several large-scale specificity shifts, but achieving consistent success and maintaining the exquisite specificity characteristic of the natural endonucleases [19, 26] are still a challenge for every approach [27–29]. A structure obtained for one of these evolved enzymes also showed extensive interface rearrangements that are not considered in standard computational design protocols [27]. All of these results indicate that there are features of the LAGLIDAG interface that are not being accurately captured by the models.

LAGLIDADG endonucleases have highly integrated interfaces, in which binding and catalysis are coordinated [19]. While this characteristic is advantageous for a gene-targeting reagent, it significantly increases the challenge of specificity modulation because there is no currently understood recognition code ([30, 31], *see* **Note 5**). This lack of a recognition code makes computation even more necessary, albeit harder, because multiple base-pair specificity switches need to be engineered as one unit, instead of being engineered separately and then recombined. Directed evolution approaches are limited in how many amino acids can be simultaneously randomized. One way to utilize the power of computational design is to identify variants with low levels of a desired feature and then use these proteins as starting points for directed evolution optimization [32–34]. Another use of computation is that it can suggest the inclusion of only certain amino acid types at each position in a protein library allows for many more positions to be concurrently explored. For example, the core positions that buttress the DNA-interacting residues and can be important for activity [35] are often excluded from libraries [11, 29] in favor of focusing on the interface residues that make direct contacts. A concerted approach is important because it is likely that a stringent alignment of the N- and C-terminal domains is required to facilitate catalysis and it is mediated by residues not in the protein–DNA interface [27, 28, 36]. There has been effectively no success at altering the specificity of the central four base pairs where catalysis occurs, presumably due to these alignment criteria and indirect readout of the DNA [37, 38], neither of which is modeled or well understood. Engineering pipelines that iterate between computational design, directed evolution, and detailed kinetic analyses are poised to discover the missing components of these computational models.

2 Materials

1. The latest release version of the Rosetta software suite (Rosetta 3.4 as of 2012) is available from http://www.rosettacommons.org and is free of charge for academics and nonprofit users. A comprehensive manual for the software is also available from the same website. For conducting protocols that are not included in the release version, the developer's version of the code must be obtained. Protocols that require these extended capabilities are noted throughout this chapter. A sponsor from a Rosetta lab is required for access to this repository, and a partial list of labs with members that can provide sponsorship or collaboration is included in **Note 6**.
2. Compiling the Rosetta code requires either an external compiling software or Python (version >= 2.2) to run the included scons.py script that runs a local version of the compiling software SCons that comes packaged with Rosetta.
3. The Rosetta software runs on multiple platforms (see manual for list). However, it is suggested that a Unix or Linux cluster be used in order to submit many runs in parallel and enhance calculation efficiency.
4. A high-resolution crystal structure (preferably <3.0 Å) of the protein of interest bound to DNA (*see* **Note 7**).

3 Methods

3.1 Standard Protein–DNA Interface Design

1. Obtain a current copy of the release version of Rosetta (*see* Subheading 2).
2. Open a terminal window (*see* **Note 8**).
3. Enter the Rosetta source directory that contains the scons.py file. Type "scons bin mode=release extras=static" to compile a production speed version of the code that can be ported to different platforms and computer systems (*see* **Note 9**).
4. If the code is going to be run on a different computer system than it was compiled, the rosettaDNA executable must be moved to that system by typing "scp ./bin/rosettaDNA.static.linuxgccrelease computerwhereitwillberun." The entire rosetta_database folder must also be moved by typing "scp -r ../rosetta_database/ computerwhereitwillberun."
5. Make a directory where the code will be run and the output collected by entering the desired location and typing "mkdir nameofdirectory" (*see* **Notes 10** and **11**).
6. Make a file that contains the arguments read by the Rosetta program with your favorite text editor (Fig. 2). The editor Vi is likely present in your Linux/Unix system. To use Vi to make the arguments file, type "vi nameofargsfile," enter insertion mode by typing "i," and then type the desired flags using Fig. 2 as a guide (*see* **Note 12**).
7. Make an XML script file (*see* **Note 13**) that contains protocol instructions given to the program through RosettaScripts [17, 39]. This file can be made by using Fig. 3 as a guide for the content and following the same Vi instructions described in **step 6** (all other files in future steps can also be made or modified with Vi).
8. The protein interface positions to be designed will be automatically calculated based on the "dna_defs" and "z_cutoff" flags that are part of the operations (TASKOPERATIONS) included in the XML file (Fig. 3). However, a type of file known as a resfile (Fig. 4) is available if the user would instead prefer to allow only a subset of amino acid types and designable positions. The addition of the line "-resfile nameoffile" to the args file (Fig. 2) will enable the resfile to override automatic detection of the interface residues. The XML script should also be modified to add the task operation "<ReadResfile name=RRF/>" and replace the use of AUTOprot with RRF in the mover. The "dna_def" option is also no longer necessary in the DnaInt operation because the target base is specified in the resfile.
9. Choose an energy function that is optimized for protein–DNA interactions [21–24] and make a file containing the necessary

```
-in:ignore_unrecognized_res # ignore anything in the pdb structure that is not recognizable
-file:s 2QOJ.pdb # input structure
-mute all # no output into an output file, skip this flag off when debugging and include for large-scale runs
-unmute protocols.dna # unmute a subset of the output if desired
-score::weights rosetta_database/scoring/weights/optimizedenergyfxn.wts # energy function for evaluating structures (see Fig. 5)
-score:output_residue_energies # include information in the pdb about the interaction energies of residues in the design
-run:output_hbond_info # include information in the pdb about the hydrogen bonding of residues in the design
-database rosetta_database # required Rosetta database, see Note 17 for useful changes to the database
-ex1 # extra rotamer sampling around chi angle 1
-ex2 # extra rotamer sampling around chi angle 2
-ex1aro::level 6 # even more extra rotamer sampling for aromatic residues around chi angle 1. This flag is recommended because aromatic residues can have large repulsion scores if the rotamer is not in the optimal position.
-ex2aro::level 6 # even more extra rotamer sampling for aromatic residues around chi angle 2
-exdna::level 4 # use DNA rotamers and include extra sampling (inclusion of this flag is highly advised for protein-DNA design)
-jd2:dd_parser # use the parser protocols
-parser:protocol XML.scriptfile # XML script (see Fig. 3)
-overwrite # if a pdb with the same name already exists in the directory where the design occurring, then overwrite the old pdb
-out:prefix design_ # an optional prefix to add to the name of designs
```

Fig. 2 Example arguments file. This file controls the parameters of the design run or specificity calculation. All writing after the # mark is a comment that is not read in by the Rosetta program

weights for each energy function component (Fig. 5). The name of the energy function is the input for the flag "-score::weights nameoffile" (Fig. 2).

10. Modify the Rosetta database to go with the optimized energy function shown in Fig. 5. The necessary changes are listed in **Note 14** [23].
11. Run code by submitting to whatever computer cluster you are using or by typing "rosettaDNA.static.linuxgccrelease @ nameofargsfile" (*see* **Notes 15** and **16**).

3.2 Assessment of Designs Using Specificity and Binding Energy Calculations

Follow instructions in Subheading 3.1 with the following described variations to the XML script (Fig. 3) and arguments files (Fig. 2). The specificity and binding energy calculations enable the user to identify the designs with the most desirable properties (*see* **Note 17**).

```
<dock_design>
 <TASKOPERATIONS>
  <InitializeFromCommandline name=IFC/> # use the information in the args file to supplement this XML
   <IncludeCurrent name=IC/> # includes the rotamers in the input structure (may not want to use)
   <RestrictDesignToProteinDNAInterface name= DnaInt base_only =1 z_cutoff =6.0 dna_defs =Z.409.GUA/> # make the target site substitution of interest (chainID.crystalposition.type) and designate the sphere of residues surrounding it that are designable and packable
     <OperateOnCertainResidues name=AUTOprot> # works with the DnaInt operation to enable residues to be chosen for design and packing if they are marked as AUTO
    <AddBehaviorRLT behavior=AUTO/>
    <ResidueHasProperty property=PROTEIN/>
   </OperateOnCertainResidues>
  </TASKOPERATIONS>
  <SCOREFXNS>
   <DNA weights=optimizedenergyfxn/> # energy function for design evaluation, this file must be put in the directory (ie, rosetta_database/scoring/weights/optimizedenergyfxn.wts)
 </SCOREFXNS>
 <FILTERS>
   <FalseFilter name=falsefilter/> # RosettaScripts has the ability to only output designs that pass a designated filter. This functionality is not being used here.
 </FILTERS>
 <MOVERS>
    <DnaInterfacePacker name=DnaPack scorefxn=DNA task_operations=IFC,IC,AUTOprot,DnaInt/>
  </MOVERS>
  <PROTOCOLS>
   <Add mover_name=DnaPack/>
  </PROTOCOLS>
</dock_design>
```

Fig. 3 Example RosettaScripts XML file. This file can be used to set up and modify Rosetta protocols. All writing after the # mark is a comment that is not read in by the Rosetta program

3.2.1 Automatic Specificity and Binding Energy Prediction Following Fixed-Backbone Design

The simplest method of specificity prediction [2, 21] is the addition of the two lines to the XML file. This method allows for multiple repacks to be done, but it is not suitable for protocols that involve any backbone movement because the backbone is optimized for the base pair originally designed for.

1. Replace *line a* with *line b* in the XML file (Fig. 3) and run the protocol exactly as described in Subheading 3.1, but with this new XML file instead of the original:

line a: <DnaInterfacePacker name=DnaPack scorefxn=DNA task_operations=IFC,IC,AUTOprot,DnaInt/>

line b: <DnaInterfacePacker name=DnaPack scorefxn=DNA task_operations=IFC,IC,AUTOprot,DnaInt binding=1 probe_specificity=3/>

2. The number following the added options refers to the number of repacks, the lowest energy of which is used in the calculations. Three is a good choice for reducing noise in the results. A repack is a search similar to the design procedure except that only the rotameric state is varied while amino acid types are fixed.
3. The calculation results are located inside the output pdb file for each design. Open the file with a text-editing program to view the data (*see* **Note 18**).

```
AUTO # all protein positions not explicitly noted are to be marked as AUTO, the same as using the AUTOprot operation
start
28 A PIKAA L # forces amino acid L at position 28 on chain A
83 A PIKAA R
-12 C NATRO # g, fixes the native rotamer
-11 C NATRO # c
-10 C NATRO # a
-9 C NATRO # g
-8 C NATRO # a
-7 C NATAA # a, fixes the native residue type, but allows different rotamers
-6 C TARGET GUA # c, target base, same as using the dna_def option, but DNA is required to be explicit in the resfile
-5 C NATAA # g
-4 C NATRO # t
-3 C NATRO # c
-2 C NATRO # g
-1 C NATRO # t
1 D NATRO # a
2 D NATRO # c
3 D NATRO # g
4 D NATRO # a
5 D NATAA # c
6 D TARGET CYT # g
7 D NATAA # t
8 D NATRO # t
9 D NATRO # c
10 D NATRO # t
11 D NATRO # g
12 D NATRO # c
```

Fig. 4 Example resfile. This file is used if specific protein positions or amino acid types need to be forced in the design run. It is an alternative to allowing the location of the target substitution to control the designable protein positions. All writing after the # mark is a comment that is not read in by the Rosetta program

3.2.2 Protocol for Specificity Calculation That Is Suitable Following Any Design Procedure

The main feature of a specificity calculation is that it is an exploration of rotameric and potentially backbone space in order to find and compare the energy of a set of given sequences. Therefore, the protocol used for the design procedure may not be optimal for doing these analyses. For example, the discreteness of rotamers is an approximation that is necessary because of computational limits when all amino acids are being considered. However, when the amino acid sequence is fixed, the number of rotamers included in the calculation can be greatly increased, and any negative effect of the approximation is lessened [23]. Flexible backbone calculations for specificity enable the protein backbone to be optimized for each particular base, reducing any energetic bias for the base pair in the crystal structure over the competing base types.

```
METHOD_WEIGHTS ref -0.3 -0.7 -0.75 -0.51 0.95 -0.2 0.8
-0.7 -1.1 -0.65 -0.9 -0.8 -0.5 -0.6 -0.45 -0.9 -1.0 -0.7 2.3 1.1 #
reference weights that are for each amino acid type

fa_atr 0.95 # attractive forces between residues
fa_rep 0.44 # repulsive forces between residues
fa_intra_rep 0.004 # repulsion within a sidechain
fa_sol 0.65 # one component of desolvation
lk_ball 0.325 # newer orientation-dependent desolvation
lk_ball_iso -0.325 # newer orientation-dependent desolvation
hack_elec 0.5 # coulombic electrostatics
fa_dun 0.56 # probability for each approximated rotamer
ref 1 # weight for the reference energies
hbond_lr_bb 1.17 # hydrogen bonding
hbond_sr_bb 1.17 # hydrogen bonding
hbond_bb_sc 1.17 # hydrogen bonding
hbond_sc 1.17 # hydrogen bonding
p_aa_pp 0.64 # probability of amino acid type given
backbone
dslf_ss_dst 0.5 # disulphides
dslf_cs_ang 2 # disulphides
dslf_ss_dih 5 # disulphides
dslf_ca_dih 5 # disulphides
pro_close 1.0 # proline ring closure
```

Fig. 5 Example energy function file. This energy function was optimized to produce high sequence recovery of protein–DNA interactions over a benchmark set of proteins [23]. All writing after the # mark is a comment that is not read in by the Rosetta program

1. Modify the XML script to fix the protein sequence of the structure being analyzed (most likely the output of a previous design calculation). In the TASKOPERATIONS section of the XML file, the operation to fix the protein sequence must be added by adding the following four lines:

```
<OperateOnCertainResidues name=ProtNoDes>
<RestrictToRepackingRLT/>
<ResidueHasProperty property=PROTEIN/>
</OperateOnCertainResidues>
```

To use this operation, the DnaInterfacePacker mover must be changed to the following:

```
<DnaInterfacePacker name=DnaPack scorefxn=DNA task_operati
ons=IFC,IC,AUTOprot,ProtNoDes,DnaInt/>
```

2. If desired, modify the arguments file to increase the number of rotamers. The addition of the flags "-ex3" and "-ex4" is a reasonable increase. Further increases can be enabled by using the "::level #" addition to any of the -ex flags. The available levels are 1–7. An advanced XML user can add the extra rotamers through the ExtraRotamersGeneric operation and complete this specificity calculation directly after design in one run.
3. Set up four separate runs, one for each base type (or more if the target has multiple base-pair substitutions. Do runs for whichever competing states are to be compared).

4. Complete a minimum of ten runs per base type for a fixed-backbone approach and at least 50 (4× or more) for any approach involving flexible backbone.
5. Collect the total_score value from inside of each pdb. The specificity can be calculated from the lowest-energy structure or from the mean or median of the energies of all structures. A comparison of all these three specificity calculations is most informative (*see* **Notes 19** and **20**).
6. The simplest way to access these values without writing a script is to execute the command "grep total_score *pdb" in the directory that contains the pdbs you are interested in analyzing.

3.3 Advanced Design Modes

3.3.1 Protein Flexibility

Follow instructions in Subheading 3.1 with the following described variations to the XML script (Fig. 3) and arguments files (Fig. 2). Protein backbone flexibility is accessible through the parser protocols and is in the release version of the code.

1. Modify the XML file to include a second mover before the standard design mover (DnaInterfacePacker). The line to add is <DesignProteinBackboneAroundDNA name=bb scorefxn=DNA task_operations=IFC,IC,AUTOprot,DnaInt type=ccd gapspan=4 spread=3 cycles_outer=3 cycles_inner=1 temp_initial=2 temp_final=0.6/>
2. Additionally, the following line must be added after the line "<Add mover_name=DnaPack/>":
<Add mover_name=bb/>
3. The DesignProteinBackboneAroundDNA enables the ccd backbone movement [40, 41]. An advanced user of RosettaScripts and the XML format could explore the protein backbone space with alternative protocols and then use those structures as input for standard design (*see* **Note 21**).
4. The diversity of the results will be significantly increased; thus, more runs are required to explore the design possibilities.

3.3.2 Multistate Design

Multistate design is a method to explicitly design for one state and against others [42, 43]. In the case of protein–DNA design, those states are the targeted bases and the alternative bases [19, 21]. This method is accessible through the parser protocols and is in the release version of the code. Follow instructions in Subheading 3.1 with the following variations to the XML script (Fig. 3).

1. Modify the XML file by replacing the standard DNA design mover with the following mover for doing multistate:
<DnaInterfaceMultiStateDesign name=msd scorefxn=DNA task_operations=IFC,IC,AUTOprot,DnaInt pop_size=20 num_packs=1 numresults=0 boltz_temp=2 anchor_offset=15 mutate_rate=0.8 generations=5/>

2. Additionally, the line "<Add mover_name=DnaPack/>" must be replaced with the line:
   <Add mover_name=msd/>
3. All of the parameters of the genetic algorithm can be varied, and the ones in the above line are testing parameters. Refer to cited literature [19, 21, 42] to identify good starting parameters for a particular design challenge.

3.3.3 DNA Flexibility

Crystal structures of engineered proteins indicate that DNA flexibility is a critical component of target site recognition [21, 27]. The DNA movement protocols in Rosetta [23, 24] are more experimental than the standard design methods and are undergoing significant development.

1. Acquire access to the developer's version of the code (*see* Subheading 2 and **Note 22**).
2. There is no RosettaScripts capability outside of the trunk version of Rosetta. Therefore there are separately compiled apps required for each protocol instead of one app with access to many movers through an XML file.
3. Compile the dna_fragment_rebuild_with_motifs app as a first step toward using DNA flexibility, or contact the people listed in **Note 22** to receive instructions or begin collaborations to access more advanced versions of the code.
4. Exact instructions and arguments files required for using this app are available in the supplemental material of reference [23].

3.3.4 High-Temp Packer

Multiple low-energy solutions exist for most protein engineering challenges. The standard design method tends to produce two or three different solutions at the most. Using computation to guide library design [14, 15] depends on having multiple designs to combine in the selection process. Flexible backbone methods can increase the number of solutions, but these solutions may not reflect the true potential movements of backbones because modeling of flexibility is a challenging problem. The high-temp packer approach increases the temperature that the algorithm driving the design process converges to and thus increases the chance of producing a design that is low energy, but not the predicted lowest energy. The energy function used in the design process may not be perfectly optimized for every design situation. Being able to produce designs that are not predicted to be the lowest energy, but that still contain high-quality contacts, is one way to alleviate the impact of an imperfect energy function on the design process. The supplemental methods of reference [23] describes the two changes required to use this method. The changes can be made to any version of Rosetta, and then the code must be recompiled.

3.3.5 Motifs

One downfall of the necessary rotamer approximation is that favorable interactions can be missed. Design procedures are limited in how large of a rotamer set can be used. One way to get around this limit is to use motifs, libraries of interactions seen in crystal structures, to increase rotamer sampling. The protocol consists of a search procedure using a greatly expanded rotamer set to see if one of these native-like interactions can be made with a target base pair. Once rotamers from the expanded set are identified, they are added to the standard rotamer set to bias the sampling for these likely favorable interactions. An energetic bonus can also be given to these rotamers to overcome potential inaccuracies in the energy function. Motif-based protocols are only available in the developer's version of the code, and their usage is described in detail in the supplemental methods of reference [23].

1. Acquire access to the developer's version of the code (*see* Subheading 2 and **Note 22**).
2. Compile the motif_dna_packer_design app. This application is available in both trunk Rosetta and in the more experimental branch of Rosetta that focuses on improving modeling of DNA flexibility.
3. Collect a library of motifs by compiling the dna_motif_collector app, downloading all protein–DNA complexes under some resolution cutoff (<2.8 is reasonable), and running the application by following the instructions in reference [23].
4. Add the line "special_rot 1.0" to the energy function (Fig. 5).
5. Add flags to the args file ([23], Fig. 2) to load in the motif library, set up cutoffs for acceptance of a motif rotamer, pick a rotamer level for the expanded motif rotamer library, and pick the energetic bonuses to try for these added rotamers.
6. If the trunk version of Rosetta is being used, add the line "-patch_selectors SPECIAL_ROT" to the args file.

4 Notes

1. DNA-interacting proteins can have either or both high activity and specificity [2]. An example of an enzyme with high activity and low specificity is DNA polymerase. Nonspecific proteins with high activity often use DNA backbone contacts to gain binding energy. Homing endonucleases are highly specific and their levels of activity vary. High-specificity proteins can also be low fidelity, meaning that they can have tolerance at some of the nucleotide positions in their target sites while maintaining an overall high level of specificity due to a long target site. Homing endonucleases tend to have low fidelity in order to maintain activity in the face of genetic drift of their target sites [5].

2. Another potential cause of high specificity is indirect readout [37, 38]. While direct readout is characterized by direct interactions with the target site, such as hydrophobic packing and hydrogen bonds, indirect readout is related to the DNA bending preferences of a target site sequence. There is some knowledge of the rules of indirect readout, but the energetics of indirect readout are just beginning to be incorporated into the Rosetta models [24], and the relationship of DNA bending to cleavage of target DNA by endonucleases is not understood.
3. Avoiding a reduction in specificity is an important consideration when engineering these reagents. Sometimes an endonuclease can maintain high levels of activity while losing interface interactions and specificity. Some target positions are nonspecific with the native enzymes because of the evolutionary pressure to maintain cleavage of a target site that is subject to genetic drift [5]. Computational redesigns at these positions can gain interface contacts and have enhanced specificity [19].
4. A rotamer is a low-energy conformation of an amino acid [22]. The computational methods rely on these discrete states to make the calculations feasible. The protocol to identify the lowest-energy design relies on a simulated annealing algorithm [17].
5. While there is no recognition code currently understood for homing endonucleases, it is possible that the rules governing specificity will be revealed as more native endonucleases are characterized. Collecting data on the specificity of many LAGLIDADG endonucleases, and analysis of their interface interactions most likely through homology modeling, is the only way to determine whether there is an understandable code that can be incorporated into the modeling.
6. David Baker (University of Washington), Philip Bradley (Fred Hutchinson Cancer Research Center), Jens Meiler (Vanderbilt), Jeffrey Gray (Johns Hopkins), Brian Kuhlman (UNC Chapel Hill), Tanja Kortemme (UCSF), Jim Havranek (Washington University of St. Louis), Richard Bonneau (NYU), Rhiju Das (Stanford), John Karanicolas (University of Kansas), Sarel Fleishman (Weizmann Institute of Science), Ora Furman (Hebrew University), Ingemar André (Lund University), and Sagar Khare (Rutgers).
7. It is possible to do design starting from a high-quality homology model instead of from a crystal structure. The model must include DNA and it can carry the DNA backbone from a starting template. This procedure requires that there is a homologue of the protein of interest that has been crystallized bound to DNA. Procedures to accomplish this work are not currently published, but they will be available to the public in the near future, and an advanced Rosetta user could accomplish such modeling with currently available tools.

8. A basic understanding of Linux/Unix commands is essential for running Rosetta. There are many available resources online, and one tutorial for a beginner user is located at the following web address: http://www.ee.surrey.ac.uk/Teaching/Unix/.
9. The mode = release command builds the release version instead of the debug version, and it is at least ten times faster than of a debugging executable. Only leave out the "mode = release" if you are developing code that needs to be debugged. The "extras = static" command means that static linking of shared libraries is done and that the code can be ported to other platforms. The only downside is that the sizes of the compiled executables are larger, but that is a worthwhile trade-off for portability. The command "-j #" can be used to parallelize the build into multiple threads if you are compiling on a multiprocessor machine (i.e., -j 20 for splitting work over 20 machines).
10. If the user plans on running parallel multiple trajectories of the same code, the output of these trajectories needs to go into different directories to avoid overwriting each other. A good strategy is to create internal directories labeled job0-job55 (or however many runs you want to complete). The command "mkdir job{0..55}" will generate those directories. Each parallel trajectory must write to a single one of these directories. The other option is to run jobs sequentially by using commands in the arguments file or by using capabilities within RosettaScripts [39]. The issue with running jobs sequentially is that it is much less time effective if the job is long. The job can be long if it is a complex protocol or if the pocket is multiple base pairs because that necessitates that more interface positions are being designed simultaneously.
11. The program GNU parallel is one (highly recommended) way to run multiple jobs in parallel on a multiprocessor system that does not have a job submission system in place. The website explaining the program is http://www.gnu.org/software/parallel/. A command to use GNU parallel to submit jobs 5 at a time and to have the results go into separate job# directories is the following: nice -19 ./bin/parallel -j 5 'cd {.}; ./bin/rosettaDNA.static.linuxgccrelease @../args>log;cd ../' ::: job* &
12. Many tutorials for using Vi are available online (i.e., http://www.infobound.com/vi.html).
13. The XML files are a part of RosettaScripts [39]. This system for protocol development is an integral part of the recent versions of Rosetta. It provides a flexible environment in which movers and operations can be recombined into different protocols without having to recompile Rosetta.

14. Change the 5th and 7th columns of the following five lines in the atom_properties.txt file (./rosetta_database/chemical/atom_type_sets/fa_standard/atom_properties.txt) to the values shown here:

    Phos P 2.1500 0.5850 -4.1000 3.5000 14.7000

    Narg N 1.7500 0.2384 -10.0000 6.0000 11.2000 DONOR ORBITALS

    NH2O N 1.7500 0.2384 -7.8000 3.5000 11.2000 DONOR ORBITALS

    Nlys N 1.7500 0.2384 -16.0000 6.0000 11.2000 DONOR

    ONH2 O 1.5500 0.1591 -5.8500 3.5000 10.8000 ACCEPTOR SP2_HYBRID ORBITALS

    Also change the fifth column of the three HC atoms in the LYS.params file to the value 0.48 from 0.33 to increase the positive charge of lysine. The LYS.params file is found here: "./rosetta_database/chemical/residue_type_sets/fa_standard/residue_types/l-caa/LYS.params"
15. If running many jobs on a multiprocessor system, always submit a single test run to confirm that all paths are correct and that all necessary files are included.
16. The number of runs that should be completed depends on how many base pairs are being mutated in the target site. The number of base pairs controls the number of interface positions that are designed (unless a resfile is used, *see* Fig. 4). As a starting point, a minimum of ten runs should be completed for a fixed-backbone standard design for a one base-pair substitution. At least 50 runs should be completed for a single base-pair pocket with flexibility (either protein or DNA). A triple base-pair pocket with backbone flexibility needs several hundred runs (300–500) to assess the full range of low-energy solutions.
17. These calculations could also be used to improve the models by comparison of results with experimental activity assays and to predict the binding sites for proteins with unknown target preferences.
18. Efficiency can be greatly increased if a script is written to pull these numbers out of each designed pdb file.
19. It is recommended that either the mean or median value of the total_energy for all structures be used for specificity prediction, rather than the score of the lowest-energy structure. The mean or median is a more accurate predictor because the protocol can generate outlier structures with energies much lower than the majority and these outliers are as likely to represent the actual energetic and structural state of the complex. This recommendation is especially true for protocols involving any amount of backbone flexibility.

20. The calculation of specificity is based on the Boltzmann distribution. The value of $k_B T$ can be changed, but a value of 1 is reasonable. The equation for calculating specificity for a guanine base pair is (2.718^0)/(2.718^0+2.178^(−Δ$E_{G\text{-}A}$)+2.178^(−Δ$E_{G\text{-}C}$)+2.178^(−ΔEG-T)).
21. Only the DesignProteinBackboneAroundDNA mover will limit protein backbone movement to around the target base pair. Other methods of protein backbone movement will require another way of designating the regions that should be flexible.
22. Contact Summer Thyme at sthyme@gmail.com or Philip Bradley at pbradley@fhcrc.org for information on the most updated branch of the developer's code needed to use DNA flexibility.

Acknowledgements

The authors would like to thank Justin Ashworth, Phil Bradley, and Jim Havranek for their vast contributions to improving protein–DNA interface design, as well as the entire RosettaCommons community for contributions to the Rosetta code base. This work was supported by the US National Institutes of Health (#GM084433 and #RL1CA133832 to DB), the Foundation for the National Institutes of Health through the Gates Foundation Grand Challenges in Global Health Initiative, and the Howard Hughes Medical Institute.

References

1. Jin X, West SM, Joshi R, Honig B, Mann RS (2010) Origins of specificity in protein–DNA recognition. Annu Rev Biochem 79:233–269
2. Ashworth J, Baker D (2009) Assessment of optimization of affinity and specificity at protein–DNA interfaces. Nucleic Acids Res 37:e73
3. Morozov AV, Havranek JJ, Baker D, Siggia ED (2005) Protein–DNA binding specificity predictions with structural models. Nucleic Acids Res 33:5781–5798
4. Stoddard BL (2005) Homing endonuclease structure and function. Q Rev Biophys 38: 39–95
5. Stoddard BL (2011) Homing endonucleases: from microbial genetic invaders to reagents for targeted DNA modification. Structure 19:7–15
6. Gao H et al (2010) Heritable targeted mutagenesis in maize using a designed endonuclease. Plant J 61:176–187
7. Windbichler N et al (2011) A synthetic homing endonuclease-based gene drive system in the human malaria mosquito. Nature 473:212–215
8. Marcaida MJ, Munoz IG, Blanco FJ, Prieto J, Montoya G (2009) Homing endonucleases: from basis to therapeutic applications. Cell Mol Life Sci 67:727–748
9. Perez EE et al (2008) Establishment of HIV-1 resistance in CD4+ T cells by genome editing using zinc-finger nucleases. Nat Biotechnol 26: 808–816
10. Takeuchi R, Certo M, Caprara MG, Scharenberg AM, Stoddard BL (2008) Optimization of in vivo activity of a bifunctional homing endonuclease and maturase reverses evolutionary degradation. Nucleic Acids Res 37:877–890
11. Chames P, Epinat JC, Guillier S, Patin A, Lacroix E, Pâques F (2005) In vivo selection of engineered homing endonucleases using

double-strand break induced homologous recombination. Nucleic Acids Res 33:e178

12. Doyon JB, Pattanayak V, Meyer CB, Liu DR (2006) Directed evolution and substrate specificity profiling of homing endonuclease I-SceI. J Am Chem Soc 128:2477–2484
13. Jarjour J et al (2009) High-resolution profiling of homing endonuclease binding and catalytic specificity using yeast surface display. Nucleic Acids Res 37:6871–6880
14. Voigt CA, Mayo SL, Arnold FH, Wang Z (2001) Computational method to reduce the search space for directed protein evolution. Proc Natl Acad Sci U S A 98:3778–3783
15. Chen MM, Snow CD, Vizcarra CL, Mayo SL, Arnold FH (2012) Comparison of random mutagenesis and semi-rational designed libraries for improved cytochrome P450 BM3-catalyzed hydroxylation of small alkanes. Protein Eng Des Sel 25:171–178
16. Khersonsky O, Röthlisberger D, Wollacott AM, Murphy P, Dym O, Albeck S, Kiss G, Houk KN, Baker D, Tawfik DS (2011) Optimization of the in-silico-designed kemp eliminase KE70 by computational design and directed evolution. J Mol Biol 407:391–412
17. Leaver-Fay A et al (2011) Rosetta3: an object-oriented software suite for simulation and design of macromolecules. Methods Enzymol 487:545–574
18. Ashworth J, Havranek JJ, Duarte CM, Sussman D, Monnat RJ Jr, Stoddard BL, Baker D (2006) Computational redesign of endonuclease DNA binding and cleavage specificity. Nature 441:656–659
19. Thyme SB, Jarjour J, Takeuchi R, Havranek JJ, Ashworth J, Scharenberg AM, Stoddard BL, Baker D (2009) Exploitation of binding energy for catalysis and design. Nature 461:1300–1304
20. Ulge UY, Baker DA, Monnat RJ Jr (2011) Comprehensive computational design of mCreI homing endonuclease cleavage specificity for genome engineering. Nucleic Acids Res 39:4330–4339
21. Ashworth J, Taylor GK, Havranek JJ, Quadri SA, Stoddard BL, Baker D (2010) Computational reprogramming of homing endonuclease specificity at multiple adjacent base pairs. Nucleic Acids Res 38:5601–5608
22. Dunbrack RL Jr, Cohen FE (1997) Bayesian statistical analysis of protein side-chain rotamer preferences. Protein Sci 6:1661–1681
23. Thyme SB, Baker D, Bradley P (2012) Improved modeling of side-chain–base interactions and plasticity in protein–DNA interface design. J Mol Biol 419:255–274
24. Yanover C, Bradley P (2011) Extensive protein and DNA backbone sampling improves structure-based specificity prediction for C2H2 zinc fingers. Nucleic Acids Res 39:4564–4576
25. Havranek JJ, Baker D (2009) Motif-directed flexible backbone design of functional interactions. Protein Sci 18:1293–1305
26. Li H, Ulge UY, Hovde BT, Doyle LA, Monnat RJ Jr (2011) Comprehensive homing endonuclease target site specificity profiling reveals evolutionary constraints and enables genome engineering applications. Nucleic Acids Res 40:2587–2598
27. Redondo P et al (2008) Molecular basis of xeroderma pigmentosum group C DNA recognition by engineered meganucleases. Nature 456:107–111
28. Takeuchi R, Lambert AR, Mak AN, Jacoby K, Dickson RJ, Gloor GB, Scharenberg AM, Edgell DR, Stoddard BL (2011) Tapping natural reservoirs of homing endonucleases for targeted gene modification. Proc Natl Acad Sci U S A 108:13077–13082
29. Grizot S, Duclert A, Thomas S, Duchateau P, Pâques F (2011) Context dependence between subdomains in the DNA binding interface of the I-CreI homing endonuclease. Nucleic Acids Res 39:6124–6136
30. Pabo CO, Nekludova L (2000) Geometric analysis and comparison of protein–DNA interfaces: why is there no simple code for recognition? J Mol Biol 301:597–624
31. Miller JC et al (2011) A TALE nuclease architecture for efficient genome editing. Nat Biotechnol 29:143–148
32. Fleishman SJ et al (2011) Computational design of proteins targeting the conserved stem region of influenza hemagglutinin. Science 332:816–821
33. Röthlisberger D et al (2008) Kemp elimination catalysts by computational enzyme design. Nature 453:190–195
34. Azoitei ML et al (2011) Computation-guided backbone grafting of a discontinuous motif onto a protein scaffold. Science 334:373–376
35. Szeto MD, Boissel SJS, Baker D, Thyme SB (2011) Mining endonuclease cleavage determinants in genomic sequence data. J Biol Chem 286:32617–32627
36. Baxter S, Lambert AR, Kuhar R, Jarjour J, Kulshina N, Parmeggiani F, Danaher P, Gano J, Baker D, Stoddard BL, Scharenberg AM (2012) Engineering domain fusion chimeras from I-OnuI family LAGLIDADG homing endonucleases. Nucleic Acids Res 40:7985–8000

37. Steffen NR, Murphy SD, Tolleri L, Hatfield GW, Lathrop RH (2002) DNA sequence and structure: direct and indirect recognition in protein–DNA binding. Bioinformatics 18: S22–S30
38. Becker NB, Wolff L, Everaers R (2006) Indirect readout: detection of optimized sequences and calculation of relative binding affinities using different DNA elastic potentials. Nucleic Acids Res 34:5638–5649
39. Fleishman SJ et al (2011) RosettaScripts: a scripting language interface to the Rosetta macromolecular modeling suite. PLoS One 6:e20161
40. Canutescu AA, Dunbrack RL (2003) Cyclic coordinate descent: a robotics algorithm for protein loop closure. Protein Sci 12: 963–972
41. Wang C, Bradley P, Baker D (2007) Protein–protein docking with backbone flexibility. J Mol Biol 373:503–519
42. Havranek JJ, Harbury PB (2003) Automated design of specificity in molecular recognition. Nature Struct Biol 10:45–52
43. Mitchell M (1996) An introduction to genetic algorithms, MIT Press

INDEX

David R. Edgell (ed.), *Homing Endonucleases: Methods and Protocols*, Methods in Molecular Biology, vol. 1123, DOI 10.1007/978-1-62703-968-0, © Springer Science+Business Media New York 2014